Nematodes as
Biological Models

VOLUME 2

Contributors

H. J. Atkinson

Alan F. Bird

R. Bolla

A. A. F. Evans

Stanley Himmelhoch

D. R. Newall

Morton Rothstein

M. R. Samoiloff

J. R. Vanfleteren

C. Womersley

K. A. Wright

D. J. Wright

Bert M. Zuckerman

Nematodes as Biological Models

VOLUME 2

Aging and Other Model Systems

Edited by

Bert M. Zuckerman

Laboratory of Experimental Biology
College of Food and Natural Resources
University of Massachusetts
East Wareham, Massachusetts

1980

ACADEMIC PRESS
A Subsidiary of Harcourt Brace Jovanovich, Publishers
New York   London   Toronto   Sydney   San Francisco

ACADEMIC PRESS, INC.
111 Fifth Avenue, New York, New York 10003

United Kingdom Edition published by
ACADEMIC PRESS, INC. (LONDON) LTD.
24/28 Oval Road, London NW1 7DX

Library of Congress Cataloging in Publication Data
Main entry under title:

Nematodes as biological models.

    Includes bibliographies and index.
    CONTENTS: v. 1. Behavioral and developmental
models.--v. 2. Aging and other model systems.
    1. Nematoda. 2. Caenorhabditis elegans.
3. Biological models. I. Zuckerman, Bert Morton,
Date [DNLM: 1. Models, Biological. 2. Nematoda.
QX203 N433]
QL391.N4N38   595.1'82   79-8849
ISBN 0-12-782402-2 (v. 2)

PRINTED IN THE UNITED STATES OF AMERICA

80 81 82 83   9 8 7 6 5 4 3 2 1

Contents

List of Contributors   **ix**

Preface   **xi**

Contents of Volume 1   **xiii**

Part 1. Aging, Nutritional, and Toxicant Testing Models

1 Nematodes as Models to Study Aging
Bert M. Zuckerman and Stanley Himmelhoch

I.	Introduction	4
II.	Aging Studies in Axenic and Monoxenic Culture	4
III.	Mass Cultures for Aging Research	6
IV.	Age Pigment	8
V.	Sex Factors in Longevity and Reproduction	11
VI.	Influence of Environment on Development and Aging	14
VII.	Behavioral Changes during Aging	14
VIII.	Longevity Increases	19
IX.	Other Aging Parameters	20
X.	Nematodes to Study Biological Aging in Outer Space	25
XI.	Summary and Future Research Directions	26
	References	26

2 Effects of Aging on Enzymes
Morton Rothstein

I.	Methods of Aging Nematodes	29
II.	Alteration of Enzymes in Aged Nematodes	36
III.	Evaluation of the Nematode Model System	44
	References	45

3   **Nematodes as Nutritional Models**
    J. R. Vanfleteren
      I.   Introduction   47
      II. Nutritional Requirements in the Basal Medium   48
      III. The Growth Factor   56
      IV. Nutrient Absorption   63
      V.   Possible Contaminants of Axenic Culture   72
      VI. Nutrition and Development   74
      VII. Conclusions and Prospects   76
           References   77

4   **Action of Chemical and Physical Agents on Free-Living
    Nematodes**
    M. R. Samoiloff
      I.   Introduction   81
      II. Methods of Assay and Probes   82
      III. Prospects   96
           References   97

Part 2. Physiology and Morphology

5   **Respiration in Nematodes**
    H. J. Atkinson
      I.   Introduction   101
      II.   Diffusion of Oxygen into Nematodes   102
      III. Oxygen and the Mitochondrion   109
      IV.   Oxygenases   115
      V.   Factors Influencing Oxygen Demand   116
      VI.   Oxygen Availability   127
      VII. Respiration in Low Oxygen Regimes   129
      VIII. Hemoglobin   132
      IX.   Anaerobiosis   136
      X.   Conclusions   137
            References   138

6   **Osmotic and Ionic Regulation in Nematodes**
    D. J. Wright and D. R. Newall
      I.   Introduction   143
      II.   Techniques   144
      III. Mechanisms in Nematode Osmoregulation and Possible
            Structures Involved in Transport Processes in Nematodes   149
      IV.   Specialized Aspects of Osmotic and Ionic Regulation in
            Nematodes   156

V.   Conclusions   161
      References   162

7   Nematode Energy Metabolism
R. Bolla
    I.   Introduction   166
    II.   Energy Storage Molecules   167
    III. Utilization of Energy Reserves   168
    IV.   End Products of Metabolism   171
    V.   Glycolysis and the Tricarboxyclic Acid Cycle   173
    VI.   Other Pathways of Energy Metabolism   178
    VII. Energy Production and Regulation of Metabolism   185
    VIII. Conclusions   189
          References   189

8   Longevity and Survival in Nematodes: Models and Mechanisms
A. A. F. Evans and C. Womersley
    I.   Introduction   193
    II.   Model Nematodes for Studying Longevity and Survival   195
    III. Physiological Aspects of Desiccation Survival   202
    IV.   Conclusions   208
          References   208

9   The Nematode Cuticle and Its Surface
Alan F. Bird
    I.   Introduction   213
    II.   Whole Cuticle Structure   214
    III. The Epicuticle   219
    IV.   Dynamic State of the Epicuticle   225
    V.   Development of the Epicuticle   232
    VI.   Summary and Conclusions   233
          References   234

10 Nematode Sense Organs
K. A. Wright
    I.   Introduction   237
    II.   Cuticular Sense Organs   239
    III. Internal Sensory Receptors   278
    IV.   Conclusions   285
          References   293

Index   **297**

List of Contributors

Numbers in parentheses indicate the pages on which the authors' contributions begin.

H. J. Atkinson (101), Department of Pure and Applied Zoology, The University of Leeds, Leeds LS2 9JT, England

Alan F. Bird (213), C.S.I.R.O. Institute of Biological Resources, Division of Horticultural Research, Adelaide 5001, South Australia

R. Bolla (165), Department of Biology, University of Missouri—St. Louis, St. Louis, Missouri 63121

A. A. F. Evans (193), Imperial College at Silwood Park, Ashurst Lodge, Sunninghill Ascot Berkshire, SLS 7DE England

Stanley Himmelhoch (3), Section of Biological Ultrastructure, Weizmann Institute, Rehovot, Israel

D. R. Newall (143), Department of Plastic and Reconstructive Surgery, The Royal Victoria Infirmary, Newcastle-upon-Tyne NE1 4LP, England

Morton Rothstein (29), Division of Cell and Molecular Biology, State University of New York at Buffalo, Buffalo, New York 14260

M. R. Samoiloff (81), Department of Zoology, University of Manitoba, Winnipeg, Manitoba R3T 2N2, Canada

J. R. Vanfleteren (47), Laboratorium voor Morfologie en Systematiek der Dieren, Rijksuniversiteit Gent, Ledeganckstraat 35, B-9000 Gent, Belgium

C. Womersley (193), Department of Zoology, University of Newcastle, Newcastle-upon-Tyne NE1 4LP, England

K. A. Wright (237), Department of Microbiology and Parasitology, Faculty of Medicine, University of Toronto, Toronto, Ontario M5S 1A1, Canada

D. J. Wright (143), Department of Zoology and Applied Entomology, Imperial College, London SW7 2BB, England

Bert M. Zuckerman (3), Laboratory of Experimental Biology, College of Food and Natural Resources, University of Massachusetts, East Wareham, Massachusetts 02538

Preface

The free-living nematode *Caenorhabditis elegans* has attained prominence as a model to study a variety of complex biological problems. Some workers believe that this organism will become the *Escherichia coli* of the metazoan world. The important early work on the nematode model was done by E. C. Dougherty, V. Nigon, and their respective colleagues between approximately 1945 and 1965. The recent surge of interest, however, is due in large part to the detailed studies on the genetics and anatomy of *C. elegans,* which began in Sidney Brenner's laboratory about 16 years ago. This work has expanded rapidly and is now proceeding in a number of institutions and universities throughout the world.

The resultant effort has brought together geneticists, cell and developmental biologists, neurologists, behavioralists, endocrinologists, toxicologists, nutritionalists, and gerontologists—all with a common interest in focusing on the nematode as a biological model in order to examine some of the most basic problems of life. A few workers, like the editor, were nematologists who developed an expanded perspective of our organism of choice.

The current volumes provide a reference source for research in which free-living nematodes have been used to examine fundamental processes in areas such as genetics, development, nutrition, toxicology, pharmacology, and gerontology. In addition to treating the work with *C. elegans,* important studies utilizing other free-living nematodes as models have also been included. Where data are lacking on free-living nematodes, pertinent information from studies of animal parasitic or plant parasitic nematodes is provided when available. With a view to obtaining the best possible coverage of this highly specialized field, the authors selected to contribute to the text are all actively engaged in research.

Volume 1 includes discussion on cell lineages, muscle development, behavior, the nervous system, control mechanisms, and genetics, with the major emphasis on *C. elegans*. However, important contributions derived primarily from studies with the animal parasitic nematode *Ascaris* and the free-living nematode *Panagrellus* are included.

Volume 2 contains discussions on free-living nematodes as biological models for pharmacologic and toxicant testing, and for studies on gerontology and nutrition. There are also several chapters on nematode physiology and morphology, so as to bring together in one reference volume information the reader will find useful to the understanding of the subject.

The significant contributions to the various areas come from studies of a large number of species, encompassing the animal parasitic, plant parasitic, and free-living nematode groups. Hopefully, one contribution of the current volumes will be to emphasize the advantages which can accrue when researchers with diverse interests focus their efforts on one nematode species. The logical choice would appear to be *C. elegans*.

Finally, I wish to acknowledge my gratitude to Margret A. Geist for her splendid efforts in assisting with the editing of the manuscripts and proofs and with the myriad of small tasks associated with the production of these volumes.

Bert M. Zuckerman

Contents of Volume 1

BEHAVIORAL AND DEVELOPMENTAL MODELS

Chapter 1. Cell Lineages and Development of *Caenorhabditis elegans* and Other Nematodes
Gunter von Ehrenstein and Einhard Schierenberg

Chapter 2. Muscle Development in *Caenorhabditis elegans:* A Molecular Genetic Approach
Janice M. Zengel and Henry F. Epstein

Chapter 3. Behavior of Free-Living Nematodes
David B. Dusenbery

Chapter 4. Neural Control of Locomotion in *Ascaris:* Anatomy, Electrophysiology, and Biochemistry
Carl D. Johnson and Antony O. W. Stretton

Chapter 5. Control Mechanisms in Nematodes
James D. Willett

Chapter 6. Genetic Analysis of *Caenorhabditis elegans*
Robert K. Herman and H. Robert Horvitz

Chapter 7. Developmental Genetics of *Caenorhabditis elegans*
Donald L. Riddle

Chapter 8. Biochemical Genetics of *Caenorhabditis elegans*
Shahid Saeed Siddiqui and Gunter von Ehrenstein

Index

part one

Aging, Nutritional, and Toxicant Testing Models

1

Nematodes as Models to Study Aging

BERT M. ZUCKERMAN

Laboratory of Experimental Biology
College of Food and Natural Resources
University of Massachusetts
East Wareham, Massachusetts 02538

STANLEY HIMMELHOCH

Section of Biological Ultrastructure
Weizmann Institute
Rehovot, Israel

I.	Introduction	4
II.	Aging Studies in Axenic and Monoxenic Culture	4
III.	Mass Cultures for Aging Research	6
	A. DNA Synthesis Inhibitors	6
	B. Heat Shock	7
	C. Size Synchrony	7
	D. Age Synchrony by Physical Separation	7
IV.	Age Pigment	8
V.	Sex Factors in Longevity and Reproduction	11
	A. Male versus Female Longevities	11
	B. Longevity of Virgin versus Nonvirgin	11
	C. Parental Age and Reproduction	12
	D. Effects of Aging on Fecundity and Reproductive Patterns	13
	E. Summary	14
VI.	Influence of Environment on Development and Aging	14
VII.	Behavioral Changes during Aging	14
	A. Movement	15
	B. Feeding	16
	C. Vulval Contractions	18
VIII.	Longevity Increases	19
IX.	Other Aging Parameters	20
	A. Electron-Dense Aggregates	20
	B. Osmotic Fragility and Specific Gravity	20
	C. Permeability Changes—Cellular	22
	D. Aging Changes of the Outer Cuticular Surface and Cuticular Permeability	22

3

X. Nematodes to Study Biological Aging in Outer Space   25
XI. Summary and Future Research Directons   26
    References   26

I. INTRODUCTION

The use of nematodes as models to study aging is currently attracting wide interest among developmental biologists. The trend is clearly indicated by the large numbers of recent papers on nematode aging cited in these volumes. Because of the rapidly expanding interest in free-living nematodes, a large amount of data has accumulated on all aspects of the biology of these organisms. Certainly there is no other group of multicellular organisms about which so much is known.

The question arises, Why was the nematode and not another small metazoan chosen as a model? This question was considered in detail in other reviews (i.e., Zuckerman, 1976a,b), and only the principal advantages need be summarized here. Briefly, these include a short life span (about 25 days for *Caenorhabditis briggsae,* small size, ease of maintainance in axenic or monoxenic culture, the relatively small number of cells that are differentiated into nervous, digestive, reproductive, and muscular systems, and, most important, the rapidly expanding body of information which is now available on the genetics, nutrition, development, and physiology of several species of free-living nematodes.

Another critical question often asked by biomedical researchers is the relevance of nematode aging to human senescence. As knowledge of molecular biology grew it rapidly became apparent that certain basic cellular processes proceed along similar paths of all living things. This chapter refers to a number of observations related to the biology and physiology of nematode aging that appear to parallel events associated with mammalian aging.

II. AGING STUDIES IN AXENIC AND MONOXENIC CULTURE

The first nematode to be cultured axenically, that is, in the absence of other organisms, was *Turbatrix aceti* (Zimmermann, 1921). Since then axenic culture has been achieved for several free-living nematode species. The method is standard for studies in which the researcher desires to eliminate all variables related to interactions with other organisms. Some workers prefer to culture nematodes on a single species of bacterium (monoxenic culture), arguing that these culture conditions more closely approximate the natural environment and that nematodes grown axenically suffer a nutritional deficit.

Most of the initial biochemical and physiological studies on nematode aging proceeded under axenic culture. However, in the past 3 years a number of researchers have chosen the monoxenic regime. The latter include the cited works of Hosono, Duggal, Suzuki, Klass, and Willett.

Croll *et al.* (1977) performed experiments with *C. elegans* which had as an objective the comparison of several aging characteristics from a single isolate grown under axenic or monoxenic conditions. The results show that large differences exist in the way nematodes develop and age under the two nutritional regimes.

Caenorhabditis elegans fed on *E. coli* attained more than twice the volume of those in axenic culture; however, 50% of the nematodes grown on bacteria died by day 12.3. By comparison, 50% of the axenically cultured nematodes survived to day 17.6. These observations suggest that nematodes cultured axenically show symptoms of starvation. Improved culture media may narrow the gap between the observed differences.

Of greater significance to the aging investigations were the findings on population survival. Until day 10 there were no observable differences between the two cultures. However, *C. elegans* in axenic culture showed a linear chronological decline in total population. In bacterial culture there was a rapid decline in numbers by day 11, and by day 14 most of the nematodes were dead. Since the nematodes were transferred to a new medium at frequent intervals during these experiments, the authors suggested that the nematodes' intestinal bacterial flora caused their premature death. This conclusion was supported by the observation that the bacteria-fed nematodes were still physiologically young at the time of death, at least in respect to some characteristics of aging such as age-pigment accumulation.

The studies cited above were on nematodes in liquid media. Other investigators were able to follow age-related changes in the cyclic nucleotide levels of *Panagrellus redivivus* grown on bacterial lawns (Willett *et al.*, 1978). These experiments showed that it is possible to demonstrate changes in biochemical pathways which occur up to midsenescence under monoxenic conditions.

The deciding factor in selecting a culture regime may be the "oldest" age level to be studied. Many nematodes in axenic culture live to a ripe old age, whereas few if any survive that long when bacteria serve as a food source. If changes that occur during late senescence are to be studied, current knowledge suggests that the experiments should proceed on axenically cultured animals.

Nutritional regime exerts a great influence on nematode aging. It seems logical that careful studies designed to define the effects of cultural conditions on nematode development should precede any large-scale effort aimed at utilizing these organisms as models to examine basic mechanisms of senescence and longevity.

III. MASS CULTURES FOR AGING RESEARCH

A technical difficulty that presents itself to the investigator who wishes to use the nematode as a model system to study aging is the problem of separating adult nematodes from their progeny. *Caenorhabditis elegans* and *C. briggsae* are sexually mature and reproductive 3½ to 5 days after hatching, and continue to lay eggs for approximately 10 more days.

Single nematodes can suffice for some types of aging experiments (i.e., Abdulrahman and Samoiloff, 1975), and other investigations may require only a few hundred individuals (i.e., Searcy *et al.*, 1976). In either case, the nematodes can be transferred manually to new media at 2- to 3-day intervals during the reproductive period, thus preventing the intermingling of parents and progeny.

However, for biochemical analyses, milligram quantities of nematodes are often needed, requiring some technique to provide a large biomass of nematodes of the same age. The several methods devised to attain age synchrony in mass cultures have served as a focus for controversy, which, in our opinion, has not been fully resolved.

A. DNA Synthesis Inhibitors

To eliminate reproduction, DNA synthesis inhibitors such as 5-fluorodeoxyuridine (FUdR), hydroxyurea, and aminopterin have been used to prevent gonad development (Pasternak and Samoiloff, 1970; Gershon, 1970; and others). This method initially involved obtaining large numbers of newly hatched larvae by glass-bead separation (Chow and Pasternak, 1969), adding the inhibitor, and then allowing the nematodes to age synchronously. This technique at first appeared to provide an elegant means for amassing large numbers of senescent organisms of the same chronological age. However, critical studies showed that inhibitors applied to newly hatched larvae not only blocked reproduction, but also retarded maturation and growth and reduced longevity (Tilby and Moses, 1975; Kisiel *et al.*, 1972, 1974).

The early published reports of nematode enzyme changes with age (Gershon, 1970; Erlanger and Gershon, 1970; Gershon and Gershon, 1970) all entailed the exposure of newly hatched larvae to one of the inhibitors referred to above. To our knowledge, the observed changes in these specific enzymes which were reported were not verified in experiments with nematodes not treated with an inhibitor. Since not all enzymes alter with age (see this volume, Chapter 2, by Rothstein), it would seem that at least some results of the earlier nematode enzyme work require confirmation. M. Rothstein (personal communication) used only older larvae in enzyme experiments in which inhibitors were used.

Recently, Mitchell *et al.* (1979) also observed adverse effects when FUdR was applied to axenic newly born larvae. However, they reported that when treatment by FUdR was delayed for several days, the drug exerted insignificant effects on development, while still effectively blocking reproduction. The conclusions drawn by Mitchell *et al.* (1979) are challenged by their own data, which show that FUdR-treated larvae lived significantly longer, moved more slowly, and had a slower bulb-pulsation rate. Judged by the preceding criteria, the present authors would not consider these FUdR-treated organisms as candidates for the study of normal aging.

Rothstein and Sharma (1978) examined three enzymes from young and old *Turbatrix aceti* that were obtained by successively screening cultures to remove newly hatched larvae. They found that all showed reduced specific activities with age. Since the results closely paralleled previous tests in which FUdR-treated organisms were used, they concluded that FUdR did not affect aging studies, at least with respect to studies of altered enzymes.

B. Heat Shock

Heat shock was used by Hieb and Rothstein (1975) to prevent reproduction of *T. aceti* with the goal of attaining age synchrony. Experiments demonstrating that heat shock alters the normal patterns of protein synthesis are discussed by Klefenz and Zuckerman (1978). At present there is no evidence that heat shock is an acceptable means for attaining nematode age synchrony.

C. Size Synchrony

Density gradient separation and successive screening are two methods used to separate nematodes into classes according to size and density. The assumption is that a nematode of a given size and density is a certain age. Klefenz and Zuckerman (1978) discuss problems inherent to aging nematodes in mass culture which indicate that under certain conditions a small nematode is not necessarily a young nematode.

Despite these objections, studies of Willett *et al.* (1978) and Willett and Bollinger (1978a,b) indicate that size synchrony is an acceptable means for segregating representative age groups of nematodes from mass cultures. In these experiments, progressive changes in cGMP/cAMP ratios and prostaglandin levels were demonstrated from seven sized groups of *Panagrellus redivivus*. Size synchrony by successive screening was used by Sharma *et al.* (1976) to examine changes in enolase from "young" and "old" *T. aceti*.

D. Age Synchrony by Physical Separation

Tilby and Moses (1975) developed an apparatus for maintaining age synchrony of a selected group of *C. elegans*. The procedure involved culturing in a continuous-flow system, with the newly born young burrowing through a screen of sufficiently small mesh to prevent passage of larger old nematodes. This system was developed only to the point where 5000 synchronously aged nematodes could be maintained at one time. We believe that relatively little effort would be required to modify this apparatus so that milligram quantities of nematodes could be obtained.

Age synchrony was also attained by successive screenings of *T. aceti* every 3–4 days (Rothstein and Sharma, 1978). This method also satisfies the criteria for age synchrony by physical means.

IV. AGE PIGMENT

The chronological accumulation of age pigment (also called lipofuscin) is characteristic of most animals, including humans (Kohn, 1971). Although the physiological significance of age pigment to the degeneration of cellular function during senescence is still uncertain, the increased accumulation of this substance is the most constant marker of advancing age in animals. In humans and small mammals, where the increase in lipofuscin has been extensively studied in heart muscle and nerve cells, it is estimated that about 90% of the volume of certain nerve cells in old animals is taken up by age pigment, a fact that indicates that the function of the cell organelles may be impaired. In rats, age-pigment accumulation is generally evaluated in 24- to 36-month-old animals. The fact that significant changes in nematode age-pigment accumulation can be demonstrated within 2 weeks provides a clear rationale for the use of nematodes to study this phenomenon.

The first work on nematode age pigment was described by Buecher and Hansen at a Gordon Conference on Biological Aging held in Santa Barbara, California, in 1970, although the results of this study were not published until much later (Buecher and Hansen, 1974). A spectrofluorometric assay, described by Fletcher *et al.* (1973), was used to quantitate and partially characterize lipofuscin extracted from *Panagrellus redivivus*. The findings showed that nematodes had more age pigment per milligram of tissue than did mouse heart tissue, and that nematode lipofuscin increased with age. Using the same procedures, Zuckerman *et al.* (abstract of a paper given at the American Aging Association Meeting, Los Angeles, 1974, published in 1978) showed that the excitation and emission spectra of age pigment from *C. briggsae* were similar to those of *P. redivivus* (Fig. 1). Klass (1977) confirmed these findings on *C. elegans*.

Electron microscope studies provided additional evidence that nematode lipofuscin is similar to that found in mammals, and revealed that the pigment granules are restricted to the intestinal epithelium (Figs. 2 and 3). These granules occur in the same place as occupied by lipid and protein droplets in the epithelium of young nematodes, strongly suggesting that they form, as does mammalian lipofuscin, as peroxidation products of lipids and protein. The morphologic appearance and location of nematode lipofuscin were first described

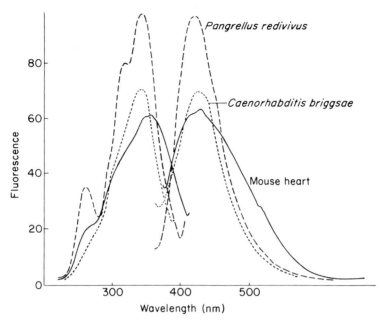

Fig. 1. Fluorescence spectra of lipid-soluble fluorochromes from *Panagrellus redivivus* and mouse heart tissue (after Buecher and Hansen, 1973) and *Caenorhabditis briggsae* (after Zuckerman *et al.*, 1978). Data from the two experiments are combined.

from *C. briggsae* (Epstein *et al.*, 1972). Histochemical tests showed that the granules contain acid phosphatase, an enzyme found within lipofuscin granules. Age pigment was also observed in electron micrographs from the intestinal epithelium of old *T. aceti* (Kisiel *et al.*, 1975) and *C. elegans* (Fig. 4). Lipofuscin granules were formed within the intestinal cells of the plant parasite *Bursaphelenchus lignicolus* during aging (Kondo and Ishibashi, 1978). Age-pigment formation is probably also a general characteristic of senescence of plant-parasitic nematodes.

Additional evidence, albeit indirect, relating to the nature of nematode lipofuscin derives from drug experiments, which suggest similar pathways of pigment formation in mammals and nematodes. Studies on *C. briggsae* showed that several drugs that retard the accumulation of age pigment in mammals (e.g., Spoerri and Glees, 1974) also produce the same effect in nematodes. These several reports included the retardation of age-pigment accumulation by vitamin E (Epstein and Gershon, 1972), centrophenoxine (Kisiel and Zuckerman, 1978) and a combination of *p*-chlorophenoxyacetic acid and dimethylaminoethanol (Zuckerman and Barrett, 1978). The latter two compounds are primary hydrolysis products of centrophenoxine.

Under the light microscope the intestinal region of young *C. briggsae* and other free-living nematode species is translucent and appears to contain

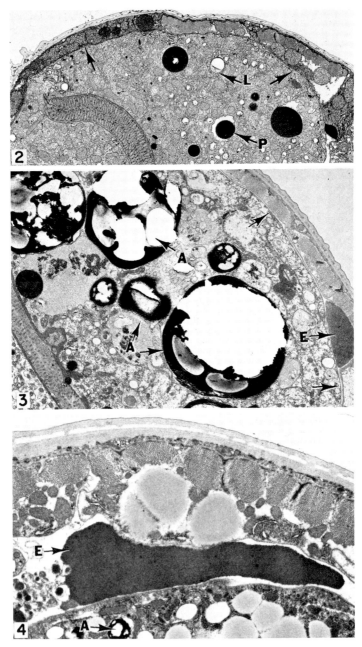

Fig. 2. *Caenorhabditis briggsae*, 6 days old. Cross section through the intestinal epithelium (arrowhead) of a young nematode showing lipid (L) and protein (P) droplets.

yellowish-green globules. As aging progresses, the intestine darkens and becomes opaque, thus allowing the progression of age-pigment accumulation to be followed by visual observation. This feature permits a rough evaluation of the effects of drugs on lipofuscin formation.

V. SEX FACTORS IN LONGEVITY AND REPRODUCTION

A common denominator in the animal kingdom is that females generally have a longer life expectancy than males (Rockstein, 1974). The gap in human longevities in the United States is greater than in many developed countries—66.5 years for males and 73.0 for females (U.S. Bureau of Census, 1976). Some workers ascribe these differences to stress, citing that the male role in society leads to greater exposure to stressful situations. However, with the exception of a few reports, studies generally show greater longevities in females of all multicellular animals. These observations suggest a genetic basis for sex longevity differences and allow for certain types of experiments that may be performed with facility using small metazoa, but only at great expense and with difficulty on higher mammals.

A. Male versus Female Longevities

All studies of nematode male–female longevities have shown that the female lived longer. In an early study on *T. aceti*, females averaged 49 days total life span whereas males averaged 48 days (Pai, 1928). Kisiel and Zuckerman (1974) reported that an average longevity of mated *T. aceti* females was 70 days and that of mated males was 64 days. The discrepancy in these results could be due to differences in cultural conditions, such as temperature and the nutrient medium. Data for mated *P. redivivus* showed an average longevity of 28 days for females and 19 days for males (Abdulrahman and Samoiloff, 1975). *Rhabditis tokai* showed a similar pattern of greater longevity of mated females than of mated males (Suzuki *et al.*, 1978a,b).

B. Longevities of Virgin versus Nonvirgin

Observations on several nematode species indicate that virgins live longer than nonvirgins. For *T. aceti*, both virgin males and females had greater longevities

---

Fig. 3. *Caenorhabditis briggsae*, 28 days old. Cross section through the intestinal epithelium (arrowhead) of an old nematode showing age-pigment granules (A) and electron-dense aggregates (E) within the pseudocoelom.

Fig. 4. *Caenorhabditis elegans*, 21 days old. Old nematode showing a large mass of aggregated substance (E) within the pseudocoelom and an age-pigment granule (A) within the intestinal epithelium.

than did their mated counterparts (Kisiel and Zuckerman, 1974). Similar findings were reported for *P. redivivus* (Abdulrahman and Samoiloff, 1975) and *Rhabditis tokai* (Suzuki *et al.*, 1978a,b). However, Duggal (1978a,b) observed that the virgin female *P. redivivus* lived longer, but virgin males had a shorter longevity.

A possible explanation for the differences in virgin–nonvirgin female longevities is that the wear and tear accompanying the reproductive process has a life-shortening effect. Statistical studies of humans in undeveloped nations where population growth is great show that women in these places have shorter average life spans than men (Beller and Palmore, 1974). A partial reason for this is that in these countries women generally bear many children, a factor that exacts a heavy toll in later life.

C. Parental Age and Reproduction

Reproduction in animals is at least under partial control of the endocrine system. During human aging there is evidence of failure of the endocrine system, accompanied by decreased responsiveness of the gonadal tissues. At the time reproduction ceases, most remaining egg cells in the ovary have died, although the reasons for this are as yet unclear. These events occur in the mid-forties, the time of female menopause (Strehler, 1977).

Only recently has knowledge of the control mechanisms that govern nematode reproduction begun to accumulate (see Chapter 5 of Volume 1), so that comparisons to age-related events associated with mammalian reproductive processes is premature. However, the limited data on reproductive changes during nematode aging suggest that nematode studies may provide relevant information on functional degeneration associated with the aging of mammalian reproductive systems.

The effects of parental senescence on succeeding generations of *C. elegans* were studied by Beguet (1972) and Beguet and Brun (1972). The fecundity of young from old parents decreased for four generations, as compared with the fecundity of young from young parents. The effect disappeared after the fourth generation (Beguet, 1972). Young from old parents matured earlier, and had a 15% reduction in the number of spermatozoa and an increase in mortality of F_2 eggs.

Klass (1977) found that the parental age and parental life span of *C. elegans* had relatively small effects on progeny life span. He concluded that progeny life span was not simply influenced by parental age or parental life span, but probably by the complex interactions of genetic and environmental factors. In our laboratory, identical series of tests were performed on *C. briggsae* (J. Lavimoniere and B. Zuckerman, unpublished data) with similar results.

D. Effects of Aging on Fecundity and Reproductive Patterns

Age-related changes in fecundity were reported for *P. redivivus* (Duggal, 1978b) and *T. aceti* (Kisiel and Zuckerman, 1974). Older females of each species produced fewer progeny than the young females. Furthermore, in *T. aceti*, the older the nematode at mating, the longer the elapsed time prior to the start of reproduction and the shorter the reproductive period (Fig. 5). These findings indicate a degeneration of gonad function with age, because the same effects occur when young males were mated with old females or old males mated with young females. Earlier, Pai (1928) reported that most reproductive cells in old *T. aceti* were dead. A similar process termed atresia occurs in human females, wherein the unexplained death of egg cells is correlated with aging (Strehler, 1977).

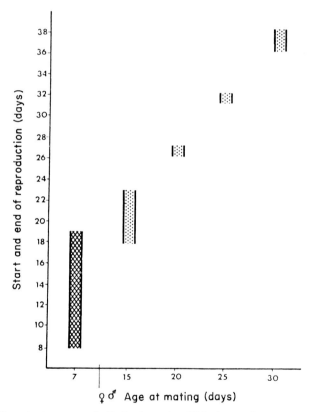

Fig. 5. Reproductive patterns in *Turbatrix aceti* at 27°C when mating occurs at different times of the life span. Mating at day 7 represents the normal reproductive pattern. (From Kisiel and Zuckerman, 1974, reproduced by permission.)

The production of a sex attractant by the female and responsiveness by the male are sex-specific age parameters that begin and end within the adult life span of *Rhabditis pellio* (Somers *et al.*, 1977). The time required for the male to migrate to a female pheromone source also increased as the nematode aged.

E. Summary

Degenerative events associated with nematode reproduction fit well in the theory of the evolutionary origins of aging. In part, this theory presupposes that evolutionary changes that preclude reproduction are of advantage to the survival of the species. The degeneration of gonad function in nematodes is accompanied by an accelerated development of other changes that characterize nematode senescence—for example, the rapid increase in osmotic fragility, age pigment, and specific gravity. The findings of the several studies cited are consistent with a theory of aging that argues that the degeneration of gonad function signals the onset of generalized aging (Gusseck, 1972).

It would appear that the nematode may provide a valid model for experimentation on the mechanisms involved in the age-related reproductive degeneration in higher animals.

VI. INFLUENCE OF ENVIRONMENT ON DEVELOPMENT AND AGING

There is conclusive evidence that nematode development and longevity are subject to control by environmental influences. Some of the many studies on free-living nematodes that established this effect showed that nutritional levels affected the fecundity of *P. redivivus* (Abdulrahman and Samoiloff, 1975), the type of nutritional substrate influenced the rate of accumulation of *C. elegans* to a food source (Hosono, 1978a,b), and temperature levels affected longevity, fecundity, growth, and movement of *C. briggsae* (Zuckerman *et al.*, 1971).

Investigations of behavioral changes during aging and longevity increases are best performed under optimal conditions with respect to factors such as nutrition and temperature. This protocol was suggested by nutritional studies which indicated that successful axenic culture requires conditions that approximate the natural environment as closely as possible (Buecher and Hansen, 1971).

Other effects of environment on aging, although pertinent to this subject, are beyond the scope of this chapter.

VII. BEHAVIORAL CHANGES DURING AGING

The musculature of free-living nematodes mediates three types of behavior: (1) movement; (2) feeding and defecation; and (3) egg-laying and copulation.

Age-related changes in these activities and the possible mechanisms involved in these changes will be considered.

A. Movement

Nematode somatic muscle cells are longitudinally oriented and are arranged in four bands separated by invaginations of hypodermal tissue. The muscle cells are attached to and lie beneath a thin layer of interchordal hypodermis. Somatic muscles control nematode movement.

A cross section through the nematode shows several nerve cords, among which are small cords that run parallel to the long axis of the muscle cells and lie within the interchordal hypodermis, approximately adjacent to the center of each muscle band. As will be discussed later, this interchordal area may play a role in the decline of movement during nematode aging.

Nematode movement has consistently been shown to slow with age. Pai (1928) recorded that movement in senile *T. aceti* was slower than in young animals. Decreases with age were noted in the numbers of sinusoidal muscle waves (Hosono, 1978a,b) and backward waves (Croll *et al.*, 1977) in *C. elegans*, head-swinging movements in *C. briggsae* (Zuckerman *et al.*, 1971), and turns, loops, and reversals in *P. redivivus* (Abdulrahman and Samoiloff, 1975).

The manner in which the slowing of nematode movement with age might parallel aging phenomena associated with muscle function in higher animals is very complex, and is not considered here in detail. Briefly, as in humans, there is the possibility that age-related changes in protein metabolism result in the degeneration of muscle constituents (Young *et al.*, 1976), an effect that could be reflected in decreased nematode movement.

Alterations in biogenic amine levels have been shown to be characteristic of aging in certain mammals (Finch, 1978), and, if such changes occur in nematodes, muscle function may be affected. Present knowledge does not allow interpretation of nematode movement decrements in terms of age-related alterations of neurotransmitter levels, but the possibility that the decrease in somatic body waves and other movement activities is due to the failure of neural control mechanisms must be considered. Findings indicating that biogenic amines are important in feeding, oviposition, and possibly in copulation (Croll, 1975b; Sulston *et al.*, 1975) support this belief.

However, with respect to nematode movement, another age-related factor is possibly of greater importance. Specifically, this is the observed physical displacement of the somatic muscle contractile elements during aging. This phenomenon has been observed from electron microscope studies of *P. silusiae* (Samoiloff and Pasternak, 1971), *C. briggsae* (Zuckerman *et al.*, 1971), and *T. aceti* (Kisiel *et al.*, 1975). During aging the nerve tissues of the interchordal hypodermis often hypertrophy to such an extent that the somatic muscles are displaced from their normal positions (Fig. 6). In some mutants it has been noted that disorganization of the body wall muscles results in aberrant movement (Hosono, 1978a,b). Muscle displacement may account for a significant part of the age-related decrement of nematode movement.

B. Feeding

Ingestion in free-living nematodes is accompanied by muscular pulsations of the esophageal basal bulb. Changes in the pulsation rate with age have been examined by several workers. The results of these studies contrast sharply, suggesting that experimental variables that can significantly influence aging studies on nematode feeding behavior must be rigorously controlled.

Croll (1975a) analyzed the feeding of *C. elegans* on *E. coli* by videotape analysis. Since pulsations of the basal bulb can proceed at a very rapid rate, this technique ensures a much higher degree of accuracy than can be achieved by visual observation through a light microscope. The observations showed that feeding occurred in two phases: one for a period of 2- to 5-min duration in which the bulb pulsations were more than 90/min, and another for the same time interval in which the pulsations decreased to 5–10/min. The worms used in these experiments were of unknown age.

Videotape analyses were also used to compare the feeding of *C. elegans* of a known age on *E. coli* with that of worms grown axenically (Croll *et al.*, 1977). The ingestion observations were on individuals for 15-min periods every 2 days, so that the "slow" and "fast" feeding rates reported previously were averaged. Large variations in bulb-pulsation rates occurred between individuals of the same age, but in general the results indicated that up to day 10 there were no consistent differences in the feeding rate (Fig. 7).

Conflicting results were obtained by D. Mitchell and J. Santelli (unpublished results). The parameters in these experiments were the same, except that observations were made for 15-sec time periods under the light microscope. A steady decline in the feeding rate was recorded from day 3 to day 10. The reason for the wide discrepancy in the findings from these two studies is not clear.

Observations by videotape on the feeding behavior of *C. elegans* in axenic culture gave more definitive results. The variations between individual feeding rates were small throughout most of these experiments (Croll *et al.*, 1977). Up to day 16 the feeding rate stayed at a constant level, after which there was a sharp decline in the bulb-pulsation frequency. Could it be that age-related changes occur in neurotransmitter levels of nerves that motivate the muscles associated with feeding?

The findings of diverse research programs suggest a rationale for the use of *C. elegans* to examine age-related behavioral changes. The neuromuscular morphology of the *C. elegans* esophageal region has been described in detail

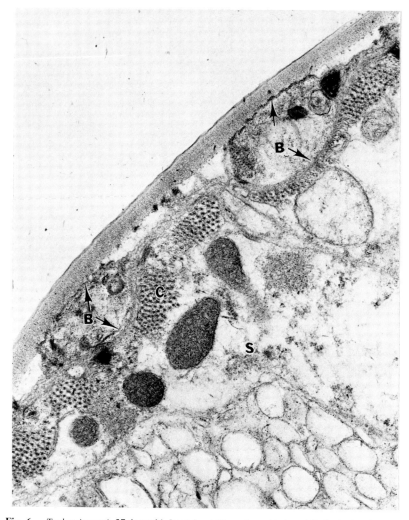

Fig. 6.   *Turbatrix aceti,* 37 days old. Interchordal hypodermal bulges (B) displacing the contractile elements (C) of somatic muscle cells (S) from their normal position contiguous to the body wall. (From Kisiel *et al.,* 1975, reproduced by permission.)

(Albertson and Thomson, 1976). The demonstration of dopamine by formaldehyde-induced fluorescence within eight esophageal neurons of *C. elegans* suggests this biogenic amine as a possible neurotransmitter for nerves that motivate the bulb musculature (Sulston *et al.,* 1975). A possible link to biological aging is provided by findings that biogenic amine levels (including dopamine and serotonin) decreased with age in certain tissues of rats (Finch,

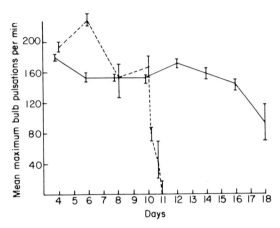

Fig. 7. Mean maximum pharyngeal bulb pulsations per minute for *C. elegans* in bacterial (broken) and axenic (continuous) culture (vertical bars ± SE). (From Croll *et al.*, 1977, *Exp. Aging Res.* **3**, 175–199, published by permission of Beech Hill Enterprises, Inc., Mt. Desert, Maine.)

1978; Meek *et al.*, 1977). The possibility of age-related changes in biogenic amine levels in *C. elegans* is a subject for future study.

The discrimination between "young" and "old" feeding activity in *C. elegans* could well provide an elaborate, yet relatively noncomplex, tool for understanding the mechanisms involved in age-related changes in biogenic amine metabolism. Conceivably, research would proceed through the use of pharmacological intervention to alter nematode behavioral activities. Manipulations such as the transformation of the feeding rate of old nematodes to that of a rate found in the young, if brought about by a drug known to alter biogenic amine levels, would provide evidence concerning the nature of changes in biogenic amine metabolism during aging.

C. Vulval Contractions

Accelerated vulval contractions are correlated with increased egg-laying in *C. elegans* (Croll, 1975b). This activity was the first sex-specific age parameter shown to begin and end within the nematode adult life span. A decrease in contractions as the female aged demonstrated that the rate of vulval–vaginal activation was age-dependent. Observations that the number of *C. briggsae* and *T. aceti* progeny decreases with age (Zuckerman *et al.*, 1971; Kisiel and Zuckerman, 1974) confirm that a degeneration of reproductive capability accompanies nematode senescence. These aging changes in nematodes would appear similar to changes in mammals, where a decreased responsivity of tissues associated with reproduction occurs with age (Strehler, 1977).

The reasons for these decrements in reproductive capacity are not clear, but several observations suggest clues to the mechanisms involved. Studies on the coordination of the contraction of the vulval musculature with other behavioral activities show that an inverse relationship exists between body movement and vulval contractions; during a period of rapid vulval–vaginal contractions, most *C. elegans* become immobile (Croll, 1975b). A similar relationship exists in feeding; during intervals of high vulval muscle contractions the pumping rate of the esophageal bulb slows (S. P. Huang, University of Massachusetts, unpublished thesis data). These data suggest a complexity of behavioral coordination and the existence of interconnecting coordinating systems in *C. elegans.*

Recent advances in the knowledge of putative nematode neurotransmitters also suggest possible bases for age-related reproductive changes. The ability to synthesize several biogenic amines that are presumptive neurotransmitters in mammals has been demonstrated in *C. briggsae* (Kisiel *et al.,* 1976) and *Phocanema decipiens* (Goh and Davey, 1976). Dopamine, and probably serotonin, were reported in several esophageal neurons, based on observations of formaldehyde-induced fluorescence (Sulston *et al.,* 1975). Significantly, no fluorescence occurred within neurons associated directly with the vulval musculature. These findings would appear contradictory to those of Croll (1975b), in which the stimulation of vulval contractions in *C. elegans* was brought about by topical applications of dopamine, 5-hydroxytryptophan, and epinephrine. This apparent conundrum could be explained by a possible role of serotonin in reproductive activities. Possibly, pools of serotonin are located in the esophageal area, and from there could pass by diffusion through the pseudocoelom to the vulval area, where contact would be made with serotonergic receptors. The presence of such receptors has not been demonstrated.

VIII. LONGEVITY INCREASES

Several studies on free-living nematodes describe induced longevity increases or physiological states under which longevity is prolonged. Currently, however, a long-lived strain of a free-living nematode, genetically distinguishable from others of its species, has not been found. Such a strain would be invaluable for the genetic evaluation of differences in body chemistry, behavior, and metabolic rates associated with longevity.

Longevity increases have been reported for free-living nematodes grown in media to which vitamin E was added. These include studies on *C. briggsae* (Epstein and Gershon, 1972), *T. aceti* (Bolla and Brot, 1975), and *P. redivivus* (Buecher and Hansen, 1974). However, no growth data are given, so that one cannot be certain that the life-span increases were not due to toxic effects of the vitamin E on growth and development.

That such toxic effects can be a factor was shown by studies on the influence of procaine on the longevity of *C. briggsae* (Zuckerman, 1974). Significant increases in longevity occurred, but only when nematodes were exposed to a concentration of the drug which reduced growth and fecundity.

Longevity increases occurred in rats and mice exposed to high concentrations of L-dopa (McNamara et al., 1978; Cotzias et al., 1974). Caenorhabditis elegans grown in a medium containing levels of L-dopa and 5-hydroxytryptophan that did not adversely affect development, did not induce significant increases in mean life span (S. P. Huang, University of Massachusetts, unpublished thesis data). Exposure of C. elegans to these two compounds also resulted in significant increases in growth as measured by volume.

At present there is no understanding of the mechanisms involved in reported longevity increases. However, the fact that they possibly occur in the absence of factors that inhibit nematode development and lengthen life span (such as hyponutrition), makes this subject of interest to gerontological research.

IX. OTHER AGING PARAMETERS

Knowledge of changes that accompany the senescence of free-living nematodes has grown since reviews by Zuckerman (1976a,b). This section describes several morphological and physiological alterations that are characteristic of the aging of free-living nematodes.

A. Electron-Dense Aggregates

Electron-dense aggregates accumulate during the aging of C. briggsae within the fluid-filled layer of the cuticle and the pseudocoelom [Kisiel et al. (1974); also shown in micrographs by Epstein et al. (1972), but not described], and in the same region in T. aceti (Kisiel et al., 1975) and C. elegans (Figs. 3 and 4). Chemical characterization has not as yet been attempted, but these aggregates physically resemble masses of cross-linked molecules that are associated with mammalian aging (Bjorksten, 1974).

Concentrations of p-chlorophenoxyacetic acid and dimethylaminoethanol (the primary hydrolysis products of the drug centrophenoxine) that brought about a reduction in accumulation of age-pigment granules in C. briggsae, also reduced the incidence of these aggregates (Zuckerman and Barrett, 1978). The mode of action of centrophenoxine in preventing cross-linkages of molecules is not known.

B. Osmotic Fragility and Specific Gravity

An increase in specific gravity and osmotic fragility characterizes the aging of many cell types, including erythrocytes (Danon and Marikovsky, 1964; and others). The same age-related changes have been noted in free-living nematodes.

The facility with which nematodes can be subjected to hypotonic shock has allowed for an evaluation of the manner in which aged nematodes lose their ability to withstand osmotic stress under a variety of experimental conditions.

Caenorhabditis briggsae (Zuckerman *et al.*, 1971) and *C. elegans* (B. M. Zuckerman, unpublished data) show increased osmotic fragility with age. In *C. briggsae* the degree to which this increase occurred was temperature-dependent. Mated female *T. aceti* also increased in osmotic fragility as they grew older, but the same change did not occur in virgin females (Kisiel *et al.*, 1975). A difference in sexual tolerance to osmotic hypotonic shock was noted for *P. redivivus:* Males were more susceptible than females of the same age (Abdulrahman and Samoiloff, 1975). Thus, reproductive factors appear important in determining the degree of age-related tolerance to osmotic shock. Why this is so is not clear.

The preceding studies were on axenically cultured nematodes. Nematodes cultured on bacteria react quite differently to osmotic stress. Neither *P. redivivus* (J. Willett, personal communication) nor *C. elegans* (B. M. Zuckerman, unpublished data) grown on *E. coli* showed significant age-related differences in osmotic fragility. A possible reason is that nematodes grown on bacteria have a shorter life span than axenically cultured nematodes. Croll *et al.* (1977) note that most *C. elegans* grown on bacteria die shortly after reproduction ceases; therefore, it is possible that the symptoms of senescence observed in axenic culture simply do not have time to develop.

Caenorhabditis briggsae increases in specific gravity with age (Zuckerman *et al.*, 1972). Possibly the conversion of low-density lipid storage materials to high-density lipofuscin is associated with this age-related change. Conflicting results were obtained with *T. aceti;* this species did not show significant differences in specific gravity between young and old nematodes (Kisiel and Zuckerman, 1974). Studies should be performed on other nematode species to determine if a specific gravity increase is a general characteristic of nematode aging.

Drugs have been reported to retard the advent of age-related increases in nematode osmotic fragility and specific gravity. Since there is substantial evidence that any environmental factor that slows nematode development also delays the onset of aging, in these studies the evaluations were made at concentrations below those that were inhibitory to nematode development. Four developmental criteria were used: growth, fecundity, duration of the reproductive period, and the interval between hatch and the start of reproduction. The drug centrophenoxine retarded both age-related increases in specific gravity and osmotic fragility (Kisiel and Zuckerman, 1978). Since the accumulation of age pigment was also decreased by the drug, it was suggested that the specific gravity increase was related to lipofuscin formation. This hypothesis receives some support from experiments with procaine in which an increased osmotic fragility was retarded, but there was no effect on specific gravity (Castillo *et al.*, 1975).

There was no evidence that procaine at noninhibitory concentrations slowed age-pigment formation.

Other possible explanations for changes in nematode specific gravity and osmotic fragility are considered in the following section.

C. Permeability Changes—Cellular

There have been few critical studies of the rate of diffusion through young and old membranes. In an investigation on human tissues, no significant differences in diffusion resistance were observed (Kirk and Laursen, 1955). Studies of age-related alterations in sensitivity of humans to drugs have not yielded conclusive results (Shuster, 1976).

There is only one report that gives evidence of age-related changes in nematode membranes, that of Pai (1928), which indicates that oocytes of old *T. aceti* stain with nuclear dyes whereas those of the young do not. This observation indicates an increased permeability of the oocyte membrane with age, and is in agreement with findings in higher animals that chemicals penetrate plasma membranes of old cells more readily and hence have a greater effect on target areas.

Work is currently under way in several laboratories to establish different cell types of *C. elegans* in tissue culture. If these efforts succeed, an excellent technique will be available for the analysis of aging of nematode cell membranes.

D. Aging Changes of the Outer Cuticular Surface and Cuticular Permeability

Bird (1971) discusses evidence that the nematode cuticle evolved from a cellular structure and, in its modified form, has retained many of the characteristics of living cells. This hypothesis provided a rationale for nematode-surface studies to determine if this structure has properties in common with animal plasma membranes, and could therefore serve as a model for examining age-related changes of biological membranes. The finding that the cuticle of *Caenorhabditis* is covered by a glycocalyx, which lies atop a layer that may be a plasma membrane (Himmelhoch and Zuckerman, 1978), resulted in a modification of the view that study of the nematode surface would yield information pertinent to membrane aging. However, the several investigations of the *Caenorhabditis* cuticular surface provided information of value to the more complete understanding of the nematode model.

During aging the outer cuticular layer of *C. briggsae* thickens, becomes more defined and, in some cases, separates from the underlying cuticular structure (Zuckerman *et al.*, 1973). The surface layer has a net negative charge that decreases in density as the nematode ages (Figs. 8 and 9, from Himmelhoch *et al.*, 1977). Since a decrease in net negative surface charge with age is a charac-

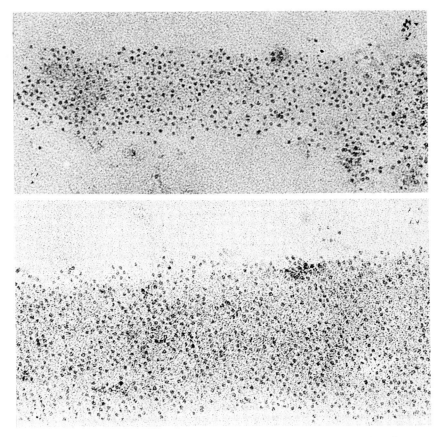

Fig. 8. (*top*) *Caenorhabditis briggsae*, 7 days old. Tangential section of the cuticle surface membrane labeled with cationized ferritin. ×150,000 (From Himmelhoch *et al.*, 1977, reproduced by permission.)

Fig. 9. (*bottom*) *Caenorhabditis briggsae*, 24 days old. Tangential section of the cuticle surface membrane labeled with cationized ferritin. The labeling density of the membrane from old nematodes was 26% less than that of membranes from young nematodes. ×150,000 (From Himmelhoch *et al.*, 1977, reproduced by permission.)

teristic of animal plasma membranes (Cook and Stoddart, 1973), the nematode surface resembles the plasma membrane in this respect.

There are other similarities; several sugars identified by lectin studies as occurring on the nematode surface—specifically glucose, galactose, mannose, and N-acetylglucosamine—are also carbohydrates common to the exposed surfaces of cell membranes (Zuckerman *et al.*, 1979). However, these surface carbohydrates in nematodes are not part of glycoprotein complexes, and the negative-charge-bearing molecules are densely packed (Zuckerman *et al.*, 1979), in contrast to the looser packing and more random arrangement of animal plasma membrane surface carbohydrates. The *C. briggsae* outer membrane is covered by a glycocalyx, probably composed of acid mucopolysaccharides (Himmelhoch and Zuckerman, 1978).

Scanning electron microscope studies of the surface of *C. briggsae* showed that the cuticle was smooth in young nematodes but wrinkled in old ones (Hogger *et al.*, 1977). Increased wrinkling and a probable decrease in elasticity, two prominent characteristics of aging skin in humans (Strehler, 1977), suggest that the aging of the protective covering of nematodes and man could proceed in similar ways.

Other investigations provided information on age-related changes in cuticle permeability. Old *C. briggsae* died when exposed to concentrations of formaldehyde which did not affect the young (Zuckerman *et al.*, 1971). This result suggests increased cuticular permeability of older nematodes, but does not eliminate the possibility that the poison was ingested. In double-labeling experiments with tritiated water, old nematodes proved to be about twice as permeable to water as were young nematodes [Fig. 10 from Searcy *et al.* (1976)]. These experiments suggest that the high internal nematode pressure is largely osmotic in origin, arising from diffusion across the cuticle. This factor probably contributes to the greater osmotic fragility of old nematodes, since water could enter

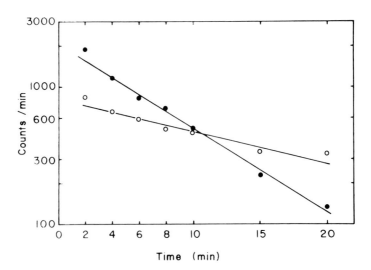

Fig. 10.   Rates of water exchange in young (O—O) and old (●—●) *Caenorhabditis briggsae* (From Searcy *et al.*, 1976, *Exp. Aging Res.* **2**, 293–301. Published by permission of Beech Hill Enterprises, Inc., Mt. Desert, Maine.)

through the cuticle faster than it could be excreted, thus causing the nematodes to burst.

X. NEMATODES TO STUDY BIOLOGICAL AGING IN OUTER SPACE

The use of free-living nematodes to look for possible changes in the aging rate in outer space was proposed at the First International Conference on Cell and Molecular Biology in Space, Toledo, Ohio, May 10–12, 1978 (B. M. Zuckerman, in press). The Toledo conference focused on proposed experiments for the first series of space-shuttle flights, originally scheduled for 1980. These experiments would be completed within 7 days and could indicate promising research directions for projected studies to be made on flights of longer duration.

The primary advantages of axenically cultured *Caenorhabditis* species for aging experiments in outer space were considered, as were the experimental approaches to be employed. First, nematodes that had just completed the reproductive phase (for *C. briggsae*, at 14 days of age) would be placed aboard the craft. The time interval, 14–21 days after hatching, would be of sufficient length to detect several changes that accompany nematode senescence. Replicated experiments would occupy little space, which is most important since the payload area would be strictly limited. Third, little special equipment would be needed. Controls at 1 *g* attained by centrifugation would parallel near-zero gravity experiments, a temperature control of ± 1°C is within the capacity of current engineering capabilities, and heavily shielded samples aboard the module would be compared with samples exposed to ambient cabin radiation levels. Similar equipment is required by many biological experiments.

The space environment encompasses several factors that could influence aging. Most obvious is the radiation effect, since exposure to cosmic radiations within the space craft is approximately five times that at sea level. High levels of ionizing radiations are known to enhance free-radical formation (Gordon, 1974), a process considered to be associated with age-pigment formation and increased molecular cross-linkage (Bjorksten, 1974). Both age pigment and cross linkages are chronological markers of advancing age in higher animals. Retrieval and examination of shielded and nonshielded samples following the recovery of the space module could provide a basis for predicting the rate of human aging in outer space with respect to these markers of senescence.

A second parameter that could be measured would be the effect of weightlessness on age-related alterations of nematode behavior. Such changes, if detected, could indirectly indicate an effect on neurotransmitter levels. Further rationale for these studies comes from Cogoli (1979) and others, who reported a multiplicity of effects of weightlessness on cell development.

There are also certain other effects on life forms noted in previous space flight experiments which cannot be explained. These are classified as unknown factors, one of which is thought to be electromagnetic forces. Another is the effect of the cosmic flux of HZE particles (known only from outer space) on man, for the total HZE environment cannot be totally duplicated on earth. Comparison of heavily shielded samples exposed to a 1 g force in the space capsule with similar samples on earth may yield clues concerning the age-related changes due to these factors.

XI. SUMMARY AND FUTURE RESEARCH DIRECTIONS

Studies on nematode aging support the contention that some events that accompany the senescence of these small worms are similar to those that occur during mammalian aging. Nematode-aging research has been under way for about 10 years. The model would appear to hold promise in several areas that are currently being intensively studied using other animal forms. Free-living nematodes are being increasingly used to examine the genetics of aging, but at this writing only fragmentary information on these studies has been published.

The large number of workers from many disciplines who have initiated research on nematode aging indicates that the nematode will play a significant role in advancing knowledge on the fundamental mechanisms of longevity and senescence in eukaryotic organisms.

ACKNOWLEDGMENTS

The authors would like to thank Dr. V. Dropkin, University of Missouri, and Ms. M. A. Geist, University of Massachusetts, for constructive criticism of this chapter.

REFERENCES

Abdulrahman, M., and Samoiloff, M. (1975). *Can. J. Zool.* **53,** 651–656.
Albertson, D., and Thomson, J. (1976). *Philos. Trans. R. Soc. London Ser. B* **275,** 300–325.
Beller, S. and Palmore, E. (1974). *Gerontologist* **14,** 373–376.
Beguet, B. (1972). *Exp. Gerontol.* **7,** 207–218.
Beguet, B., and Brun, J. (1972). *Exp. Gerontol.* **7,** 195–206.
Bird, A. (1971). "The Structure of Nematodes." Academic Press, New York.
Bjorksten, J. (1974). *In* "Theoretical Aspects of Aging" (M. Rockstein, ed.), pp. 43–59. Academic Press, New York.
Bolla, R., and Brot, N. (1975). *Arch. Biochem. Biophys.* **169,** 227–236.
Buecher, E., and Hansen, E. (1971). *Annu. Meet. Arizona Acad. Sci., 15th Abstr.* 261.
Buecher, E. and Hansen, E. (1974). *IRCS Libr. Compend.* **2,** 1595.
Castillo, J., Kisiel, M., and Zuckerman, B. (1975). *Nematologica* **21,** 401–407.

Chow, H., and Pasternak, J. (1969). *J. Exp. Zool.* **170**, 77–83.

Cogoli, A. (1979). *Proc. Int. Conf. Cell Mol. Biol. Space, 1st* 1978 (in press).

Cook, G., and Stoddart, R. (1973). "Surface Carbohydrates of the Eukaryotic Cell." Academic Press, New York.

Cotzias, G., Miller, S., Nicholson, A., Maston, W., and Tang, L. (1974). *Proc. Natl. Acad. Sci. U.S.A.* **71**, 2466–2469.

Croll, N. (1975a). *J. Zool.* **176**, 159–176.

Croll, N. (1975b). *Can. J. Zool.* **53**, 894–903.

Croll, N., Smith, J., and Zuckerman, B. (1977). *Exp. Aging Res.* **3**, 175–199.

Danon, D., and Marikovsky, Y. (1964). *J. Lab. Chim. Med.* **64**, 668–674.

Duggal, C. (1978a). *Nematologica* **24**, 257–268.

Duggal, C. (1978b). *Nematologica* **24**, 269–276.

Erlanger, M., and Gershon, D. (1970). *Exp. Gerontol.* **5**, 13–19.

Epstein, J., and Gershon, D. (1972). *Mech. Ageing Dev.* **1**, 257–264.

Epstein, J., Himmelhoch, S., and Gershon, D. (1972). *Mech. Ageing Dev.* **1**, 245–255.

Finch, C. (1978). *Int. Congr. Gerontol. Proc., 11th Abstr.* pp. 133–134.

Fletcher, B., Dillard, C., and Tappel, A. (1973). *Anal. Biochem.* **52**, 1–9.

Gershon, D. (1970). *Exp. Gerontol.* **5**, 7–12.

Gershon, H., and Gershon, D. (1970). *Nature (London)* **227**, 1214–1217.

Goh, S., and Davey, K. (1976). *Tissue & Cell* **8**, 421–435.

Gordon, P. (1974). *In* "Theoretical Aspects of Aging" (M. Rockstein, ed.), pp. 61–81. Academic Press, New York.

Gusseck, D. (1972). *Adv. Gerontol. Res.* **4**, 105–166.

Hieb, W., and Rothstein, M. (1975). *Exp. Gerontol.* **10**, 145–153.

Himmelhoch, S., and Zuckerman, B. (1978). *Exp. Parasitol.* **45**, 208–214.

Himmelhoch, S., Kisiel, M., and Zuckerman, B. (1977). *Exp. Parasitol.* **41**, 118–123.

Hogger, C., Estey, R., Kisiel, M., and Zuckerman, B. (1977). *Nematologica* **23**, 213–216.

Hosono, R. (1978a). *Exp. Gerontol.* **13**, 31–36.

Hosono, R. (1978b). *Dobutsugaku Zasshi* **87**, 191–195.

Kirk, J., and Laursen, T. (1955). *Ciba Found. Collog. Ageing* **1**, 69–75.

Kisiel, M., and Zuckerman, B. (1974). *Nematologica* **20**, 277–282.

Kisiel, M., and Zuckerman, B. (1978). *Age (Omaha Nebr.)* **1**, 17–20.

Kisiel, M., Nelson, B., and Zuckerman, B. (1972). *Nematologica* **18**, 373–384.

Kisiel, M., Himmelhoch, S., and Zuckerman, B. (1974). *Exp. Parasitol.* **36**, 430–438.

Kisiel, M., Castillo, J., Zuckerman, L., Zuckerman, B., and Himmelhoch, S. (1975). *Mech. Ageing Dev.* **4**, 81–88.

Kisiel, M., Deubert, K., and Zuckerman, B. (1976). *Exp. Aging Res.* **2**, 37–44.

Klass, M. (1977). *Mech. Ageing Dev.* **6**, 413–429.

Klefenz, H., and Zuckerman, B. (1978). *Age (Omaha, Nebr.)* **1**, 60–67.

Kohn, R. (1971). "Principles of Mammalian Aging" Prentice Hall, New York.

Kondo, E. and Ishibashi, N. (1978). *Int. Congr. Gerontol. Proc., 11th,* Abstr., p. 16.

McNamara, M., Miller, A., and Libby, G. (1978). *Age (Omaha, Nebr.)* **1**, 83–84.

Meek, J., Bertilsson, L., Cheney, D., Zsilla, G., and Costa, E. (1977). *J. Gerontol.* **32**, 129–131.

Mitchell, D., Stiles, J., Santelli, J., and Sanadi, D. (1979). *J. Gerontol.* **34**, 28–36.

Pai, S. (1928). *Z. Wiss. Zool. Abt. A.* **131**, 293–344.

Pasternak, J., and Samoiloff, M. (1970). *Comp. Biochem. Physiol.* **33**, 27–38.

Patel, T., and McFadden, B. (1978). *Nematologica* **24**, 51–62.

Rockstein, M. (1974). *In* "Theoretical Aspects of Aging" (M. Rockstein, ed.), pp. 1–10. Academic Press, New York.

Rothstein, M., and Sharma, H. (1978). *Mech. Ageing Dev.* **8**, 175–180.

Samoiloff, M., and Pasternak, J. (1968). *Can. J. Zool.* **46**, 1019–1022.

Searcy, D., Kisiel, M. and Zuckerman, B. (1976). *Exp. Aging Res.* **2**, 293–301.

Sharma, H., Gupta, S., and Rothstein, M. (1976). *Arch. Biochem. Biophys.* **174**, 324–332.

Shuster, L. (1976). *In* "Special Review of Experimental Aging Research. Progress in Biology" (M. Elias, B. Eleftheriou and P. Elias, eds.), pp. 55–70. EAR, Inc. Bar Harbor, Maine.

Somers, J., Shorey, H., and Gaston, L. (1977). *J. Chem. Ecol.* **3**, 467–474.

Spoerri, P. E., and Glees, P. (1974). *Mech. Ageing Dev.* **3**, 131–155.

Strehler, B. (1977). "Time, Cells and Aging." Academic Press, New York.

Sulston, J., and Horvitz, R. (1977). *Dev. Biol.* **56**, 110–156.

Sulston, J., Dew, M., and Brenner, S. (1975). *J. Comp. Neurol.* **163**, 215–226.

Suzuki, K., Hyodo, M., Ishii, N., and Moriya, Y. (1978a). *11th Inter. Congr. Gerontol. Proc. 11th*, Abstr., p. 15.

Suzuki, K., Hyado, M., Ishii, N., and Moriya, Y. (1978b). *Exp. Gerontol.* (in press).

Tilby, M., and Moses, V. (1975). *Exp. Gerontol.* **10**, 213–223.

U.S. Bureau of Census (1976). Curr. Pop. Rep. No. 59. Washington, D.C.

Willett, J., and Bollinger, J. (1978a). *Age (Omaha, Nebr.)* **1**, 75.

Willett, J., and Bollinger, J. (1978b). *Age (Omaha, Nebr.)* **1**, 158.

Willett, J., Rahim, I., and Bollinger, J. (1978). *Age (Omaha, Nebr.)* **1**, 75.

Young, V., Winterer, J., Munro, H., and Scrimshaw, N. (1976). *In* "Special Review of Experimental Aging Research. Progress in Biology." (M. Elias, B. Eleftheriou and P. Elias, eds.), pp. 217–252. EAR, Inc., Bar Harbor, Maine.

Zimmermann, A. (1921). *Rev. Suisse Zool.* **28**, 357–379.

Zuckerman, B. (1974). *In* "Theoretical Aspects of Aging" (M. Rockstein, ed.), pp. 177–186. Academic Press, New York.

Zuckerman, B. (1976a). *In* "The Organization of Nematodes" (N. Croll, ed), pp. 211–241. Academic Press, New York.

Zuckerman, B. (1976b). *In* "Special Review of Experimental Aging Research. Progress in Biology" (M. Elias, B. Eleftheriou and P. Elias, eds.), pp. 429–451. EAR, Inc., Bar Harbor, Maine.

Zuckerman, B. M. (1979). *Proc. Int. Conf. Cell Biol. Space, 1st* (in press).

Zuckerman, B., and Barrett, K. (1978). *Exp. Aging Res.* **4**, 133–139.

Zuckerman, B., Himmelhoch, S., Nelson, B., Epstein, J., and Kisiel, M. (1971). *Nematologica* **17**, 478–487.

Zuckerman, B., Nelson, B., and Kisiel, M. (1972). *J. Nematol.* **4**, 261–262.

Zuckerman, B., Himmelhoch, S., and Kisiel, M. (1973). *Nematologica* **19**, 109–112.

Zuckerman, B., Fagerson, I., and Kisiel, M. (1978). *Age (Omaha, Nebr.)* **1**, 26–27.

Zuckerman, B., Kahane, I., and Himmelhoch, S. (1979). *Exp. Parasitol.* **47**, 419–424.

2

Effects of Aging on Enzymes

MORTON ROTHSTEIN

Division of Cell and Molecular Biology
State University of New York at Buffalo
Buffalo, New York 14260

I. Methods of Aging Nematodes 29
  A. Introduction 29
  B. Use of FUdR to Maintain Synchronous Cultures of *Turbatrix aceti*
    and *Caenorhabditis elegans* 30
  C. Use of Repeated Screening Procedures 34
  D. Use of Heat Treatment 35
  E. Comparison of *Turbatrix aceti* and *Caenorhabditis elegans* for Use
    in Aging Studies 36
II. Alteration of Enzymes in Aged Nematodes 36
  A. Criteria for Detection of Altered Enzymes 36
  B. Heat Sensitivity Patterns of Altered Enzymes 38
  C. Comparison of Nematode and Mammalian Enzymes 38
  D. Unaltered Enzymes 39
  E. Comparison of Results with Mammalian Enzymes 39
  F. Changes Found in Old Enzymes 39
  G. Possible Mechanisms of Formation of Altered Enzymes 40
  H. Protein Turnover in Nematodes 41
  I. Conclusions 43
III. Evaluation of the Nematode Model System 44
  References 45

I. METHODS OF AGING NEMATODES

A. Introduction

The practical use of free-living nematodes for aging studies must overcome two problems. Not only must cultures begin with organisms of a similar age, but

NEMATODES AS BIOLOGICAL MODELS
VOLUME 2

also reproduction must be prevented, or synchrony will be lost and the aging cultures will become contaminated with newborn organisms and will eventually revert to typical "mixed" cultures. The problem of obtaining uniformly small organisms to start cultures has been solved by the use of screens for *Turbatrix aceti* (Hieb and Rothstein, 1975) and the hatching of isolated egg masses for *Caenorhabditis elegans* (Mitchell *et al.*, 1979; Gandhi *et al.*, 1979). Subsequent reproduction is prevented by the use of the DNA inhibitor fluorodeoxyuridine (FUdR) (Gershon, 1970; Hieb and Rothstein, 1975; Hosono, 1978; Mitchell *et al.*, 1979), or by culturing the organisms at elevated temperatures (Hieb and Rothstein, 1975). Another practical method for the aging of *T. aceti* is the use of a repeated screening process that periodically removes small (young) organisms from the aging cultures (Rothstein and Sharma, 1978).

B. Use of FUdR to Maintain Synchronous Cultures of *Turbatrix aceti* and *Caenorhabditis elegans*

Gershon (1970) first developed the use of inhibitors of DNA synthesis for preventing reproduction during the aging of *T. aceti*. This author used a partially defined medium containing liver extract, according to the original formulation of Rothstein and Cook (1966). Various concentrations of FUdR and aminopterin were tested. Both were effective in providing a greater than 90% inhibition of normal DNA synthesis at concentrations of 100 μg/ml. Hydroxyurea also was reported to be effective at 500 μg/ml. The author settled upon using FUdR at a concentration of 100 μg/ml, which resulted in a 96% inhibition of DNA synthesis. Uridine was added concomitantly (100 μg/ml), presumably to inhibit formation of fluorouracil. The final procedure selected was as follows. Cultures of *T. aceti* were passed through columns of glass beads from which the smallest (youngest) organisms emerged first. These young organisms were then aged in the presence of FUdR and uridine, each at 100 μg/ml. At this concentration, the organisms increased in size to 1.4 mm in length in 20 days, no reproduction occurred, and the life span of individual organisms was identical to that of untreated organisms.

Hieb and Rothstein (1975) subsequently modified the FUdR procedure of Gershon (1970) to accommodate larger scale preparations and the use of a chemically defined medium (Rothstein, 1974). Large numbers of young worms were separated by a screening process using stainless steel Robusta filter cloth (Tobler, Ernst and Traber, Inc./Kressilk Products, Inc.). The mesh size that proved to be most suitable was 720/140, with a 12- to 15-μm nominal pore size and an 18- to 21-μm absolute pore size. The screen was soldered between two bronze rings precisely machined for flatness. The diameter of the screens used at first was 47 mm, fitting exactly into a disposable plastic Sterifil Filtration System (Millipore

Filter Corp.). A silicone rubber gasket acted as a seal between the bottom ring and the apparatus. The whole unit is autoclavable.

Later, larger screens were built into a 100-mm ring machined to replace the filter support of the Sartorius Membrane Filter Apparatus (SM 163 08). The exposed screen area has a diameter of 80 mm. The use of this particular screening material is very effective in that two washes equal to the culture volume result in a yield of about 70% of the nematodes small enough to pass through the filter. Screening material made with other weaves is less effective. The maximal size of the organisms passing through the screen is 1100 μm. However, the vast majority is in the range of about 370–700 μm. The oldest organisms that pass through the screen are estimated, based on the size of worms grown individually, to be 4 days old. The bulk of the screened worms are much younger. Typical mixed cultures of T. aceti contain few large organisms, only about 15% being above the screen cutoff size.

There is a 2- to 3-day safety factor after screening before reproduction begins. Therefore, the screened cultures are allowed to incubate for at least 24 hr before FUdR is added. A suitable concentration, using the defined medium, is 100 μg/ml of FUdR plus 150 μg/ml of uridine. In Roux bottles, each containing 50 ml of medium at 30°C, the organisms grow to an average size of 1400 μm in 13 days, the majority being between 1200 and 1600 μm. Thus, the aging worms, on an average, grow reasonably well in the presence of FUdR. In untreated cultures, because of rapid reproduction, most of the population is "young"—that is, below 800 μm in size.

After testing various types of vessels, it was found that the most efficient use of incubator space was obtained by the use of stationary Roux bottles, each containing 50 ml of sterile culture medium. An unfortunate limitation of the system is that aging populations cannot be maintained at more than 20,000–30,000 worms/ml. If a greater concentration is used, the organisms simply die back to this level. On the assumption that the difficulty was related to lack of oxygen (although this is not a problem in non-aging cultures), the cultures were shaken during aging. The results were unsatisfactory since the young organisms, although they survive well, do not grow in size. Turbatrix aceti, under normal conditions, does not reproduce well when shaken (Hieb and Rothstein, 1975).

Hieb and Rothstein (1975) found that the half-life of cultures grown in Roux bottles is 26 days, in good agreement with the 25 days reported by Gershon (1970). For individually grown organisms, the former authors reported a value of 50 days compared to the 25 days estimated by Gershon (1970). The difference may lie in the use of different media (defined medium versus one supplemented with liver extracts). It is of interest that the antioxidant α-tocopherolquinone has been reported to increase markedly the life span of C. briggsae and T. aceti (Epstein and Gershon, 1972).

Recently, several papers have appeared dealing with the use of *C. elegans* for aging studies. Klass (1977) examined a number of biological and environmental factors. Included were the effects of parental age, temperature, and food concentration (*E. coli*) on life span and egg production. Croll *et al.* (1977) examined age-related behavioral changes and size changes of axenically and monoxenically grown organisms. Hosono (1978) grew *C. elegans* on *E. coli* and found that growth was normal in the presence of 2 m*M* FUdR + 2 m*M* uridine, but was suppressed at higher concentrations. The treated nematodes laid eggs at a somewhat reduced rate, but these failed to hatch.

In a more comprehensive study, Mitchell *et al.* (1979) observed that addition of FUdR, if delayed until *C. elegans* reaches sexual maturity, permits normal postmaturational development but completely inhibits egg hatching. Thus, the organisms can be maintained in synchronous culture by the addition of the reagent to the cultures. Initial synchrony was obtained by isolating egg masses from axenic cultures or from monoxenic cultures. It was found that in axenic culture, 0.4 m*M* FUdR applied 3 days after hatching had no effect on the growth of the organisms. If added after 2 days, a modest inhibition resulted. In a monoxenic medium, parallel results were obtained at 2–5 days (no effect on growth) and 1.8 days (modest effect). If applied to newborn larvae, severe stunting resulted. These findings are in agreement with the work of Kisiel *et al.* (1972), who found that FUdR severely inhibited the growth of *C. briggsae* and *T. aceti,* when applied to first-stage larvae.

Mitchell *et al.* (1979) found that the mean survival rate of *C. elegans* in axenic culture increased by the addition of 0.4 m*M* FUdR, being 29 versus 21 days for the controls. Croll *et al.* (1977) reported a value of 17.6 days, but these authors used a crude medium rather than the nearly defined medium used by Mitchell *et al.* (1979). In a monoxenic medium using *E. coli,* mean survival was similar with or without FUdR, being 16 and 15 days, respectively, at 20°C. Klass (1977) reported 14.5 days and Croll *et al.* (1977) 12.3 days for cultures grown on *E. coli.* The use of 6 m*M* FUdR did not dramatically shorten the life span of nearly mature organisms, and even 50 m*M* FUdR only shortened the mean life span to 12.2 days. The results reported for axenic culture are corrected for *endotokia matricida*—the hatching of eggs while still in the adult. The use of FUdR eliminates this condition, since eggs do not hatch.

D. H. Mitchell and J. Santelli (private communication) report that 0.030 m*M* FUdR may be added directly to newly hatched larvae of *C. elegans* grown in monoxenic culture. The organisms develop into fully functional adults that grow to normal length and produce normal numbers of sperm, oocytes, and fertilized eggs in a normal-appearing reproductive system. However, the eggs do not hatch, thus permitting synchronous cultures. Lower levels of FUdR (0.01 m*M*) were less effective. At 0.1 m*M*, the organisms matured more slowly and were slightly smaller. At 0.4 m*M*, growth was sharply inhibited.

Unfortunately, even at 0.030 mM, FUdR cannot be added to freshly hatched *C. elegans* in axenic medium since abnormal developmental effects are observed unless the addition is delayed until near maturity.

The above results explain the differences in the findings of Kisiel *et al.* (1972) and Hieb and Rothstein (1975). The former authors found that FUdR caused damage and prevented the adequate growth of larvae of *C. briggsae* and *T. aceti*, whereas Hieb and Rothstein (1975) observed good growth of *T. aceti*. The method devised by Hieb and Rothstein (1975) starts with organisms that are on average much older than first-stage larvae. In addition, their procedure uses an extra 1- to 2-day delay before FUdR is added. Thus, the drug is being applied to mature or nearly mature nematodes. Moreover, it is clear from the above results with axenic versus monoxenic cultures and from media containing live versus heat-killed *E. coli* (Hosono, 1978), that the composition of the culture medium plays a substantial role in the effects of FUdR. Thus, the differences in results can be readily reconciled. On the other hand, Gershon (1970) treated small larvae of *T. aceti* with FUdR and reported good growth of the organisms. The contradiction between these results and those of Kisiel *et al.* (1972) remains unexplained.

Gandhi *et al.* (1979) described a mass aging technique for *C. elegans* in axenic medium. The organisms were first roughly synchronized by passage through a column of glass wool, which yielded only small larvae. These organisms produced a culture containing more than 100,000 eggs/ml. The eggs were obtained by flotation on 20% sucrose and provided newly hatched larvae which were separated from unhatched eggs by filtration. FUdR at 0.025 mM was added just before the organisms reached maturity. Eggs were subsequently laid but did not hatch. According to Gandhi *et al.* (1979), this procedure can provide synchronous worms in the gram range. An inconvenience is the changing of the medium every 3–6 days to maintain its original pH. However, the authors indicate that this procedure is not necessary.

Assuming that FUdR affects aging nematodes in a relatively subtle manner, it may still play a dramatic role in altering intermediary metabolism. After all, inhibition of DNA synthesis, even if that is the only effect of the drug, may have major metabolic consequences. Thus, one is justified in asking if nematodes aged in the presence of FUdR can serve as a suitable system for biochemical study. In this regard, two important metabolic changes have so far been reported to occur in aging *T. aceti:* (1) an age-related alteration of enzymes (Reiss and Rothstein, 1975; Gupta and Rothstein, 1976; Sharma *et al.*, 1976) and (2) an age-related slowing of protein turnover (Prasanna and Lane, 1979; Sharma *et al.*, 1979). Both studies yielded nearly identical results when carried out with *T. aceti* aged in the presence of FUdR and aged by a repeated screening process (see Section I,C) which periodically removes newly born organisms from aging cultures. That such a delicately balanced process as protein turnover should be unaffected by the drug is indeed reassuring. Nonetheless, studies carried out with FUdR-aged nematodes should have an external control. Either aging manifestations discovered in FUdR-treated worms should be present in other animals—e.g., altered enzymes in rodents—or they should be shown to be present in nematodes aged by the screening procedure.

C. Use of Repeated Screening Procedures

Repeated or continuous screening of nematode cultures can be used to remove newly born organisms and thus maintain the synchrony of aging cultures. Tilby and Moses (1975) devised a procedure in which *C. elegans* larvae were grown to adult size and then transferred to the upper surface of a stainless steel wire mesh with a pore size that retains only adult organisms (nominally 13 μm). An enclosed apparatus was designed in such a way that the medium level came just up to the mesh. Small organisms that passed through the mesh were washed away by a flow of medium. Provision was made for occasional flooding of the screen with fresh medium to help newly born organisms pass more readily through the screen. The number of young worms contaminating the aging cultures on top of the screen appears to be very small. The 50% survival was reported as 58 days. This is much longer than for *C. elegans* grown in bulk cultures in flasks (Gandhi *et al.*, 1979).

The problem with this screening procedure is that it is not amenable to large-scale production of nematodes. The maximum population density attained was 280 adults/ml. Moreover, Mitchell *et al.* (1979) point out that a substantial number of young *C. elegans* hatch within the uteri of adults and are thus not removed by filtration. Biochemical studies in which old organisms are homogenized could be distorted by the presence of young tissue.

A screening procedure devised by Rothstein and Sharma (1978) for aging *T. aceti* appears to work well and provides adequate amounts of tissue for biochemical work. Mass cultures grown in defined axenic medium were screened through stainless steel filter cloth machined to fit into a 100-mm Sartorius Membrane Filter Apparatus (see Section I,B). The mesh size was 670 × 120, permitting passage of organisms up to 1.2 mm in length. The organisms remaining on top of the screen were transferred to flasks containing fresh medium. These organisms were subsequently screened every 3 or, at most, 4 days, thus removing young worms born in the interim. The size distribution of worms on top of the screen shifted continuously toward larger sizes and by the seventh and eighth screenings consisted nearly entirely of large worms (approximately 1.5–2.0 mm in length). Checks made late in the process (sixth screening and later) of the size distribution of the cultures just before screening revealed a pattern consisting of only small and large worms with no intermediate sizes. This result showed that worms transferred to the medium from the previous screening were not contaminated with young organisms, since these would show up as intermediate sizes in the 3-day interim before the next screening.

The procedure does not permit one to pinpoint the exact ages of the screened organisms, since the initial screening isolates organisms with a range of ages. However, approximations can be made based on the following reasoning. Initial cultures before the first filtration are permitted to increase in population for 14 days. Thus, the organisms old enough to be retained by the first screening cannot be older than this (the small number in the original inoculum can be ignored). Since the screens will not retain organisms of less than 6 days of age, and given the normal population distribution for cultures of *T. aceti* in the axenic medium (Hieb and Rothstein, 1975), the great bulk of the starting organisms can be assumed to be between 7 and 12 days old. Each screening adds 3 days to their age, less the effect of the oldest organisms on top of the screens dying more frequently than the youngest ones. Thus, after six screenings, the organisms probably average 25–27 days rather than the theoretically possible 30 days. Of course, experiments using the first screening and the sixth or seventh screening will provide a basis for a "young" versus "old" comparison without concern for actual age. Those organisms passing through the screens are clearly very young.

It is interesting to note that rates of enolase synthesis and degradation in worms screened seven times closely match the rates in worms aged for 20–22 days in the presence of FUdR (Sharma *et al.*, 1979). Moreover, the alteration found in enolase was similar in *T. aceti* from the sixth screening and from 25-day-old organisms (Rothstein and Sharma, 1978).

The yield of organisms after eight screenings is 1.7% of those retained after the first screening (0.5% of the population originally screened). Since the starting cultures contain about 350,000 worms/ml, and 1000 ml of medium was utilized in the experiments reported, about 1.6×10^6 aged organisms were obtained. However, the procedure can readily handle much larger numbers.

D. Use of Heat Treatment

Hieb and Rothstein (1975) used increased incubation temperatures to prevent reproduction of *T. aceti*, and thus maintained the synchrony of cultures during aging. After an initial screening to obtain a young inoculum (see Section I,B), the organisms were placed in 50 ml of medium in Roux bottles for 24 hr at 30°C and then incubated at 36°C. The inhibiting temperature is critical. At 35°C, reproduction occurs; at 37°C, the organisms die quite rapidly. The mean survival of *T. aceti* at 36°C is only 16 days and the organisms grow quite slowly, most attaining less than about 1300 μm by 12 days. It is interesting to note, however, that even after prolonged periods at 36°C, cultures of *T. aceti* will reproduce if the temperature is dropped to 30°C.

Mitchell *et al.* (1979) note that the use of a heating procedure is unsuitable for

C. elegans since that organism develops abnormally at temperatures required to inhibit reproduction.

As an alternative to the use of FUdR or screening procedures, the use of elevated temperatures would seem to offer little potential. Nonetheless, altered enzymes, shown to be present in *T. aceti* aged in the presence of FUdR, were also found in *T. aceti* aged by the high-temperature procedure (Reiss and Rothstein, 1975).

E. Comparison of *Turbatrix aceti* and *Caenorhabditis elegans* for Use in Aging Studies

There are a number of practical advantages in the use of *T. aceti* for aging studies. For example, the culture medium has a starting pH of 3.2, which discourages growth of bacteria even if contamination should occur. Moreover, the culture medium is completely defined, whereas for maximal growth *C. elegans* requires the addition of a fraction from soy peptone (Rothstein and Coppens, 1978), or must be grown in the presence of *E. coli*. Another factor of great practical value is that very few dead worms are found in the old cultures of *T. aceti*. One may speculate that perhaps because of the acid pH of the medium dead organisms simply dissolve on account of lysosomal action. Other factors favoring *T. aceti* are the substantial amount of biochemical information available and the existence of a routine procedure for aging the organisms, which has been successfully used for over 7 years. Furthermore, an alternative aging procedure, based on a repeated screening technique rather than use of FUdR (see Section I,C), is available for the organisms, thus permitting a check on biochemical results obtained from worms aged in the presence of the drug. Negative factors for *C. elegans* are the possible formation of dauer larvae in liquid medium and the strong tendency for eggs to hatch within the adult. This condition is reported by Mitchell *et al.* (1979) to afflict 50–90% of the population in axenic medium versus 10% in monoxenically grown organisms.

On the other hand, a great deal of genetic, neurobiological, and behavioral research is currently being performed on *C. elegans*. Therefore, if future research on aging is to be linked to any of these areas, this organism should provide a valuable model system.

II. ALTERATION OF ENZYMES IN AGED NEMATODES

A. Criteria for Detection of Altered Enzymes

When altered enzymes are found in old organisms, the most obvious change in properties is a loss of specific activity. However, measurements of activity cannot simply be carried out in crude homogenates because a lowered activity per milligram of protein or per milligram of DNA does not distinguish between two explanations for the results: either fewer enzyme molecules are present or there is a lowered catalytic ability but a normal number of molecules. One way around this problem is to prepare antiserum to the young enzyme in question and determine, by immunotitration, the amount of cross-reacting material present per unit of enzyme activity in preparations from both young and old organisms. If more antiserum is utilized per unit of activity for old preparations, the assumption can be made that more enzyme molecules are required per unit of activity. One may argue that young antiserum may merely react less effectively with old enzyme, but that result itself is proof that the old enzyme is altered. A second procedure is to compare the specific activity of pure enzyme preparations from young and old animals. Preferably both procedures should be used.

In addition to a change in specific activity, several altered enzymes show a difference in sensitivity to heat. Of the enzymes tested, isocitrate lyase (IL) (Reiss and Rothstein, 1975), enolase (Sharma *et al.*, 1976), and aldolase (Reznick and Gershon, 1977) from the free-living nematode *T. aceti*, and superoxide dismutase (SOD) (Reiss and Gershon, 1976) and phosphoglycerate kinase (PGK) (Sharma *et al.*, 1980) from rat liver and muscle, respectively, show substantial changes in this parameter. An indication of the danger of comparisons made in crude homogenates is the fact that aldolase in homogenates of both young and old *T. aceti* show identical sensitivity to heat (Zeelon *et al.*, 1973), whereas in pure preparations, the old enzyme is more stable.

Other properties that have been shown to differ in young and old enolase from *T. aceti* are stability to heat, uv spectra, CD spectra, and the rate of degradation by the proteolytic enzyme trypsin (Sharma and Rothstein, 1978a).

No significant differences have been observed for any young–old pairs of enzymes by gel electrophoresis, isoelectric focusing, immunodiffusion, or by measurements of K_m or K_I. Enough evidence is available in the literature to make it conclusive that these procedures do not detect altered enzymes in aged animals. However, differences are found in some of these characteristics in certain mutant forms of enzymes such as glucose-6-phosphate dehydrogenase, where there is a change in amino acid sequence (Kahn *et al*, 1976).

As mentioned above, a test for the existence of an altered enzyme is the titration of old enzyme with antiserum prepared to young enzyme. If the old enzyme is catalytically less active than the young enzyme, it will require more molecules per unit of activity, and this situation will be reflected in the immunotitration. If the antigenic sites differ, more young antiserum may be required to precipitate old enzyme, even if the specific activities of the young and old enzymes are the same. In the case of enolase from *T. aceti*, both young and old antisera cross-react with young and old enzyme, but the young–young and old–old pairs bind most effectively (Sharma and Rothstein, 1978b).

B. Heat Sensitivity Patterns of Altered Enzymes

The heat sensitivity patterns of old enzymes are highly individual. For example, old nematode PGK is unchanged; old IL shows a biphasic pattern, one component being heat sensitive and the second component being parallel to the line given by young enzyme; old enolase shows a biphasic pattern, but in this case, both components are more rapidly denatured than the young form of the enzyme. Old rat liver SOD gives similar results. Nematode aldolase, which also gives a biphasic pattern, is unusual in that the old enzyme is the more stable form.

We see, then, that although all altered enzymes so far discovered have a lower catalytic ability (indeed, this measurement is the means of detecting such enzymes), heat sensitivity shows several patterns. These patterns may be interpreted as follows. If the old form of an enzyme has a lowered specific activity, it must result from one of three situations: (a) the old enzyme consists of molecules, all of which are altered; (b) it consists of a mixture of normal plus inactive molecules; or (c) it consists of molecules with a range of activities. Based upon the heat sensitivity patterns, one can surmise that old IL consists of partially active molecules (the heat sensitive component) and normal molecules. Old enolase and aldolase consist of two populations of altered molecules, as does old rat liver SOD. PGK consists of normal plus inactive molecules. The latter would not affect the heat sensitivity pattern but would lower the specific activity of the preparation. Alternatively, the altered enzyme may retain its original heat stability.

C. Comparison of Nematode and Mammalian Enzymes

Five enzymes have been purified from young and old *T. aceti* and the properties of the young–old pairs were compared. The enzymes are IL (Reiss and Rothstein. 1975), PGK (Gupta and Rothstein, 1976), aldolase (Reznick and Gershon, 1977), enolase (Sharma *et al.,* 1976), and triosephosphate isomerase (TPI) (Gupta and Rothstein, 1976). Of these, the first four show substantial differences when isolated from young versus old organisms, but TPI is unchanged.

Thus far, three pure enzymes have been reported to be altered in old rodents. These are SOD from old rat liver (Reiss and Gershon, 1976), aldolase from mouse muscle and heart (Chetsanga and Liskiwskyi, 1977) and PGK from rat muscle (Sharma *et al.,* 1980). The characteristics of these altered enzymes differ from the related young enzymes in ways similar to those isolated from *T. aceti.* Old SOD, for example, shows a biphasic heat curve and lowered specific activity, and requires an increased amount of antiserum per unit of activity. Mouse muscle and heart aldolases show differences in heat stability and specific activity.

D. Unaltered Enzymes

In addition to the altered enzymes mentioned above, several enzymes have been reported to be unchanged with age. In mammalian species these consist of creatine kinase (CK) and aldolase from human muscle (Steinhagen-Thiessen and Hilz, 1976), aldolase from rat liver (Anderson, 1976; Weber *et al.*, 1976), and enolase from rat muscle and rat liver (Rothstein *et al.*, 1980). The first three enzymes were examined in crude homogenates, but immunotitration with anti-sera prepared to pure enzyme showed no change in the specific activity of the old preparations. Although the respective heat sensitivity patterns of aldolase from young versus old human muscle and aldolase from young versus old rat liver were not identical, the evidence supports the conclusion that there is no substantial age-related change. Unequivocal proof that old enzymes are not necessarily altered has been obtained (Rothstein *et al.*, 1980) by comparing pure enolase from rat muscle and liver: For both enzymes, the young versus old pairs were identical. In *T. aceti*, pure TPI (Gupta and Rothstein, 1976) isolated from young and old organisms was shown to be identical by a number of criteria, including specific activity, heat sensitivity, and immunotitration.

E. Comparison of Results with Mammalian Enzymes

From the above, it may be concluded that the phenomenon of age-related changes in proteins in nematodes and mammals is analogous. That is, altered enzymes and unaltered enzymes have been reported in both systems. The alteration of properties reported for mammalian enzymes appears to parallel that reported in nematodes: they have a lowered specific activity, are more sensitive to heat, and require more antiserum per unit of activity. K_m and behavior on polyacrylamide gels or after isoelectric focusing are unchanged. Since the nature of the changes, at least so far as has been determined, appears to be the same for nematode and rodent enzymes, it seems likely that the mechanism of alteration is identical in both types of organisms. Thus, it should be advantageous to use the nematode as a model system for studying this aspect of aging. Undoubtedly the use of the organisms for study of other age-related changes will also be rewarding.

F. Changes Found in Old Enzymes

It is clear from the isoelectric focusing of several enzymes (SOD, aldolase, enolase) that there is no change in the charge of altered enzymes. The technique should readily detect the loss of a single amide group. Therefore, deamidation, phosphorylation, or acylation cannot be involved in the alteration process. The amino acid analysis of enolase shows no methylated derivatives of lysine or histidine, so that methylation is unlikely to be a factor. Proteolysis of the enzyme has been ruled out by showing that the N-terminal and carboxy terminal groups of young and old enolase are identical. Moreover, the four cysteine residues are all reduced in both young and old enolase, so that oxidation of SH groups is not responsible for the observed alterations.

The conformation of old enolase differs from that of the young enzyme as shown by CD spectra, uv spectra and the titration of exposed tryptophan and tyrosine groups (Sharma and Rothstein, 1978a). The spectral differences disappear in 6-M guanidine, suggesting that there are no differences in sequence, at least in areas near the aromatic amino acid residues. Although the possibility of errors has been the subject of much debate, as will be seen below, the evidence weighs heavily against this idea insofar as proteins are concerned.

G. Possible Mechanisms of Formation of Altered Enzymes

Studies of altered enzymes, along with other related information, have placed sharp limits on the options available to explain the formation of altered enzymes. The idea of an "error catastrophe" is hardly credible, and even the possibility that proteins become altered because of one or two amino acid substitutions is highly unlikely. If errors in transcription (e.g., faulty RNA polymerase) or translation (due to faulty synthetases or ribosomes) occur, then all proteins should possess errors in sequence. Clearly, TPI from *T. aceti* as well as enolase from rat muscle and rat liver are unaltered. In addition, there is good evidence that several enzymes are unaltered in old organisms (see Section II,D). Also, immunotitrations have shown that six different enzymes are unaltered in leukocytes from old (more than 80 years) compared to young (20–30 years) subjects (Rubinson *et al.*, 1976). Similarly, SOD in the red cells of old humans is identical to that in cells of young subjects (Joenje *et al.*, 1978). These results prove that old people can make error-free proteins in systems requiring continuous synthesis. This information, along with that available from the studies of young versus old enolase, virtually eliminates sequence errors as the reason for altered enzymes.

The alternative to sequence changes as the cause of altered enzymes is the postsynthetic modification of proteins. From the studies carried out with enolase, changes caused by proteolysis, methionine sulfoxide formation, deamidation, phosphorylation, methylation, or acylation are not responsible for the alterations observed (Sharma and Rothstein, 1978a). Only a few logical possibilities remain. One of these is that old enzymes are merely subtly denatured forms of young enzymes. Rothstein and co-workers (Reiss and Rothstein, 1974; Rothstein, 1975; Rothstein, 1977; Sharma and Rothstein, 1978b) have proposed a theory based upon this concept and have provided considerable evidence in support of the idea. The proposal suggests that the formation of altered enzymes results from the slowing of protein turnover in old organisms. Under these conditions, enzymes would have an increased ''dwell time'' in the cells and there would be ample opportunity for a subtle denaturation to take place, either enzymatically or kinetically. With both protein synthesis and proteolysis slowed, intermediate products (altered enzymes) would accumulate.

According to the hypothesis, young enzyme is synthesized normally and subsequently becomes altered. Direct evidence in favor of this hypothesis is the finding that young enolase from *T. aceti* can be converted in vitro to a product similar to old enolase (Sharma and Rothstein, 1978a). If young enolase is passed three times through DE-52, an ion exchanger, the specific activity of the recovered enzyme drops from 1200 units/mg to 700 units/mg. Moreover, the heat lability pattern changes from monophasic to biphasic, although it does not exactly match that of old enolase. Antiserum prepared to the thrice-columned young enzyme has a greater affinity for old than for young enzyme (Sharma and Rothstein, 1978b). These three changes in properties do not prove that young enzyme has become identical to old enzyme. However, they certainly show that enolase can be changed by treatment in vitro to something similar to old enzyme without sequence changes or exposure to agents that bind covalently.

Further evidence for the hypothesis is the finding that inactive enolase, which is absent in homogenates of young *T. aceti,* steadily accumulates with increasing age of the organisms (Sharma and Rothstein, 1978b). Moreover, the repeated passage of young enolase through DE-52 results in cross-reacting material emerging from the column in the position where inactive enolase is found after the chromatography of old enzyme. The inactive material from recolumned young enolase is immunologically identical to the inactive material found in old (but not young) homogenates (Sharma and Rothstein, 1978a).

Thus, at least for enolase from *T. aceti,* there is considerable evidence supporting the idea that old enzymes result from a slight denaturation of young enzymes. Results obtained with other enzymes, altered or unaltered, can also be explained on this basis (Rothstein, 1977; Sharma and Rothstein, 1978b).

The hypothesis rests upon the idea of a slowed protein turnover in aged organisms. Although the situation has not been clearly resolved in higher animals, protein turnover, as shown in the next section, does slow dramatically with age in *T. aceti.*

H. Protein Turnover in Nematodes

The study of protein turnover in free-living nematodes serves three purposes: (1) support or negation of the hypothesis that protein turnover is involved in the formation of altered enzymes; (2) determination of the effect of age on protein turnover in nematodes; and (3) the provision of basic information involving this aspect of nematode metabolism.

Only recently have reports appeared dealing specifically with protein metabolism in free-living nematodes. The organisms appear to make excellent models for this type of study. Incorporation of [^{35}S]methionine into protein is linear for at least 30 hr. After either a 12- or 24-hr labeling period, washing the organisms and placing them in medium containing 40 mM unlabeled methionine results in rapid cessation of isotope incorporation (2–4 hr). Decay of protein activity is essentially linear thereafter (Prasanna and Lane, 1979).

The rate of degradation of soluble proteins in T. $aceti$ labeled with [^{35}S]methionine decreases dramatically with age. The half-life increased from 25 hr in young (2-day-old) to 269 hr in old (20-day-old) organisms, with intermediate values at intermediate ages. K_D values dropped tenfold, from 28.3 $\times$ 10^{-3}/hr to 2.58 $\times$ 10^{-3}/hr over the same range of ages. A similar age-related drop in K_D values was observed for nematodes aged at 36°C. As noted in Section I,D, this temperature prevents reproduction and permits aging of the nematodes without using FUdR. Thus, these results suggest that the use of this drug in the medium is not the cause of the age-related slowing of protein turnover.

A more detailed study of protein turnover was carried out by Sharma et $al.$ (1979) using [^{3}H]leucine. In addition to studies with soluble proteins, the synthesis and degradation rates of pure enolase in aging T. $aceti$ were determined. The enzyme was precipitated from labeled homogenates with antiserum prepared to the pure enzyme. It was shown that 95% of the label in the resulting precipitate was in enolase.

The synthesis and degradation rates for both soluble proteins and the single protein enolase were determined in T. $aceti$ aged by both the use of FUdR and the repeated-screening procedure (Section I,C) which avoids the use of the drug. From both aging techniques, the respective synthesis rates and K_D values were similarly depressed with age, proving that use of FUdR had little or no effect on protein metabolism. The experiments again demonstrated a dramatic age-related slowing of protein turnover. FUdR was used to ascertain that the half-life of soluble protein for young T. $aceti$ (5 days) was 73 hr and for enolase was 58 hr. The values increased to 163 and 161 hr, respectively, at 22–30 days of age. The large spread of ages in the older organisms was due to the extended sampling time. The values for the half-life of soluble proteins and enolase after three screenings were 76 and 109 hr, respectively, and for eighth-screen organisms, 228 and 226 hr, respectively. The ages of the screened worms are not precisely determinable, but are estimated to be 10–16 days for the third screening and about 15 days older for the later screening.

The fact that pure enolase is degraded at the same rate as soluble proteins in aged T. $aceti$ makes it clear that the altered enzyme present in old organisms is not degraded with greater rapidity than are "normal" proteins. Enzymes altered by the incorporation of amino acid analogs have been shown to yield a faster-than-normal rate of degradation.

The above results clearly indicate that old *T. aceti* synthesize and break down very little protein. Although the absolute values are subject to problems of reutilization of labeled amino acids, this consideration cannot be of major consequence, since the half-life of aldolase was calculated to increase from 40 hr in young animals to 200 hr in old organisms, using cycloheximide (Zeelon *et al.*, 1973) to prevent reincorporation of labeled amino acids.

The slowing of protein synthesis in *T. aceti* is in agreement with results from the relatively few in vitro experiments that have been performed in cell-free preparations from aging rats and mice. These preparations show a reduced ability to incorporate labeled amino acids into protein. Microsomal preparations from old mice (30 months) were less able to incorporate [^{14}C]phenylalanine into protein, compared to similar preparations from 5-month-old animals (Mainwaring, 1969). Similar results were obtained with labeled valine (Layman *et al.*, 1976). Lack of synthetases or of tRNA is not involved, as cell sap from both young and old animals worked equally well in conjunction with microsomal or ribosomal preparations from young animals. Moreover, [4,5-^{3}H]leucyl-tRNA gave less incorporation of label into protein when incubated with old compared to young ribosomal preparations (Britton and Sherman, 1975).

If the above information can be applied to the nematode system, the observed slowing of protein turnover is due not to elements such as lack of tRNA or synthesizing enzymes but to problems associated with the ribosomes. In this respect, it is interesting that Wallach and Gershon (1974) reported that there is a substantial reduction in the number of polysomes in old *T. aceti*.

It is clear that in the nematodes the regulation of protein turnover, whatever its mechanism, causes slowing of both synthesis and breakdown in a consistent manner with increasing age. Unfortunately, there is available at present only inferential evidence for an age-related slowing of protein turnover in higher animals.

I. Conclusions

The use of FUdR to obtain synchronized cultures of *T. aceti* initially gave rise to concern that the use of the drug could be responsible for the changes in enzymes observed in old organisms. However, the fact that altered enzymes had been found in old rodents strongly suggested that aging and not FUdR was the causative agent. Direct proof that FUdR was not responsible was reported by Rothstein and Sharma (1978). They obtained aged nematodes by the repeated screening process (Section I,C) and isolated pure enolase from these organisms. The enzyme proved to be altered, possessing a specific activity of 700 units/mg versus 1200 units/mg for young enzyme. Moreover, immunotitration with antiserum prepared to young enolase showed that the old enzyme had a lowered activity per unit of antiserum. The results matched exactly those obtained with pure old enolase from organisms aged in the presence of FUdR. The additional finding that protein turnover in *T. aceti* is similar whether or not FUdR is used (Section II,H) provides assurance that use of the drug is innocuous insofar as current metabolic studies are concerned.

III. EVALUATION OF THE NEMATODE MODEL SYSTEM

Although the nature of the change in altered enzymes and the mechanism of their formation are not yet unequivocally established, studies utilizing *T. aceti* have contributed substantially toward the solution of these problems. In fact, there are a number of distinct advantages to using the nematode system for studies of aging, providing, of course, that the parameters being investigated are widespread effects of aging and not limited to the nematodes. Among the attributes of the nematode system is the ability to grow the organisms in routine fashion in axenic, defined medium. Thus, one may manipulate individual components (e.g., vitamins or amino acids) or examine the effects of hyponutrition. Addition of isotopes to the medium permits metabolic studies such as those on protein turnover discussed above. Moreover, the physical environment can be manipulated so that the effects of temperature or low oxygen tension on the life span or metabolism can be determined. Added to this is the fact that the organisms are easily handled, either in bulk culture or as individuals.

A great advantage of using nematodes for aging studies is of course their relatively short life span. Old *T. aceti* can be obtained in 22–28 days, providing great flexibility in designing experiments. By starting aging cultures of nematodes at short intervals, one may have the security of knowing that if an experiment fails or needs revision, more aged organisms will shortly be at hand. Contrast this with the problems of maintaining a supply of 30-month-old rats or mice, especially in the case of investigators who cannot maintain their own large colonies of aging rodents. Even more valuable is the capability of utilizing nematodes in an aging continuum, rather than at two, or at most three, points (e.g., 12 and 24 months), as is typical for the rodent systems.

The availability of a clear medium makes the isolation of the organisms by filtration or centrifugation a simple task. Isolation of pure enzymes has proved particularly easy compared to working with rats or mice, probably because there are fewer contaminating proteins in the nematodes. Certainly, ample amounts of tissue can be obtained without difficulty. For mixed cultures, which are used to evaluate procedures, about 2 g of organisms are obtained per 100 ml of medium. Yields of old organisms are smaller because of the limitation of using only

20,000 worms/ml during FUdR treatment. The yield of old worms is about 200 mg/100 ml of medium, taking into consideration the attrition caused by aging. However, adequate if not generous amounts of tissue can be obtained using a single large incubator.

There are, of course, disadvantages to the nematode system. One is that homogenates of whole organisms contain material from several cell types. However, the bulk of tissue is composed of muscle cells, with intestinal cells making up most of the remainder. Other cell types such as nerve cells are minor constituents. An advantage, from one point of view, is that the cells in adult organisms are all postmitotic. This fact guarantees that in an old nematode all the cells are old.

One must be certain that the aging characteristic being investigated in nematodes also applies to higher forms. Otherwise, one might be studying aging in nematodes rather than aging as a common phenomenon. The limitation of nematodes in this sense is not yet clear. Altered enzymes, the formation of age pigments, and perhaps protein turnover as studied in *T. aceti* all seem to apply to rodents. It is to be hoped that future studies dealing with areas such as gene regulation and regulation of protein synthesis and protein degradation will be found to be equally related, making the nematode system of continuing value for studies of aging. In this regard, considerable potential for aging studies exists in the rapidly expanding use of *C. elegans* for genetic, behavioral, and neurobiological studies.

In conclusion, it may be stated that the nematode model system has already played an important role in the study of aging and gives every promise of making valuable contributions in the future.

REFERENCES

Anderson, P. J. (1976). *Can. J. Biochem.* **54**, 194–196.

Britton, G. W., and Sherman. F. G. (1975). *Exp. Gerontol.* **10**, 67–77.

Chetsanga, C. J., and Liskiwskyi, M. (1977). *Int. J. Biochem.* **8**, 753–756.

Croll, N. A., Smith, J. M , and Zuckerman, B. M. (1977). *Exp. Aging Res.* **3**, 175–189.

Epstein, J., and Gershon, D. (1972). *Mech. Ageing Dev.* **1**, 257–264.

Gandhi, S., Santelli, J., Mitchell, D. H., Stiles, J. W., and Sanadi, D. R. (1979). *Mech. Ageing Dev.* (in press).

Gershon, D. (1970). *Exp. Gerontol.* **5**, 7–12.

Gupta, S. K., and Rothstein, M. (1976). *Biochim. Biophys. Acta* **445**, 632–644.

Hieb, W. F., and Rothstein, M. (1975). *Exp. Gerontol.* **10**, 145–153.

Hosono, R. (1978). *Exp. Gerontol.* **13**, 369–374.

Joenje, H., Frants, R. R., Arwert, F., and Eriksson, A. W. (1978). *Mech. Ageing Dev.* **8**, 265–267.

Kahn, A., Esters, A., and Habedank, M. (1976). *Hum. Genet.* **32**, 171–180.

Kisiel, M. J., Nelson, B., and Zuckerman, B. M. (1972). *Nematologica* **18**, 373–384.

Klass, M. R. (1977). *Mech. Ageing Dev.* **6**, 413–429.

Layman, D. K., Ricca, G. A., and Richardson, A. (1976). *Arch. Biochem. Biophys.* **173**, 246–254.
Mainwaring, W. I. P. (1969). *Biochem. J.* **113**, 869–878.
Mitchell, D. H., Stiles, J. W., Santelli, J., and Sanadi, D. R. (1979). *J. Gerontol.* **34**, 28–36.
Prasanna, H. R., and Lane, R. (1979). *Biochem. Biophys. Res. Commun.* **86**, 552–559.
Reiss, U., and Gershon, D. (1976). *Eur. J. Biochem.* **63**, 617–623.
Reiss, U., and Rothstein, M. (1974). *Biochem. Biophys. Res. Commun.* **61**, 1012–1016.
Reiss, U., and Rothstein, M. (1975). *J. Biol. Chem.* **250**, 826–830.
Reznick, A. Z., and Gershon, D. (1977). *Mech. Ageing Dev.* **6**, 345–353.
Rothstein, M. (1974). *Comp. Biochem. Physiol. B* **49**, 669–678.
Rothstein, M. (1975). *Mech. Ageing Dev.* **4**, 325–338.
Rothstein, M. (1977). *Mech. Ageing Dev.* **6**, 241–257.
Rothstein, M., and Cook, E. (1966). *Comp. Biochem. Physiol.* **17**, 683–692.
Rothstein, M., and Coppens, M. (1978). *Comp. Biochem. Physiol. B* **61**, 99–104.
Rothstein, M., and Sharma, H. K. (1978). *Mech. Ageing Dev.* **8**, 175–180.
Rothstein, M., Coppens, M., and Sharma, H. K. (1980). *Biochim. Biophys. Acta, in press.*
Rubinson, H., Kahn, A., Boivin, P., Schapira, F., Gregori, C., and Dreyfus, J. C. (1976). *Gerontology* **22**, 438–448.
Sharma, H. K., and Rothstein, M. (1978a). *Biochemistry* **17**, 2869–2876.
Sharma, H. K., and Rothstein, M. (1978b). *Mech. Ageing Dev.* **8**, 341–354.
Sharma, H. K., Gupta, S. K., and Rothstein, M. (1976). *Arch. Biochem. Biophys.* **174**, 324–332.
Sharma, H. K., Prasanna, H. R., Lane, R., and Rothstein, M. (1979). *Arch. Biochem. Biophys.* **194**, 275–282.
Sharma, H. K., Prasanna, H. R., and Rothstein, M. (1980). *J. Biol. Chem., in press.*
Steinhagen-Thiessen, E., and Hilz, H. (1976). *Mech. Ageing Dev.* **5**, 447–457.
Tilby, M. J., and Moses, V. (1975). *Exp. Gerontol.* **10**, 213–223.
Wallach, Z., and Gershon, D. (1974). *Mech. Ageing Devel.* **3**, 225–234.
Weber, A., Gregori, C., and Schapira, F. (1976). *Biochim. Biophys. Acta* **444**, 810–815.
Zeelon, P., Gershon, H., and Gershon, D. (1973). *Biochemistry* **12**, 1743–1750.

3

Nematodes as Nutritional Models

J. R. VANFLETEREN

Laboratorium voor Morfologie en Systematiek der Dieren
Rijksuniversiteit Gent
Ledeganckstraat 35, B-9000 Gent, Belgium

I. Introduction 47
II. Nutritional Requirements in the Basal Medium 48
   A. Chemically Defined Basal Medium 48
   B. Nutritional Requirements in Defined Basal Medium 49
III. The Growth Factor 56
IV. Nutrient Absorption 63
V. Possible Contaminants of Axenic Culture 72
VI. Nutrition and Development 74
VII. Conclusions and Prospects 76
   References 77

I. INTRODUCTION

The use of nematodes as model organisms for the study of metazoan organization assumes a detailed knowledge of all facets of their biology. This chapter covers the nutritional aspects of the nematode model as revealed by studies on *Caenorhabditis briggsae* and *Caenorhabditis elegans* and a few other nematode species that have been established in axenic culture. The term axenic, which means "free of other organisms" was proposed by Dougherty (1960) in preference to other terms such as "sterile," "aseptic," and "pure" that would be rather ambiguous in this respect. Since most of the nematodes under discussion are bacteria feeders and some of them fungal-feeding species, it is essential that axenic culture conditions are rigidly maintained when studying nematode requirements.

This chapter appraises the present state of nematode nutrition, and refers

47

particularly to those nematodes that are being currently used as model organisms for basic research of behavior, development, and aging. Detailed information on the culture methods and specific diets of these and other nematodes in axenic culture may be found in a previous review (Vanfleteren, 1978).

II. NUTRITIONAL REQUIREMENTS IN THE BASAL MEDIUM

A. Chemically Defined Basal Medium

It is clear that the proper media for studying nematode nutrition should be entirely composed of known chemicals. Only then could adding or omitting components be evaluated with relative ease. Unfortunately, holidic media, as Dougherty (1960) called them, were not available until recently, and much pioneer work was done with chemically undefined (oligidic) media or with media consisting of a defined basal portion and a less defined supplement—e.g., a tissue extract, so-called meridic media.

The search for a suitable defined basal medium led to the formulation in 1963 of a mixture designated EM1 (Sayre *et al.*, 1963), which was based on the amino acid ratios found in *Escherichia coli*, the best associate of *C. briggsae* in monoxenic culture. Two alterations thereof are still in common use. One is medium No. 75, which is a modification of Sayre's medium by Buecher *et al.* (1966) and has become widely known as *C. briggsae* Maintenance Medium (CbMM, see Table I). The other is Rothstein's (1974) modification of Sayre's medium. It is designated EM and differs from CbMM by containing DL-alanine (15 mg/liter) and lacking pantetheine, calcium folinate, and niacinamide. Myoinositol and choline are more concentrated in this medium (by approximately tenfold). Pyridoxal phosphate (7.5 mg/liter) replaces the three forms of the vitamin used in CbMM.

Although both CbMM and EM were developed primarily for the continuous culture of *C. briggsae*, they have proved equally suitable for the cultivation of several other nematode species in axenic culture, among which the rhabditids *Caenorhabditis elegans, Panagrellus redivivus,* and *Turbatrix aceti,* the tylenchids *Aphelenchus avenae* and *Aphelenchoides rutgersi,* and the insect-parasitic nematodes *Neoaplectana carpocapsae* DD-136 and *Neoaplectana glaseri* are the most thoroughly studied species (Table II). The rhabditid nematodes listed are free-living nematodes that feed on bacteria. *Aphelenchus avenae* and *A. rutgersi* are fungal-feeding organisms. They are often, perhaps incorrectly (Nicholas, 1975), called plant-parasitic nematodes, mainly because they bear a stylet and belong to a group that is almost entirely composed of stylet-feeding, plant-parasitic species. In contrast, *N. carpocapsae* and *N. glaseri* are true parasites, but their infectious larvae transmit bacteria to their insect hosts, which are killed, and the nematodes feed on the bacteria and the putrefying tissues of the insects.

TABLE I
Caenorhabditis briggsae **Maintenance Medium**[a,b]

Component	mg/liter	Component	mg/liter
$CaCl_2 \cdot 2\,H_2O$	220.50	L-Serine	788.00
$CuCl_2 \cdot 2\,H_2O$	6.50	L-Tyrosine	272.00
$MnCl_2 \cdot 4\,H_2O$	22.20	L-Phenylalanine[d]	180.00
$ZnCl_2$	10.20	Glutathione, reduced	204.00
KH_2PO_4	1225.50	N-Acetylglucosamine	15.00
Potassium citrate $\cdot\,H_2O$	486.00	Cyanocobalamine	3.75
$Fe(NH_4)_2\,(SO_4)_2 \cdot 6\,H_2O$	58.80	Folinate (Ca)	3.75
Magnesium citrate $\cdot$ 5 H_2O	915.00	Niacinamide	7.50
KOH	—[c]	Pantetheine	3.75
D-Glucose	1315.00	Pantothenate (Ca)	7.50
L-Arginine	975.00	Pyridoxal phosphate	3.75
L-Histidine	283.00	Pyridoxamine 2 HCl	3.75
L-Lysine-HCl	1283.00	Pyridoxine HCl	7.50
L-Tryptophan	184.00	Riboflavine-5'-PO_4(Na) $\cdot$ 2 H_2O	7.50
L-Phenylalanine	623.00	Thiamin HCl	7.50
L-Methionine	389.00	p-Aminobenzoic acid[e]	7.50
L-Threonine	717.00	Biotin[e]	3.75
L-Leucine	1439.00	Niacin[e]	7.50
L-Isoleucine	861.00	Pteroylglutamic acid[e]	7.50
L-Valine	1020.00	DL-Thioctic acid[e]	3.75
L-Alanine	1395.00	Choline dihydrogen citrate	88.50
L-Aspartic acid	1620.00	Myo-inositol	64.50
L-Cysteine HCl $\cdot\,H_2O$	28.00	Adenosine-3'-(2')-phosphoric acid $\cdot\,H_2O$	365.00
L-Glutamate (Na) $\cdot\,H_2O$	550.00	Cytidine-3'-(2')-phosphoric acid	323.00
L-Glutamine	1463.00	Guanosine-3'-(2')-$PO_4Na_2 \cdot H_2O$	363.00
Glycine	722.00	Uridine-3'-(2')-phosphoric acid	324.00
L-Proline	653.00	Thymine	126.00

[a] Reproduced, with permission, from *Annu. Rev. Phytopathol.* **16**, 131–157. © 1978, by Annual Reviews, Inc.

[b] After Buecher *et al.* (1966). Commercially available from GIBCO Bio-Cult, Paisley, Scotland or GIBCO, Grand Island, New York.

[c] As needed for adjustment to pH 5.9 ±0.1.

[d] In replacement for an amount of tyrosine, because of the solubility limitations of the latter.

[e] Made up first as concentrated stock solutions in 5% (wt/vol) triethanolamine.

Thus, although specific adaptations have developed, the nematodes under study appear not too far removed nutritionally from an unspecialized diet and this may be the basis of common nutritional properties.

B. Nutritional Requirements in Defined Basal Medium

Both CbMM and EM contain many constituents that are not really needed. Progress has been made toward a minimum essential medium, however, which

TABLE II

Comparative conditions for Axenic Culture of Nematodes[a]

Species	Reproductive habit[b]	Basal medium[c]	Growth-promoting supplement[c]	Temperature (°C)	References
C. briggsae, C. elegans	H	SP (3–4%) + YE (1–3%)	HLE (5–15%)	20–23	Tomlinson and Rothstein (1962); Cryan et al. (1963); Vanfleteren and Roets (1972); Rothstein (1974)
T. aceti	B	EMSP + S SP (3–4%) + YE (1–3%) + HAc (2–3%) EMS + HAc (4%)	Hb or Mb (0.5 mg/ml) HLE (5–15%) Mb or Hb (0.5 mg/ml)	28–30	Rothstein and Cook (1966); Vanfleteren and Roets (1972); Rothstein (1974); Hieb and Rothstein (1975)
P. redivivus (=P. silusiae)	B	SP (3–4%) + YE (1–3%) EMS + ethanol (1%) CbMM	HLE (10–15%) Mb (0.5 mg/ml)	25	Cryan et al. (1963); Rothstein and Cook (1966); Rothstein and Coppens (1978)
A. avenae	P	CbMM	Horse serum (10%) + CEE (25%) CEE (10%)	23	Hansen et al. (1970) Myers et al. (1971)
A. rutgersi	P	SP (3%) + YE (2%) CbMM M-10	Horse serum (10%) + CEE (25%) CEE (20%)	23	Buecher et al. (1970b) Thirugnanam (1976)
N. carpocapsae DD-136	B	SP (3%) + YE (3%) + dextrose (0.7%)	HLE (5–10 mg/ml)	20–25	Buecher and Hansen (1971)
Czechoslovakian	B	CbMM	HLE (10 mg/ml)	20	Hansen et al. (1968)
N. glaseri	B	SP (3%) + YE (3%)	FYE (20%, dialyzed)	23	Lower and Buecher (1970)

[a] Additional data in Vanfleteren (1978).

[b] H, Self-fertilizing hermaphrodite with occasional males; B, bisexual; P, parthenogenetic.

[c] Some of the culture media listed have been compiled from many data, omitting minor components. Thus the few references given are only indicative. CEE, Chick embryo extract, aseptically prepared supernatant from chick embryos (about 10 days old) in water, clarified by low centrifugation; EMS, chemically defined basal medium (EM) containing 50 μg/ml sterols; EM may be replaced by CbMM; EMSP, EM minus all amino acids, but including 4% soy peptone; FYE, fresh yeast extract, supernatant from baker's yeast homogenate in 0.05 M potassium phosphate buffer pH 7, sterilized by filtration through Millipore filter 0.30 μm; HAc, glacial acetic acid; Hb, hemoglobin; HLE, heated liver extract, the supernatant from liver homogenate in water, that has been heated at 53°C for 6 min, centrifuged and sterilized by Seitz filtration; Mb, myoglobin; M-10, chemically defined basal medium, designed for the axenic culture of A. rutgersi; S, 50 μg/ml sterols; SP, soy peptone; YE, yeast extract.

by definition would be composed of nutritional requirements only. According to Sayre *et al.* (1963), nutrients that cannot be synthesized at all may be called absolute nutritional requirements, whereas those that are synthesized, but at rates insufficient to permit normal unimpaired development, are designated limited nutritional requirements. The occurrence of the latter group precludes tracer experiments using radioactive precursors as a useful tool for ascertaining nutrients as nonessential, and conclusive proof should always await deletion assays. Still other nutrients are not required, but their inclusion in the culture medium may have a beneficial effect.

1. Assay Methods

One way to evaluate nematode growth involves the accurate determination of the maturation time or F_1 generation time (F_1 time), which is expressed as the number of days or hours required for freshly hatched juveniles to mature and to produce new juveniles. This type of assay (larval assay method) was developed by Dougherty *et al.* (1959) for use with *C. briggsae*. Briefly, adult nematodes are transferred through a series of dishes containing buffer solution and are allowed to lay eggs in the final solution. These hatch and are immediately transferred to small culture tubes containing a thin film of test medium (e.g., 0.25–0.3 ml in 10 × 75-mm culture tubes). Obviously, one juvenile per culture tube would yield inconsistent results since a varying percentage of the newly hatched juveniles are "laggards," i.e., unable to grow in permissive conditions (Dougherty *et al.*, 1959). On the other hand, too many juveniles per tube would produce a positively biased response since the method measures the first appearance of offspring, which of course are generated by the fastest-developing individuals. Statistical analysis suggested that three juveniles per tube would give reliable results with hermaphrodites such as *C. briggsae* (Lower *et al.*, 1966). More juveniles (e.g., 10) are recommended for work with bisexual species to ensure the presence of both sexes.

Eggs produced in dense populations of *C. briggsae* and *C. elegans* stick together, and hatching is strongly delayed. Washing of these so-called egg masses by serial transfer through buffer solution is a very simple task, and hatching will be complete after standing overnight. Numerous first-stage juveniles will result, and these may be used for assay (Vanfleteren, 1973). A still faster but less reliable method (see the above comments) involves the direct transfer of washed egg masses into the culture medium. The observed maturation rate must then be reduced by 6–12 hr to compensate for the time required for hatching.

Limiting media may stimulate the original juveniles to mature and to produce offspring, but will fail to support further development. Thus a 2- to 3-week observation period is advisable, and the appearance of a second generation (F_2 time), the number of progeny after a specified time interval, and the production of egg masses by the resulting population eventually provide additional information on the adequacy of the medium.

An alternative method involves the periodic counting of a sample that has been taken from growing cultures and is known as mass-culture assay (Rothstein and Nicholas, 1969). This method gives reliable results when applied to repeated subcultures. This is necessary to preclude false positive results because of initial growth at the expense of stored reserves (Rothstein, 1974). This method may be used preferentially to check media for their capability of sustaining dense populations (Rothstein and Coppens, 1978).

2. Vitamin Requirements

The omission of thiamin, riboflavine, folic acid (pteroylglutamic acid), niacinamide, pantothenic acid, and pyridoxine from basal medium supplemented with 10% chick embryo extract (CEE) impaired normal growth of *C. briggsae,* suggesting that these B vitamins are essential nutrients (Dougherty *et al.,* 1959; Nicholas *et al.,* 1962). The requirement for folic acid also became evident upon addition of its antagonist aminopterin (Dougherty and Hansen, 1957; Vanfleteren and Avau, 1977). Folic acid is generally required for thymine biosynthesis (Vanfleteren and Avau, 1977) and for the breakdown of histidine (Lu *et al.,* 1974). A biotin requirement was demonstrated similarly by adding avidin, a protein that inhibits the uptake of biotin, but whose effect is reversed by excess biotin (Nicholas and Jantunen, 1963). Nicholas (1975) found no evidence for a vitamin B_{12} requirement in an otherwise complete medium, but Lu *et al.* (1976) demonstrated that this vitamin and folic acid were required when homocysteine was replaced for methionine in the basal medium. As mentioned above, Rothstein (1974) reduced the redundancy in the vitamin complement and deleted calcium folinate and pantetheine with no ill effects.

3. Amino Acid Requirements

The systematic omission of amino acids from CbMM supplemented with low levels of yeast ribosomes showed that the following amino acids were required to sustain the reproduction of *C. briggsae:* arginine, histidine, lysine, tryptophan, phenylalanine, methionine, threonine, leucine, isoleucine, and valine (Vanfleteren, 1973). These were classified as dietary essentials, although limited biosynthesis of most of these amino acids had been demonstrated by adding labeled precursors (Nicholas *et al.,* 1960; Rothstein and Tomlinson, 1961, 1962; Rothstein, 1963, 1965). Alanine, aspartic acid (erroneously mentioned as asparagine in the original paper), cysteine, glutamic acid, glutamine, glycine, proline, serine, and tyrosine could be deleted with impunity (Vanfleteren, 1973). It was recently confirmed by deletion experiments using a fully chemically defined medium (J. R. Vanfleteren, unpublished observations) that these amino acids are nonessential for both *C. briggsae* and *C. elegans* (Table III). It is apparent from Table III that *C. elegans* (N_2) grows consistently more slowly than *C. briggsae* under identical conditions. Interestingly, the same trend is always observed in axenic culture regardless of the medium used. Perhaps the rate of some metabolic activity is lower in *C. elegans*. Eventually, this might explain why multiple deletion of nonessential amino acids delays reproduction in *C. briggsae* by 2–3 days, but has no noticeable effect on *C. elegans* (Table III).

Nutritional physiology is one of the characteristics that differentiates *C. elegans* and *C. briggsae* in addition to genetic, biochemical, and morphological characteristics (Friedman *et al.*, 1977). However, the hermaphrodites of both species cannot be distinguished from each other by simple inspection, and much confusion has arisen about the identity of the species under study. In order to clear up this embarassing situation as much as possible, several laboratory strains of putative *Caenorhabditis briggsae* and their respective identification are listed in Table IV. In this chapter, the author has adopted the notation *C. "briggsae"* referring to work reportedly done with *C. briggsae* but actually dealing with *C. elegans*.

The amino acid requirements of *N. glaseri* (Jackson, 1973) are very similar to those of *C. briggsae*, probably reflecting a comparable feeding habit. *Aphelenchoides rutgersi*, which feeds on fungi, is nutritionally further removed from *Caenorhabditis*. The appropriate amino acid levels were different and reproduction failed when isoleucine, leucine, methionine, phenylalanine, threonine, his-

TABLE III

Effect of Omitting Nonessential Amino Acids from Holidic Medium[a] on the Generation Time of *Caenorhabditis briggsae* and *Caenorhabditis elegans*[b]

Amino acid omitted	pH of the medium	*C. elegans*	*C. briggsae*
L-Alanine	5.5	6.9 ± 0.1	5.3 ± 0.3
L-Aspartic acid	5.4	9.3 ± 0.3	5.1 ± 0.1
L-Cysteine	5.4	9.6 ± 0.7	5.3 ± 0.3
L-Glutamine	5.4	8.8 ± 0.4	5.0 ± 0.3
L-Glutamic acid	5.3	9.8 ± 0.3	5.9 ± 0.4
Glycine	5.5	9.4 ± 0.6	6.5 ± 0.2
L-Proline	5.5	7.4 ± 0.2	5.4 ± 0.2
L-Serine	5.4	7.4 ± 0.4	5.3 ± 0.1
L-Tyrosine	5.5	9.0 ± 0.6	5.3 ± 0.1
Control (complete medium)	5.3	7.9 ± 0.2	4.9 ± 0.1
MEM (all nonessential amino acids omitted)	5.4	7.8 ± 0.3	7.3 ± 0.4

[a] CbMM + 50 μg/ml sterols dissolved in 10 mg/ml Tween 80 + 50 μg/ml acid-precipitated hemin chloride.

[b] Mean values ± SE of four replicates. The generation time is measured as the time (in days) from the inoculation of three newly hatched juveniles to the production of first-stage juveniles of the next generation.

TABLE IV

Identity of *Caenorhabditis* Strains Currently Used as Nutritional Models

Supposed identity	Source	Species identification
C. briggsae	Department of Nutritional Sciences, University of California, Berkeley	C. elegans[a]
C. briggsae	Division of Cell and Molecular Biology, State University of New York, Buffalo	C. elegans[a]
C. briggsae	Laboratory of Experimental Biology, University of Massachusetts, East Wareham	C. briggsae[a]
C. briggsae	Laboratory of Morphology and Systematics, State University of Ghent, Ghent (Belgium)	C. briggsae[b]
C. elegans N₂	MRC Laboratory of Molecular Biology, Cambridge University Medical School, Cambridge (England)	C. elegans

[a] Species identification performed by Friedman *et al.* (1977).
[b] Derived from the above strain (Laboratory of Experimental Biology, East Wareham) in 1970.

tidine, tryptophan, lysine, and cysteine, cystine, or cysteic acid were deleted from the basal medium. When serine, glycine, alanine, proline, hydroxyproline, valine, and arginine were deleted, reproduction occurred at significantly reduced rates and was not affected only when one of either glutamine or glutamic acid and asparagine or aspartic acid was deleted (Balasubramanian and Myers, 1971; Myers and Balasubramanian, 1973).

4. Lipid-Related Requirements

The nematodes tested show a nutritional requirement for sterol. This was first demonstrated in *N. carpocapsae* DD-136 (Dutky *et al.*, 1967) and subsequently in *C. "briggsae," C. elegans* (both Bristol and Bergerac strains), *T. aceti,* and *P. redivivus* (Hieb and Rothstein, 1968; Cole and Dutky, 1969; Lu *et al.*, 1977). Strong evidence indicates that *Nippostrongylus brasiliensis,* which is a parasite of mammals, requires sterol compounds in its diet (Bolla *et al.*, 1972). Very small amounts of sterol will effectively sustain nematode growth and reproduction: about 0.1 μg/ml for *T. aceti* and slightly more than 1 μg/ml for *C. "briggsae."* One or two generations may be completed in a sterol-free medium at the expense of stored reserves. This might indicate that sterol biosynthesis is not completely defective in these organisms; however, it appears that the required levels of sterols cannot be synthesized by these organisms (see Volume 1, Chapter 5).

Both CbMM and EM contain no sterols because these media were devised before a requirement for sterols was demonstrated. This can be met by adding sterols (e.g., cholesterol, 7-dehydrocholesterol, ergosterol, β-sitosterol, stigmasterol) individually or as a mixture at 50 μg/ml sterols. The author has adopted the notations CbMMS and EMS in reference to sterol-supplemented CbMM and EM, respectively. Sterols are usually added from stock concentrations made in Tween 80 to overcome solubility problems. Tween comprises about 1.3 mg/ml in the final medium (Hieb and Rothstein, 1968; Rothstein, 1974; Lu *et al.*, 1978).

Recently Lu *et al.* (1978) reported on the growth-promoting activity of several lipid-related compounds in C. *"briggsae."* The addition of sodium oleate or sodium stearate to a medium consisting of CbMMS, 1.3 mg/ml Tween 80, and 50 μg/ml cytochrome *c* boosted the population growth to considerably higher levels. About 0.1 mg/ml was required but 1 mg/ml gave better results. As expected, Tween 80 (polyoxyethylene sorbitan monooleate, 5–20 mg/ml) and Tween 85 (polyoxyethylene sorbitan trioleate, 1–10 mg/ml) proved effective, but were less active than oleate on a molar basis. J. R. Vanfleteren (unpublished observations) repeated the experiment with sodium oleate at 1 mg/ml using *C. briggsae* and *C. elegans* and obtained no beneficial effect. The culture medium contained 10 mg/ml of Tween 80, provided along with sterols which were routinely dissolved at 5 mg/ml in undiluted Tween. Thus, the final amount of oleate available to the nematodes was probably supraoptimal. The fact that Vanfleteren (1974) obtained better growth of *C. briggsae* in holidic medium consisting of CbMMS and 50 μg/ml of acid-precipitated hemin than other investigators (Pinnock *et al.*, 1975) may be partly due to the beneficial effect of the higher level of Tween 80.

Biosynthesis of oleic acid and polyunsaturated fatty acid has been demonstrated in *C. "briggsae," T. aceti,* and *P. redivivus* (Rothstein and Gotz, 1968; Rothstein, 1970), and since a source of oleate (Tween 80) is inevitably included in a defined medium, it is not clear at present whether fatty acids are either nonessential stimulatory nutrients or are to be classified as nutritional essentials. Nature is more versatile than any classification, however, and the effect of fatty acids may well be somewhere in between. The free-living marine nematode *Rhabditis marina* was reported to be still more dependent on fatty acid supplementation of the culture medium than were the soil-inhabiting nematodes studied (Tietjen and Lee, 1975).

In addition to fatty acids, several short-chain carbon compounds reportedly proved stimulatory, e.g., ethanol (4 mg/ml), *n*-propanol (4 mg/ml), and potassium acetate (5 mg/ml) (Lu *et al.*, 1978). This is not unexpected since it has been shown that free-living nematodes possess the glyoxylate pathway to synthesize other cell components using 2-carbon fragments (Rothstein and Mayoh, 1964a, 1964b, 1965, 1966). *Turbatrix aceti* probably utilizes 2-carbon units in preference to glucose. The addition of ethanol (Hansen and Cryan, 1966) to the culture medium was beneficial, but acetic acid gave much better results (Rothstein and Cook, 1966). The effect was not just one of lowering the pH, since HCl was unable to replace the acetic acid. The latter compound is now being currently added at 3-4% in the culture medium of *T. aceti*. Ethanol added at 1% or acetic acid at 0.5% improved the medium for *Panagrellus silusiae* (Rothstein and Coppens, 1978). *Panagrellus silusiae* has been synonymized with *P. redivivus* (Hechler, 1971).

5. Carbohydrate and Heme Requirements

Lack of carbohydrate reduced reproduction in *C. briggsae*, *C. elegans*, and a few other nematode species markedly, but *P. redivivus* was less affected (Hansen and Buecher, 1970; Petriello and Myers, 1971; Roy, 1975). Glucose and trehalose were utilized by all the species tested, but other sugars gave more inconsistent results. Since the experimental media used permitted only poor growth of the nematodes, it is the view of the author that firm conclusions should not be drawn from these data.

Besides sterols, a heme source must be added to CbMM to make it an effective medium for the indefinite culture of *Caenorhabditis* and other nematodes as well. Heme is the essential nutrient that was supplied along with growth factor, but whose identity remained unknown for many years. Holidic medium consisting of CbMM or EM to which sterols and hemin chloride are added will support the growth and reproduction of *C. briggsae*, *C. elegans*, *T. aceti* (plus 3-4% acetic acid), and *P. redivivus* through repeated subculture, though the final population levels obtained remain relatively low (Vanfleteren, 1974; Rothstein and Coppens, 1978). At 250 μg/ml, hemin provides for good growth of *C. briggsae* and *C. elegans* (J. R. Vanfleteren, this chapter). Lower concentrations of hemin, e.g., 50 μg/ml, may give inconsistent results unless supplied in the proper precipitated form (Vanfleteren, 1974; Rothstein and Coppens, 1978).

Chemically defined medium is expensive and time-consuming to prepare. For the simple purposes of maintaining stock cultures and growing axenic nematodes for biochemical study, the amino acid and vitamin complement of the basal medium may be replaced by soy peptone (SP) and yeast extract (YE), respectively, with sterols being provided as impurities in these preparations (Table II).

III. THE GROWTH FACTOR

All efforts to obtain reproduction of *C. briggsae* in defined medium failed unless an unknown requirement for maturation and reproduction was supplied. This requirement was originally met by some heat-labile proteinaceous component (destroyed by autoclaving), which was best provided by certain tissue extracts, e.g., chick embryo extract (CEE) and liver extract (LE) (Dougherty *et al.*, 1950). Of these, LE was easily prepared and attempts were soon made to identify the active material by biochemical purification methods. Biological activity was recovered in the fraction coming down at 20-40% (LPF-B) saturation with ammonium sulfate, but most activity was retained in the fraction precipitated at between 40 and 60% saturation (LPF-C) (Dougherty, 1953). This fraction became even more active when diluted in defined medium and frozen prior to assay. The effect slowly disappeared after thawing but was restored by refreezing, and it was thought that it might interact with some medium components (Hansen *et al.*, 1961).

Further fractionation of LPF-C on hydroxyapatite yielded a partially purified growth factor, which also showed a pattern of freeze activation (Sayre *et al.*, 1961). In addition, a similar effect could be obtained in a number of ways, including dialysis against low phosphate, lowering the pH of the medium to 5.6, addition of the sucrose polymer ficoll, and heating at 37°C for 24 hr or longer (Hansen *et al.*, 1964; Buecher *et al.*, 1966).

Subfractions derived from lamb liver growth factor had identical amino acid compositions but different specific activities, and all of these could be freeze-activated. This result led Sayre *et al.* (1967) to believe that growth factor existed in multiple aggregated states of a smaller subunit and that freeze activation was related to conformational changes within the molecule.

However, it appeared difficult to reconcile this idea of a specific protein with the observation that growth factor could be prepared from a variety of vertebrate tissues and resides in other growth-promoting preparations such as CEE and bacteria as well. It soon became clear that the essential process of activation involved the controlled precipitation of nutritionally active material rather than conformational changes within the active molecules. Thus Buecher *et al.* (1969b) found that activation generally resulted in increased turbidity and that a complete medium containing ficoll-activated growth factor lost its activity upon filtration. This was further substantiated by experiments in which particles, e.g., latex particles, were added to the culture medium, demonstrating that most biological activity was confined to the precipitable portion of the medium as long as this contained growth factor.

As mentioned above, it has been known for a long time that growth-factor preparations were heat-labile, i.e., autoclaving under standard conditions (121°C for 15 min) destroyed the growth-promoting activity, although this result was variable. Likewise, *E. coli* cells became ineffective upon autoclaving (Nicholas *et al.*, 1959). It was observed, however (Hieb and Rothstein, 1970), that defined medium alone did stimulate the development of newly hatched juveniles to the size of advanced fourth-stage juveniles or small adults as long as it contained autoclaved *E. coli* cells and that numerous offspring were produced when as little as 10 μg/ml of a protein fraction from liver (HLE; for preparation see Table II, Footnote *c*) was added to this medium. It thus appeared that autoclaved *E. coli* cells retained part of their activity and that an essential nutrient required for reproduction was provided by HLE. Attempts at further purification of the active material led to the isolation of a red fraction, which showed the absorption spectrum of oxymyoglobin or oxyhemoglobin. Subsequent experiments proved that pure hemoglobin, myoglobin, cytochrome *c,* and hemin chloride effectively replaced the liver fraction, whereas nonheme proteins such as β-lactoglobulin, bovine serum albumin, and soluble casein were not effective (Hieb *et al.,* 1970).

Eventually the conclusion was reached that growth factor supplied at least three components to the medium: (1) a sterol (since CbMM contained no sterols and still a requirement for sterols had been demonstrated; see Section II, B,4), (2) a heme moiety, and (3) a third component that remained a mystery since it was somehow clearly involved in the activation effect and appeared heat labile and heat stable at the same time. For example, it was well known that the active material in HLE was heat labile, but myoglobin, which corresponded best to the active fraction isolated, proved heat stable (120°C for 8 min). Hemoglobin and cytochrome *c* were also heat stable.

New data became available when Buecher and Hansen (1969) reported on the growth-promoting activity of freshly prepared yeast extract. An alternative to liver growth factor called yeast factor was prepared from baker's yeast as the fraction precipitating at 70% ammonium sulfate saturation. Yeast extract and yeast factor could be activated in much the same way as could liver preparations, but were most active at pH 7.4. Most interestingly, fresh yeast extract yielded a very active pellet when centrifuged at 30,000 *g* for 30 min; the pellet consisted of partially denatured ribosomes (Lower and Buecher, 1970; Buecher *et al.,* 1969a, 1970a). Clearly, the above results were again antithetical to the idea of a specific protein structure and emphasized the importance of particulate material.

Bearing this knowledge in mind, Buecher *et al.* (1970c, 1971) supplied several commercially available proteins as far as possible in precipitated form to defined medium containing hemin chloride. Among these, human γ-globulin was the most effective, and α- and β-globulin, β-lipoprotein, bovine serum and egg albumin, soy peptone, tobacco mosaic virus protein, and pork insulin were slightly active. However, even the best substitute was much less active than tissue extracts. Nevertheless, these experiments proved that autoclaved *E. coli* cells were not actually required as a supplement to a defined medium that contained suitable, pure proteins. They suggested that, in a sense, biological activity was correlated with the rate of uptake of these nutrients by the nematodes.

From the above, it is clear that the active component provided by growth factor, besides sterols and heme, cannot be a discrete protein. Whether it was a question of a specific amino acid balance or of high levels of amino acids seemed unlikely when considering the fact that very small amounts of protein (10 μg/ml or less) of the most active preparations effectively supported the growth and reproduction of *C. briggsae* in defined medium. Could the active component be a small peptide occurring in all active proteins? The answer was reached by Hansen *et al.* (1971), who found that pure carbohydrate such as glycogen stimulated the growth and reproduction of *C. briggsae* in a hemin-containing medium.

Sterols were provided as impurities of the glycogen, but there was no protein to account for the biological activity. *Caenorhabditis briggsae* was maintained in this medium (CbMM + 100 mg/ml glycogen + 8 μg/ml hemin) through 10 serial subcultures, and mass cultures of *C. elegans* under continuous aeration (Buecher and Hansen, 1971) increased 600-fold in a peptone–yeast basal medium supplemented with glycogen at 10 mg/ml and hemin at 10 μg/ml. Most interestingly, biological activity was consistently confined to the precipitated portion of the medium.

Thinking further along this line, Vanfleteren (1974) postulated that growth factor contributed only two dietary essentials, sterols and heme, to the basal medium, and that the third portion of growth factor was nothing more than a convenient vector of the heme component, which must be supplied in particulate form. If so, any material should prove effective as long as it was available in particulate form and effectively carried heme, either as part of a molecule (heme protein) or by binding hemin chloride.

Vanfleteren (1975a) confirmed that ribosomal preparations from yeast, *E. coli,* mammalian liver, and HeLa cells showed high biological activity, which he ascribed to contaminating cytochromes that became adsorbed to the ribonucleoprotein (RNP) particles during the isolation procedure. Accordingly, RNP particles from *E. coli* washed with 1 *M* salt to remove contaminating protein were inactive. There was probably no heme protein present in the preparation from pea seedlings, which might explain its inactivity. This preparation appeared to become somewhat activated upon the addition of hemin chloride.

Several pure heme proteins were subsequently tested. Assuming that the presence of particulate heme-containing material would improve uptake by the nematodes, whereas excessive denaturation might rapidly have the adverse effect, Vanfleteren (1975b) experimented with a number of methods for the controlled precipitation of heme proteins. As a result, pure cytochrome *c* was complexed with the phospholipid phosphatidylethanolamine; catalase, myoglobin, and hemoglobin were precipitated at their respective isoelectric points, and horseradish peroxidase was assayed as the peroxidase–antiperoxidase immunoprecipitate. All of the above preparations stimulated the growth and reproduction of *C. briggsae* in a sterol-containing defined basal medium, although cytochrome *c* proved less effective than the other heme proteins. Gentle precipitation methods failed with yeast L-lactate dehydrogenase and cytochrome *c* reductase, two crude cytochrome-containing preparations. They were assayed after acid precipitation but were inactive, perhaps because heme was removed during the activation treatment. These preparations became very active upon adding hemin chloride.

In addition to the above experiments, Vanfleteren (1976) also tested several artificial heme-containing supplements. In one series of experiments, iron-binding proteins were carefully precipitated in the presence of hemin chloride on the assumption that hemin might be adsorbed on the protein through interaction with the ferric moiety of hemin. It was realized that the subsequent precipitation treatments might well interfere with that (eventual) interaction, however, so the aggregated material was diluted with and assayed in CbMMS to which 10 μg/ml of hemin chloride was added to enhance the eventual adsorption of hemin on the surface of the particles formed. Ferritin (precipitated at its isoelectric point or with acetone), apoferritin (acetone precipitated), and transferrin (tested as transferrin–hemin–antitransferrin immunoprecipitate) proved biologically very active in the presence of hemin. Bovine serum albumin, egg white, and conalbumin were assayed as coagulates made after hemin was added. These stimulated the growth and reproduction of newly hatched juveniles in CbMMS, but failed to support the continued culture of *C. briggsae*.

Similar experiments involved the use of nonproteinaceous carriers of hemin and led to the development of a chemically completely defined medium for the cultivation of *C. briggsae* (Vanfleteren, 1974). Egg lecithin was dispersed ultrasonically and sterilized by membrane filtration (0.30 μm pore size). Hemin chloride was allowed to interact with the phospholipid particles before addition to CbMMS. At concentrations of 1 mg/ml lecithin and 50 μg/ml hemin, first-stage juveniles of *C. briggsae* matured and gave rise to two or three generations. Alternatively, hemin chloride was precipitated from stock solution (1 mg/ml in 0.1 *N* KOH) at a low pH and was resuspended in CbMMS at pH 5.0–5.4. Holidic medium thus prepared supported the growth and reproduction of *C. briggsae* through repeated subculture, although the population densities obtained remained low. CbMMS to which 3% acetic acid and 50 μg/ml of hemin chloride were added proved equally suitable for serial subculture of *T. aceti*.

However, acid-precipitated hemin was found to give various results. Research workers at the State University of New York, Buffalo (M. Rothstein, personal communication), and at the University of California, Berkeley (Pinnock *et al.*, 1975), obtained but poor growth of C. *"briggsae"* with particulate hemin. J. R. Vanfleteren (unpublished results), upon repeating the experiments several times, observed that the activity of various preparations of acid-precipitated hemin varied considerably, but repeated subculturing was consistently attained, provided that the hemin particles were neither too large (unfavorable) nor too fine (they redissolved too rapidly), and that they did not contribute too much salt to the culture medium. The higher levels of Tween 80 used by Vanfleteren (see Section II, B,4) possibly produced an additional beneficial effect.

The variable results obtained with acid-precipitated hemin and the fact that tissue extracts consistently promoted faster and much more profuse growth than nonproteinaceous supplements thus reinforced the search for nutritionally active low molecular weight peptides. One such peptide seemed to be available from soy peptone, since neither myoglobin (Mb), hemoglobin (Hb), nor cytochrone *c* (Cyt *c*) was able to sustain repeated subculture of C. *"briggsae"* in autoclaved

(121°C, 7 min) defined medium (EMS) unless soy peptone was added (Rothstein, 1974). *Turbatrix aceti*, however, grew readily in EMS supplemented with Mb or Hb only, and failed on Cyt c. Thus the idea of a specific peptide requirement led to the unlikely situation in which one had to admit that *C. "briggsae,"* unlike *T. aceti*, would require another factor present as an impurity in Hb and Mb.

Vanfleteren (1978), in trying to find a more likely explanation for the effectiveness of the above media, examined some essential features such as precipitate formation and absorbance in the Soret region. He pointed out that the ineffective media provided heme in unfavorable form, i.e., either as a solute or as heat-denatured material. The incorporation of soy peptone yielded active but also turbid media, so Vanfleteren (1978) suggested that soy peptone probably improved the uptake of solubilized heme proteins. The culture medium (EMS + Mb + HAc) recommended for repeated subculture of *T. aceti* was clear, but had a maximum absorbance at 400–402 nm, which is typical of free hemin, but not of Mb. Thus coordinatively bound heme (e.g., in Hb and Mb) is apparently removed by autoclaving in an acid medium. The free hemin formed is insoluble in this medium and probably forms colloidal aggregations (Falk, 1964); these may be readily taken up by the nematodes. In contrast, Cyt c contains covalently bound heme and is unaffected under identical conditions. Cyt c is completely dissolved in an EMS/Cyt c/HAc medium and this might explain the poor activity observed with this medium.

The fractionation of soy peptone yielded a series of increasingly purified components, all of which enhanced the population densities of *C. "briggsae"* and *C. elegans* sharply when added to EMS. The presumptively pure material was called "nutritional factor" (Rothstein and Coppens, 1978) and was tentatively identified as a small peptide containing lysine, histidine, glutamic and aspartic acids, arginine, serine, glycine, and proline. Most interestingly, there was no reduction in the amount of material needed as purity increased, and nutritional factor itself was required at high levels, suggesting that still other components in soy peptone were at least partly active and that nutritional factor acted as a nutrient rather than as a specific requirement. Whatever its role, the nutritionally active material prepared from soy peptone was a good growth stimulant for *C. "briggsae"* and *C. elegans*, supporting a considerable increase in population densities. However, it was much less effective with *P. silusiae* and was not absolutely required by either of these species (Table V).

A beneficial effect of peptides on the population growth of *C. "briggsae"* was also reported by Pinnock *et al.* (1975). These authors reported that a medium consisting of CbMM, 50 μg/ml β-sitosterol, and 50 μg/ml Cyt c would not support the continued growth and reproduction of *C. "briggsae"* unless casein, casamino acids (acid hydrolysate of the former), casitone (enzymatic hydrolysate), or soy peptone was added. Several defined di- and tripeptides were then

TABLE V

Growth of C. "briggsae," C. elegans, and P. silusiae in Various Media

Species	Medium[a]	Initial population (nematodes/ml)	Final population[b] (nematodes/ml)	Reference
C. "briggsae"	CbMMS + Hm	400	500 (15)	Pinnock and Stokstad (1975)
C. "briggsae"	CbMMS + Hm + Leu-Phe	400	820 (15)	Pinnock and Stokstad (1975)
C. "briggsae"	CbMMS + Hm + cationic guar	400	15,900 (15)	Pinnock and Stokstad (1975)
C. "briggsae"	EMS + Hm	500	16,000 (9)	Rothstein and Coppens (1978)
C. "briggsae"	EMS + Hm + N	500	130,000 (9)	Rothstein and Coppens (1978)
C. "briggsae"	CbMMS + Cyt c	400	640 (15)	Pinnock and Stokstad (1975)
C. "briggsae"	CbMMS + Cyt c + Leu-Phe	400	12,100 (15)	Pinnock and Stokstad (1975)
C. "briggsae"	CbMMS + Cyt c + cationic guar	400	33,000 (15)	Pinnock and Stokstad (1975)
C. "briggsae"	CbMMS + Cyt c	500–900	6,000–12,000 (29)	Lu et al. (1978)
C. "briggsae"	EMS + Mb	500	36,000 (9)	Rothstein and Coppens (1978)
C. "briggsae"	EMS + Mb + N	500	186,000 (9)	Rothstein and Coppens (1978)
C. elegans	EMS + Hm	500	28,000 (9)	Rothstein and Coppens (1978)
C. elegans	EMS + Hm + N	500	190,000 (9)	Rothstein and Coppens (1978)
P. silusiae	EMS + Hm	500	17,000 (10)	Rothstein and Coppens (1978)
P. silusiae	EMS + Hm + Et	500	30,000 (10)	Rothstein and Coppens (1978)

[a] Cationic guar (2 mg/ml): CbMMS, defined basal medium (CbMM) containing 50 μg/ml sterols; Cyt c, cytochrome c (50 μg/ml); EMS, defined basal medium (EM) containing 50 μg/ml sterols; ET, ethanol fraction (4 mg/ml), a growth-promoting peptide fraction from soy peptone; Hm, hemin chloride (50 μg/ml); Leu-Phe, L-leucyl-L-phenylalanine (1 mg/ml); Mb, myoglobin (0.5 mg/ml); N, neutral fraction (4 mg/ml), a growth-promoting peptide fraction from soy peptone that yields pure nutritional factor after additional purification.

[b] Figures in parentheses, number of days.

tested for their growth-promoting ability and some, especially L-leucyl-L-phenylalanine, proved to be effective supplements. Pinnock *et al.* (1975) interpreted this as an argument against the idea that facilitation of heme uptake was the primary function of growth factor and they suggested that the contribution of protein itself was essential. The latter hypothesis was not substantiated by further experimental work (Pinnock and Stokstad, 1975), however, which revealed that nonproteinaceous material like cationic guar (a sugar polymer) was equally effective when added as a supplement to a CbMMS/Cyt *c* medium. Unlike L-leucyl-L-phenylalanine, cationic guar supported the population growth of *C. "briggsae"* in a CbMMS/hemin medium (Table V). Possibly, L-leucyl-L-phenylalanine derived its activity partly from lowering the solubility of Cyt *c,* besides providing amino acids in a more favorable form (Vanfleteren, 1978). As a matter of fact, the CbMMS/Cyt *c* medium contributes all the nutrients required by *C. "briggsae,"* and later work performed at the University of California, Berkeley (Lu *et al.,* 1978), confirmed that this medium alone supported poor but continued population growth of *C. "briggsae."*

In summary, it may be said that growth factor contributes only two dietary essentials, sterol and heme, to the culture medium. Protein, although not absolutely required, has as its primary role improving the uptake of heme by the nematodes, and also has a general beneficial effect, possibly by improving the amino acid balance of the medium or providing amino acids in a form that is more readily utilized.

IV. NUTRIENT ABSORPTION

From the foregoing it is clear that growth factor derives its activity mainly from the facilitation of the uptake of heme. Precipitated material is more readily utilized than solutes. The beneficial effect of activation, which is a form of controlled precipitation, clearly points to that direction. Heat-precipitated growth factor is usually inactive, although the essential nutrients provided—i.e., sterol and heme—are sufficiently heat stable. Moreover, heat treatment produces inconsistent results. As an example, live *E. coli* cells provide for excellent growth of *C. "briggsae,"* but autoclaved cells are not effective unless an additional heme source is added (Hieb and Rothstein, 1970). Similarly, tissue extracts become inactive upon autoclaving. Yet Cyt *c,* Hb, and Mb—the common active components of these extracts—appear more resistant to heat when added to the culture medium individually (Rothstein, 1974). In addition, a coagulate of bovine serum albumin and hemin chloride that proved effective when formed at 60°C was totally inactive when heated to 100°C for 5 min (Vanfleteren, 1978).

Realizing that the activity of growth factor was centered about the availability of heme to the nematodes, Vanfleteren (1974, 1975a, 1975b, 1976, 1978) postulated that part of the food, including heme, was taken up by the intestinal cells as small particles by a supposed process of pinocytosis, unfortunately less correctly referred to as phagocytosis in the earlier papers. Preferential uptake of heme compounds by a process of pinocytosis, if correct, is in good agreement with what the author called "the proper particulate form" of growth-promoting material, any reverse effect being unfavorable (e.g., complete dissolution of the active material, but also any other alteration making it unsuitable for uptake by pinocytosis). One good example is heat-denatured protein, which forms coarse precipitates that can probably withstand the grinding action of the nematode bulb and, because of their size, are unlikely to be accessible to pinocytosis. Thermal denaturation is a gradual and complex process that is readily influenced by the experimental conditions, and this might explain the inconsistent results observed with heat-denatured growth-promoting material.

Although the preferential utilization of precipitated heme is indisputable, it has also become clear that soluble heme is not quite unavailable to the nematodes. Tilby and Moses (1975) reported that defined medium supplemented with cholesterol (100 μg/ml) and solubilized Hb (500 μg/ml) supported good growth of *C. elegans,* although better growth was obtained with old medium, in which a precipitate appeared on standing. However, Vanfleteren (1975b) was unable to obtain the maturation of *C. briggsae* juveniles with Hb dissolved at 1 mg/ml in CbMMS, and he observed that the growth response of this nematode to CbMMS supplemented with decreasing amounts of Hb and Mb sharply decreased as soon as the precipitated heme proteins redissolved, which occurred quickly at below 100 μg of protein/ml. Vanfleteren (1978), therefore, assumed that either physiological differences between *C. briggsae* and *C. elegans* were involved or that the growth-promoting activity observed by Tilby and Moses (1975) resulted from the formation of small protein aggregates, invisible to the naked eye.

Vanfleteren (1978) also suggested that the biological activity of cationic guar and the various peptide supplements discussed above derived part of their activity from forming complexes with the soluble heme sources added so as to make them more available to pinocytosis. The earlier observation (Pinnock *et al.,* 1975; Pinnock and Stokstad, 1975) that unsupplemented CbMMS/ Cyt *c* medium was ineffective (Table V) strengthened this idea.

More recently, however, experiments performed at the University of California, Berkeley (Lu *et al.,* 1978; Cheng *et al.,* 1979) provided evidence that, in contrast to the former reports, unsupplemented CbMMS/Cyt *c* medium consistently supported a slow but continued population growth of *C. "briggsae"* (Table V). The appearance of particles in this medium, in the opinion of the author, is out of the question; since this point was crucial, the experiment was repeated by the author. The initial results were very similar to those reported by Lu *et al.,* (1978) and Cheng *et al.* (1979), but the nematodes died out in the third subculture. Both *C. briggsae* and *C. elegans* were assayed and behaved similarly (J. R. Vanfleteren, unpublished observations).

To explore this question further, the author also reexamined the biological activity of clear medium composed of CbMMS and Hb or Mb. Remarkably, several preparations had to be discarded since the heme proteins tended to precipitate spontaneously when diluted in CbMMS. Preparations that were slightly turbid or opalescent lost a considerable amount (up to 50% and over) of their protein content upon filtration through 0.30-μm pore-size membranes, indicating that hardly noticeable precipitates actually may contribute large amounts of readily available heme to the nematodes, and that membrane sterilization of complete media may lead to unexpected losses of heme protein. In any case, a supposedly particle-free medium consisting of CbMMS and Hb or Mb at about 1 mg/ml was finally reached and proved a good culture medium for *C. briggsae* and *C. elegans* (J. R. Vanfleteren, unpublished observations).

The question of why soluble Hb at 1 mg/ml produced inconsistent results remains unclear. The above experiments were performed with Hb preparations purchased from Serva and ICN. The Hb used in earlier experiments (Vanfleteren, 1975b) was obtained from Calbiochem (2× crystalline) and showed no noticeable (e.g., as revealed by light scattering) tendency to particle formation in the medium. As previously mentioned, this medium was inactive, but the addition of latex particles (0.234 μm in diameter and added at about 5×10^9/ml) stimulated sparse but consistent growth and reproduction of *C. briggsae* (Vanfleteren, 1975b).

The problem of the inconsistent growth-promoting activity observed with dissolved heme proteins became even more complex when Rothstein and Coppens (1978) obtained relatively slow but continued growth of *C. "briggsae," C. elegans,* and *P. silusiae* in clear EMS/hemin chloride medium (Table V). The stimulatory effect of dissolved hemin was maintained through repeated subcultures.

Clearly, this result was in strong conflict with earlier unsuccessful attempts to grow *Caenorhabditis* in a clear CbMMS/hemin medium (Pinnock *et al.,* 1975; Pinnock and Stokstad, 1975; Vanfleteren, 1978). The present author has, therefore, repeated the experiment with *C. briggsae* and *C. elegans.* The larval-assay method was used, but egg masses were transferred directly to the culture tubes after being extensively washed in buffer, each tube thus receiving 20–50 eggs. These hatched and developed no further than the second to fourth juvenile stage throughout the observation period (21 days) in CbMMS containing 50 μg/ml of solubilized hemin. In one tube out of three inoculated with *C. elegans,* very few worms reached maturity in 11 days, and these produced very few offspring; the offspring themselves failed to develop any further. Essentially the same results were observed with *C. briggsae.* With acid-precipitated hemin at 50 μg/ml, maturation was reached within 4.3 ± 0.1 (*C. briggsae*) and 6.5 ± 0.2 (*C. elegans*) days, and nematode populations of approximately 1000 worms formed in otherwise identical conditions (J. R. Vanfleteren, unpublished observations).

When supplied at higher concentrations, solubilized hemin consistently proved stimulatory, however, and at the highest concentrations tested (Table VI), egg masses and second-generation juveniles were produced. Noticeably, solubilized hemin was slightly more active at 50 μg/ml than mentioned above when it was made by serial dilution from the mixtures that were more concentrated, although again it was by no means satisfactory (J. R. Vanfleteren, unpublished results).

From the foregoing, it is clear that soluble heme sources can be utilized by the nematodes. However, soluble heme must be supplied at considerably higher levels, say tenfold, to be effective on a comparative basis. Lower concentrations will inevitably produce largely fluctuating results since any stored heme or any eventual trace of aggregated heme will bear a noticeably stimulatory effect on nematode growth and reproduction in an otherwise complete medium. In addition, it would now appear that nematodes definitely have undergone some nutritional change since their first establishment in axenic culture some 30 years ago, and may have become adapted to soluble food.

It is clear also that the mechanism by which heme compounds are absorbed in the intestinal cells remains the key to a better understanding of nematode nutritional behavior. Whatever the nature of this mechanism, it must explain the

TABLE VI

Dose Response of *C. briggsae* and *C. elegans* to Solubilized Hemin

Species	Concentration[a] (μg/ml)	Maturation time[b] (days)	Population at 14 days[b]
C. briggsae	500	4.3 ± 0.1	$7,000 \pm 850$
C. briggsae	250	4.3 ± 0.1	$10,600 \pm 1,250$
C. briggsae	100	4.5 ± 0.1	$1,600 \pm 100$
C. briggsae	50	6.8 ± 0.1	90 ± 10
C. elegans	500	5.3 ± 0.1	$4,500 \pm 600$
C. elegans	250	5.7 ± 0.1	$3,900 \pm 300$
C. elegans	100	7.3 ± 0.3	$1,450 \pm 150$
C. elegans	50	$11 - 14$	120 ± 10

[a] Hemin chloride from stock solution (10 mg/ml in 0.1 N KOH) to defined basal medium containing 50 μg/ml sterols (CbMMS). Lower concentrations of hemin were made by serial dilution with CbMMS.

[b] Triplicate tubes containing 0.3 ml of medium inoculated with approximately 50 eggs. Maturation time was measured as the time from inoculation to the first appearance of juveniles of the next generation, reduced by 0.5 days to compensate for the time needed for hatching. Mean values $\pm$ SE are given.

general finding that heme protein provides heme in a more readily available form than free hemin, and it must account for the preferential uptake of precipitated forms of heme protein or hemin, with the exception made, for example, for heat-denatured (heme) proteins. The hypothesis that the concentration of nutritive material as a precipitate might be necessary to provide the nutrient in a more concentrated form (Buecher *et al.*, 1969b; Lu *et al.*, 1978) offers no plausible explanation for the failure of heat-denatured supplements and is less likely to occur.

Recently, it was proposed that the growth-promoting activity of either particulate protein or precipitated hemin was linked to the physical presence of particles in the medium, rather than being due to the precipitated form of protein or heme per se (Cheng *et al.*, 1979). This idea was based on the observation that several particulate compounds such as Celite, glass powder, tin oxide, diamond powder, and Microcrystalline Cellulose, Solka Flok, and Cellophane Spangles (three cellulose products) consistently stimulated the population growth of *C. "briggsae"* in CbMMS/Cyt *c* (50 μg/ml) medium, although the population increases reported varied considerably. Interestingly, poor adsorbents such as diamond powder and large-size particles, which were unable to enter the gut, proved still active. The present author agrees with the conclusion by the authors that the physical presence of particles in the culture medium is somehow stimulatory.

By which mechanism does this stimulation occur? J. R. Vanfleteren (unpublished observations) observed the feeding movements of the bulbar apparatus in the presence and absence of latex particles. A single *C. elegans* egg mass was washed and allowed to hatch in CbMMS/hemin (50 μg/ml), and observations were made in a cold room. As the nematodes were cooled from room temperature to about 5°C, they became increasingly paralyzed, but it took several minutes before the feeding movements became irregular and stopped. During that period, bulb action proceeded at essentially the same high rate whether latex particles (0.234-μm diameter) were added or not: Pumping took place at two to three cycles per second initially and then slowed down.

Thus the stimulatory activity of (inert) particles in the medium is not likely to be due to the increased ingestion of food. Besides, it is not clear how faster ingestion of soluble nutrients such as hemin or Cyt *c* might enhance the uptake of these by the intestinal cells. Moreover, the loss of activity by heat-denatured supplements once more precludes the physical presence of particulate material in the medium as the sole mechanism that determines the effectiveness of an otherwise complete medium.

Thus, we must consider again that the growth-promoting activity of precipitated versus soluble forms must be determined somehow by the mechanisms by which nutrients are absorbed in the intestine.

It might be anticipated that precipitation of a solute actually results from charge-alteration. Thus J. R. Vanfleteren (unpublished observations) examined whether the surface charge of a heme protein possibly influenced the absorption by the intestinal cells. Cyt *c* from *Pseudomonas aeruginosa* (strain 1896) was obtained from Van Beeumen, Laboratory of Microbiology, Ghent. This cytochrome is unusual in being acidic (p*I* near 4.5). This and Cyt *c* from horse heart (p*I* near 10) were added at 50 μg/ml to CbMMS and were assayed for their growth-promoting activity on *C. briggsae* and *C. elegans*. When the pH of the culture medium was 5.9, horse-heart cytochrome can be supposed to carry a large positive charge, whereas the bacterial cytochrome can be expected to be negatively charged. No significant difference in growth-promoting activity was observed; perhaps the final populations reached were somewhat larger with the bacterial cytochrome. More heme was provided along with this cytochrome at the supplied level (50 μg/ml) since it is smaller (MW about 9000) than horse-heart Cyt *c* (MW about 12,000).

The hypothesis that particulate nutrients are readily taken up by some process of pinocytosis may be reconsidered and adapted to cover all the observations discussed. Central to this is the assumption that the formation of pinocytotic vacuoles is a basic property of the intestinal cells of those nematodes that normally feed on bacteria. The size of the particulate material in the intestine is thought to be a main factor determining effective uptake by pinocytosis; for example, large particles would be precluded. Heat-denatured proteins usually form copious precipitates, which doubtless are ingested but may withstand crushing in the bulb. This might explain why heat destroys the activity of proteinaceous growth factors increasingly with time. Too-small particles, on the other hand, e.g., completely solubilized macromolecules, would fail to stimulate the formation of pinocytotic vacuoles. They would become absorbed rather erratically, thus supplying insufficient amounts of heme, unless they were present in substantially higher levels (i.e., higher levels would compensate for poor absorption).

On the other hand, molecular aggregation may occur more frequently with increasing concentration and may enhance the formation of pinocytotic vacuoles. Thus one might suspect that soluble (heme) protein, unlike precipitated forms, may support fast growth initially (that is, by stimulating newly hatched juveniles to develop and produce offspring) and may fail thereafter so that further development is suddenly impaired. This is exactly what happens frequently with solubilized (heme) protein supplements. A typical example is presented in Fig. 1, showing the *C. elegans* populations produced in CbMMS supplemented with 50 μg/ml of hemin and 1 mg/ml of solubilized and acid-precipitated ferritin, respectively. Initial growth was faster in the clear medium (maturation times of about 4 and 4.5 days, respectively), but the numerous offspring produced did not develop any further. When precipitated ferritin was used, growth was not impaired and a

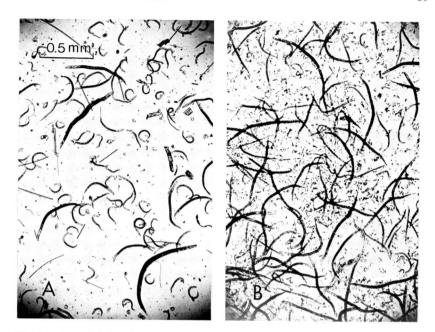

Fig. 1. *Caenorhabditis elegans* populations produced in CbMM supplemented with 50 μg/ml sterol mixture, 50 μg/ml hemin chloride, and 1 mg/ml ferritin. Culture media were inoculated with egg masses and incubated at 21°C for 2 weeks. (A) Solubilized ferritin. The population consists mainly of adults and early-stage juveniles. (B) Acid-precipitated ferritin. Normal population comprising all stages of development.

normal population was produced, consisting of worms in all stages of development (J. R. Vanfleteren, unpublished observations).

The biological activity of particles, including inert particles in the culture medium, has been attributed to their capability of adsorbing heme compounds or free hemin (Vanfleteren, 1978). In view of the recent results with various particulate materials discussed above (Cheng *et al.*, 1979), the more likely explanation is that the physical presence of particles stimulates the formation of pinocytotic vesicles in the intestinal cells by a supposed neural regulation. The existence of such a mechanism in nematodes that feed on particulate nutrients (i.e., bacteria feeders) seems quite conceivable.

It is improbable that heme protein molecules would pass through the intestinal cell membranes by simple diffusion. Free hemin is much smaller (MW of hemin chloride, 652), but the simple observation that it is much less active than heme proteins on a molar base suggests that it also does not pass through.

From the information presented, we are forced to conclude that cavital digestion of proteins must be very limited, if it occurs at all, in the bacteria-feeding

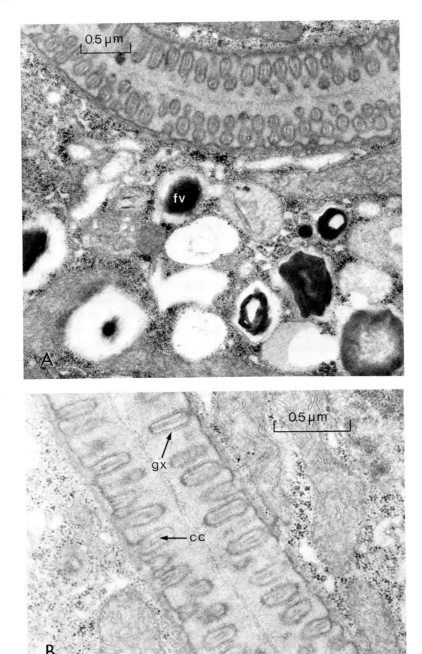

species studied. Proteolytic activity in the intestinal lumen would be hardly compatible with almost all the observations on nutrient uptake discussed. It may be concluded that the intracellular digestion of protein is a characteristic property of bacteria-feeding nematodes. It is probable that in those fungal-feeding species investigated and in the insect parasites *N. glaseri* and *N. carpocapsae* DD-136 a similar requirement for sterols and heme exists, but much less is known about the patterns of the uptake of these nutrients by these species. *Aphelenchus avenae* and *A. rutgersi* require comparatively higher levels of protein and utilize soluble forms of protein by preference, suggesting that the mechanisms by which proteins are absorbed may be quite different in stylet-bearing nematodes (Hansen *et al.*, 1968; Buecher *et al.*, 1970b; Hansen *et al.*, 1970; Lower and Buecher, 1970; Buecher and Hansen, 1971; Myers *et al.*, 1971; Thirugnanam, 1976; reviewed by Vanfleteren, 1978).

Ultrastructurally, it should be possible to visualize nutrient uptake by pinocytosis and the subsequent intracellular digestion if it is occurring. In order to investigate this, two *C. elegans* egg masses were washed in the usual way, transferred to each of two culture tubes containing CbMMS, and left to hatch overnight. Acid-precipitated ferritin was next added to one of the tubes at 1 mg/ml, and the worms were allowed to feed during 1 hr. After this period the worms were fixed in glutaraldehyde and osmium tetroxide and were routinely embedded in ERL. Unstained sections revealed the occurrence of large electron-dense bodies in the cytoplasm of several intestinal cells. No such bodies were seen in the gut cells of the control nematodes. Conceivably these formations corresponded to food vacuoles filled with ferritin, which might be locally digested. The high iron content (23%) of ferritin might account for the observed electron density in the unstained sections. Sections stained in lead citrate and uranyl acetate showing putative food vacuoles can be seen in Fig. 2A, and an enlargement of the microvillous border is given in Fig. 2B. Rather unexpectedly, no pinocytotic vesicles were seen. It is possible that they may have disappeared during fixation, or, alternatively, the feeding period was too long (J. R. Vanfleteren, unpublished observations).

Experiments are now in progress in our laboratory to trace the uptake of Hb using the diaminobenzidine (DAB) reaction for selective staining. The very preliminary results presently available suggest an unusual pattern of protein transfer along the hollow central core of the microvilli. More experimental work is needed for valid conclusions to be drawn, however.

Fig. 2. Electron micrograph of a section through the intestine of a first-stage juvenile of *C. elegans*, fed on acid-precipitated ferritin (1 mg/ml) for 1 hr. (A) Putative food vacuoles (fv) filled with ferritin occur within the cytoplasm. ×42,000. (B) Enlargement of the microvillus border. Note the glycocalyx (gx) and the hollow, central core (cc) associated with each microvillus. ×66,000.

V. POSSIBLE CONTAMINANTS OF AXENIC CULTURE

A main problem in studying nematode nutrition by adding or withdrawing nutrients is the risk of unsuspected contamination. Nematode culture medium is rich enough to stimulate the growth of abundant microorganisms. In general, bacterial contamination is easily distinguished by the appearance of cloudiness in the medium, odor, and sudden death of the worms. Contamination with fungi or yeasts proceeds less dramatically and can normally be detected under the low power of a dissecting microscope. Nevertheless, where possible, it is advisable that all experimental cultures, even after an experiment has been ended, be incubated for at least 3–4 weeks and carefully inspected for contamination before being discarded.

It is better, of course, although much more time-consuming, that sterility checks be run as a matter of routine. Tight sterility testing should be performed using two different temperatures (e.g., 20° and 37°C for at least 2 weeks) and a series of different microbiological media, e.g., brain–heart infusion (fastidious microorganisms), thioglycollate (anaerobic microorganisms), and Sabaroud dextrose (fungi and yeasts). As a matter of fact, nutrient agar or brain–heart infusion alone have been found to be sufficient to detect the usual contaminants that propagate in nematode cultures (Dougherty et al., 1959; Rothstein, 1974).

Suitable techniques for axenization have been reviewed elsewhere (Rothstein and Nicholas, 1969; Nicholas, 1975). Egg-laying nematodes can be axenized with relative ease. The eggs will tolerate surface sterilization with substances such as 1% merthiolate (Cryan et al., 1963), 25% glutaraldehyde (Murfitt et al., 1976), or 0.4 M NaOH (Patel and McFadden, 1978). The particular resistance of the soil-inhabiting rhabditid nematodes to antibiotics (e.g., penicillin, 1000 units/ml; streptomycin, 1 mg/ml and amphotericin B, 25 μg/ml) may be helpful.

Slow-growing bacteria have been reported by Casida (1965) to occur abundantly in soil. They form very small colonies on agar slants and usually provoke no turbidity in a liquid medium, and there is a real possibility that such microorganisms may have accidentally been introduced into axenic nematode cultures. Such contaminants might remain unsuspected in the absence of tight sterility checks and might bias the experimental results considerably. The coexistence of nematodes and contaminating microorganisms in the same liquid medium is rather exceptional, but does occur. Thus J. D. Willett (personal communication) isolated a yellow, slow-growing bacterium from apparently healthy cultures of *P. redivivus,* which increased the population densities of the nematode by a considerable amount. The bacteria grew well on brain–heart infusion agar and on agar medium containing yeast extract, malt extract, peptone, and dextrose, but no growth occurred on the other bacteriological media tested.

The author has found no reports on the occurrence of mycoplasmas in supposedly axenic culture. Mycoplasmas are a microbial form of life which normally parasitize eukaryotic cells, but which are able to replicate in a cell-free environment. Unlike bacteria, they lack a rigid cell wall and are therefore resistant to antibiotics that inhibit cell-wall synthesis. They are also smaller than bacteria, the minimal reproduction unit being in the range near 300 nm, and may escape retention by current membrane filtration (Tully, 1977). Mycoplasmas require fastidious growing conditions, and their occurrence in liquid medium is not necessarily revealed by turbidity. It may take 2 weeks or more for the small typically fried-egg-shaped colonies to appear on agar slants.

Probably most mycoplasma species would not propagate under the nematode cultural conditions. A few species, e.g., *Acholeplasma laidlawii,* are likely candidates for infecting nematode cultures, however, especially those maintained on a soy peptone/yeast basal medium supplemented with crude tissue extracts. *Acholeplasma laidlawii* is a common contaminant of bovine and equine serum and might possibly be introduced along with HLE. The mycoplasma growth medium consists of 3% brain–heart infusion, 2.5% fresh yeast extract, and 20% horse serum (Taylor-Robinson, 1977). When considering that a medium consisting of 1–3% yeast extract, 3–4% soy peptone, and 10% HLE is being widely used for keeping nematode stock cultures or as an experimental medium, and that *A. laidlawii* grows readily at room temperature, we must be on the alert for possible mycoplasmal contamination.

For this reason, the author (unpublished observations) has carefully examined his stock cultures of *C. briggsae* and *C. elegans* for the possible presence of mycoplasmas. Samples were taken from all stock cultures and combined in a culture tube. About 0.1-ml volumes were then inoculated into 2 ml of mycoplasma broth (Gibco Bio-Cult) and onto 5 ml of mycoplasma agar medium, respectively. Duplicate sets were inoculated at room temperature and at 36°C. The broth cultures were subcultured to an agar medium after a week. The agar cultures were examined under a dissecting microscope (12–50×) for microbial growth during 3 weeks, all with negative results. Samples taken from the *C. "briggsae"* Berkeley (obtained from N. C. Lu) and Buffalo (obtained from M. Rothstein) strains also proved negative for microbial growth. The test was not entirely tight since it was not possible to inoculate duplicate sets in a nitrogen (5% CO_2 in N_2) atmosphere, but it is improbable that preferentially anaerobic prokaryotes would be a nuisance in nematode cultures. It is recommended, however, that HLE no longer be used and that it be replaced by safer supplements such as hemoglobin.

Other organisms that might bias the experimental results considerably if present in axenic culture are nematode parasites (Croll and Matthews, 1977). Introduction of these, with exception made for viruses, is rather theoretical provided that the usual aseptic precautions are taken. Predaceous fungi are readily detectable by visual inspection, but infection with sporozoans might become manifest only by a considerable increase of dead worms. Too high a percentage of dead organisms in young cultures should arouse suspicion. The presence of viruslike particles in nematodes (Foor, 1972; Zuckerman *et al.*, 1973) has been discussed, but no report has been made on the eventual effect of viruses on nematode health, so this question must await further study.

VI. NUTRITION AND DEVELOPMENT

Despite emerging knowledge of nematode nutrition, the best growth of nematodes generally realized in axenic culture has become very close, though not identical, to growth realized in the presence of bacteria. Maturation of *C. briggsae* takes approximately 3 days in monoxenic association with *E. coli* and 4–5 days in CbMMS or soy peptone/yeast extract basal medium supplemented with 500 μg/ml Hb or 10% HLE. For *C. elegans* these data are 3.5 and 4.5–5.5 days, respectively. One exception was reported in a paper, where faster maturation of *C. elegans* occurred in axenic rather than in monoxenic culture (Croll *et al.*, 1977). Adults of both *C. briggsae* and *C. elegans* are consistently wider and more voluminous (Hansen *et al.*, 1964; Croll *et al.*, 1977) and eventually longer (Hansen *et al.*, 1964) when grown on *E. coli* cells, but this difference may be nullified by the addition of lipids to the culture medium (Lu *et al.*, 1978).

As a matter of fact, the observed differences between axenically and monoxenically reared nematodes are in no way considerable, with the exception of differences in longevity. Both *C. briggsae* and *C. elegans* live significantly longer in axenic culture than in bacterial culture. It has been suggested that shorter longevity in bacterial culture is caused by toxins given off by the bacteria (Hansen *et al.*, 1964; Croll *et al.*, 1977). Although this is quite possible, it is also true that the entire developmental cycle generally is reported as proceeding faster in monoxenically reared nematodes. Work is in progress in our laboratory to examine what developmental event, if any, determines the onset of aging by altering the nutritive ingredients at distinct stages of nematode development.

Nutritional cues somehow trigger the regulatory mechanisms that govern nematode development. Food supply, for example, is one of the poorly understood cues of dauer larva formation (Yarwood and Hansen, 1969; Cassada and Russell, 1975; Klass and Hirsh, 1976). Nutritive ingredients, notably steroids, and poly(A)biosynthesis would stimulate fourth-stage juveniles of *P. silusiae* to develop into adults (Samoiloff, 1978), and there is much reason to believe that heme is involved in reproduction in a similar way. The high levels of heme required for reproduction may also indicate that it is incorporated into the eggs. Such incorporation was reported to occur in *Ascaris lumbricoides* eggs (Smith and Lee, 1963).

Malnutrition eventually induces morphological variations. J. R. Vanfleteren (unpublished observations) observed that adults of *C. elegans* and *C. briggsae*

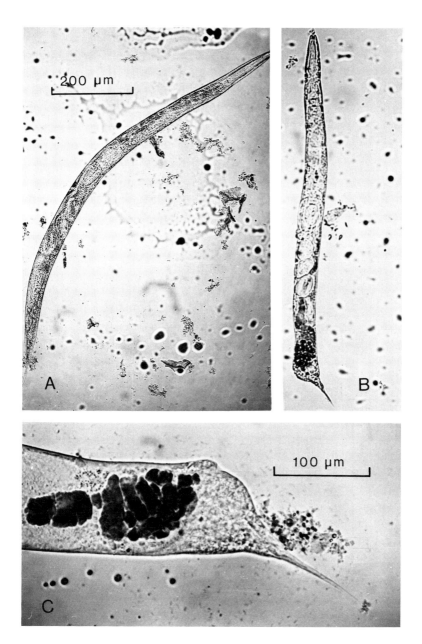

Fig. 3. Light micrographs of adult *C. elegans* reared in CbMM containing sterols (50 μg/ml) and hemin chloride (50 μg/ml) and to which 1 mg/ml of acid-precipitated ferritin (A) or acid-precipitated cationized ferritin (B) was added. (C) Higher magnification of the posterior part of a specimen grown on cationized ferritin, showing the swelling near the anal opening and the accumulation of ferritin in the hindgut.

grown in CbMMS to which were added 50 μg/ml of hemin chloride and 1 mg/ml of acid-precipitated ferritin were shorter when normal ferritin was replaced by cationized ferritin (Miles Laboratories; Figs. 3A and 3B). A swelling was seen just posterior to the anal opening and the hindgut was filled with large brownish aggregates, apparently consisting of ferritin (Fig. 3C). The development of a swelling immediately posterior to the anal opening also occurred in nematodes reared axenically in CbMM to which a bacterial growth factor, derived from a *Flavobacterium* species, was added (Kisiel *et al.*, 1969). The true nature of the above interactions is not yet understood.

VII. CONCLUSIONS AND PROSPECTS

It is the hope of the author that nutritional studies performed on selected nematodes will contribute to a better understanding of the regulatory mechanisms governing nutrient uptake and assimilation in the lower metazoans. Doubtless, this knowledge will ultimately prove useful in the study of the nutritional behavior of parasitic nematodes and perhaps other parasitic helminths as well. Thus *Haemonchus contortus,* a parasitic nematode of sheep, developed to the beginning of the fourth molt in a medium designed for the axenic culture of *C. briggsae* (Hansen *et al.*, 1966), and this nematode and the cestode *Hymenolepis nana* have been successfully cultured in interchanged media (Hansen and Berntzen, 1969). On the other hand, nutritional and environmental studies performed on *Strongyloides fülleborni* demonstrated that the requirements of the rhabditoid phase of this parasitic nematode were those of a parasite and not of a free-living nematode (Hansen *et al.*, 1975).

The availability of a chemically defined medium for the axenic culture of nematodes may stimulate further elucidation of the nutritional requirements of these organisms and may eventually lead to the development of a minimum essential medium. This may further lead to the isolation and characterization of nutritionally defective mutant nematodes. It may also contribute to further study of the biosynthetic abilities of these nematodes by experiments in which various precursors of essential nutrients are added to the medium and examined for their growth-promoting properties in order to discover which step is missing or which pathway is altered in nematodes.

The very particular pattern of nutrient, especially protein, uptake by the free-living nematodes discussed certainly warrants further studies on the digestive physiology of these organisms using appropriate histochemical techniques. The question of which mechanism, if any, might mediate between food perception and the rate of nutrient uptake by the intestinal cells must also await further investigation.

Finally, the availability of suitable media for the axenic culture of nematodes that have been deemed suitable as model organisms for the study of particular properties of multicellular organisms has the advantage of eliminating possible interaction with other organisms. In particular, studies on hormonal regulation and aging may be performed on axenically cultured nematodes by preference. The effect of altering nutritional ingredients on the aging rate in axenic culture may eventually contribute to a better understanding of the mechanisms governing physiological aging.

ACKNOWLEDGMENTS

The author is sponsored by the Belgian National Fund for Scientific research (NFWO). Special thanks are due to Dr. G. De Maeyer and Dr. P. Grootaert with regard to the electron micrographs, Dr. G. Haspeslagh for taking the light micrographs, and Dr. J. Van Beeumen for providing *Pseudomonas aeruginosa* cytochrome *c*. Thanks are also due to Drs. Friedman, Lu, Rothstein, and Willett for communicating results.

REFERENCES

Balasubramanian, M., and Myers, R. F. (1971). *Exp. Parasitol.* **29**, 330–336.

Bolla, R. I., Weinstein, P. P., and Lou, C. (1972). *Comp. Biochem. Physiol. B.* **34**, 487–501.

Buecher, E. J., and Hansen, E. L. (1969). *Experientia* **25**, 656.

Buecher, E. J., and Hansen, E. L. (1971). *J. Nematol.* **3**, 199–200.

Buecher, E. J., Hansen, E. L., and Yarwood, E. A. (1966). *Proc. Soc. Exp. Biol. Med.* **121**, 390–393.

Buecher, E. J., Hansen, E. L., and Gottfried, T. (1969a). *Nematologica* **15**, 619–620.

Buecher, E. J., Perez-Mendez, G., and Hansen, E. L. (1969b). *Proc. Soc. Exp. Biol. Med.* **132**, 724–728.

Buecher, E. J., Hansen, E. L., and Gottfried, T. (1970a). *J. Nematol.* **2**, 93–98.

Buecher, E. J., Hansen, E. L., and Myers, R. F. (1970b). *J. Nematol.* **2**, 189–190.

Buecher, E. J., Hansen, E. L., and Yarwood, E. A. (1970c). *Nematologica* **16**, 403–409.

Buecher, E. J., Hansen, E. L., and Yarwood, E. A. (1971). *J. Nematol.* **3**, 89–90.

Casida, L. E., Jr. (1965). *Appl. Microbiol.* **13**, 327–334.

Cassada, R. C., and Russell, R. L. (1975). *Dev. Biol.* **46**, 326–342.

Cheng, A. C., Lu, N. C., Briggs, G. M., and Stokstad, E. L. R. (1979). *Proc. Soc. Exp. Biol. Med.* **160**, 203–207.

Cole, R. J., and Dutky, S. R. (1969). *J. Nematol.* **1**, 72–75.

Croll, N. A., and Matthews, B. E. (1977). "Biology of Nematodes," pp. 179–184. Blackie, Glasgow and London.

Croll, N. A., Smith, J. M., and Zuckerman, B. M. (1977). *Exp. Ageing Res.* **3**, 175–189.

Cryan, W. S., Hansen, E., Martin, M., Sayre, F. W., and Yarwood, E. A. (1963). *Nematologica* **9**, 313–319.

Dougherty, E. C. (1953). *J. Parasitol.* **39**, 371–380.

Dougherty, E. C. (1960). *In* ''Nematology'' (J. N. Sasser and W. R. Jenkins, eds), pp. 297–318. Univ. of North Carolina Press, Chapel Hill, North Carolina.

Dougherty, E. C., and Hansen, E. L. (1957). *Anat. Rec.* **128**, 541–542.

Dougherty, E. C., Raphael, J. C., and Alton, C. (1950), *Proc. Helminthol. Soc. Wash.* **17**, 1–10.

Dougherty, E. C., Hansen, E. L., Nicholas, W. L., Mollett, J. A., and Yarwood, E. (1959). *Ann. N.Y. Acad. Sci.* **77**, 176–217.

Dutky, S. R., Robbins, W. E., and Thompson, J. V. (1967). *Nematologica* **13**, 140.

Falk, J. E. (1964). ''Porphyrins and Metalloporphyrins. Their General, Physical and Coordination Chemistry and Laboratory Methods,'' p. 115. Elsevier, Amsterdam.

Foor, W. E. (1972). *J. Parasitol.* **58**, 1065–1070.

Friedman, P. A., Platzer, E. G., and Eby, J. E. (1977). *J. Nematol.* **9**, 197–203.

Hansen, E. L., and Berntzen, A. K. (1969). *J. Parasitol.* **55**, 1012–1017.

Hansen, E. L., and Buecher, E. J. (1970). *J. Nematol.* **2**, 1–6.

Hansen, E. L., and Cryan, W. S. (1966). *Nematologica* **12**, 138–142.

Hansen, E. L., Sayre, F. W., and Yarwood, E. A. (1961). *Experientia* **17**, 32–33.

Hansen, E. L., Buecher, E. J., and Yarwood, E. A. (1964). *Nematologica* **10**, 623–630.

Hansen, E. L., Silverman, P. H., and Buecher, E. J., Jr. (1966). *J. Parasitol.* **52**, 137–140.

Hansen, E. L., Yarwood, E. A., Jackson, G. J., and Poinar, G. O. (1968). *J. Parasitol.* **54**, 1236–1237.

Hansen, E. L., Buecher, E. J., and Evans, A. A. F. (1970). *Nematologica* **16**, 328–329.

Hansen, E. L., Perez-Mendez, G., and Buecher, E. J. (1971). *Proc. Soc. Exp. Biol. Med.* **137**, 1352–1354.

Hansen, E. L., Buecher, E. J., and Yarwood, E. A. (1975). *Exp. Parasitol.* **38**, 161–166.

Hechler, H. C. (1971). *J. Nematol.* **3**, 227–237.

Hieb, W. F., and Rothstein, M. (1968). *Science* **160**, 778–779.

Hieb, W. F., and Rothstein, M. (1970). *Arch. Biochem. Biophys.* **136**, 576–578.

Hieb, W. F., and Rothstein, M. (1975). *Exp. Gerontol.* **10**, 145–153.

Hieb, W. F., Stokstad, E. L. R. and Rothstein, M. (1970). *Science* **168**, 143–144.

Jackson, G. J. (1973). *Exp. Parasitol.* **34**, 111–114.

Kisiel, M., Nelson, B., and Zuckerman, B. M. (1969). *Nematologica* **15**, 153–154.

Klass, M., and Hirsh, D. (1976). *Nature (London)* **260**, 523–525.

Lower, W. R., and Buecher, E. J. (1970). *Nematologica* **16**, 563–566.

Lower, W. R., Hansen, E., and Yarwood, E. A. (1966). *J. Exp. Zool.* **161**, 29–35.

Lu, N. C., Hieb, W. F., and Stokstad, E. L. R. (1974). *Proc. Soc. Exp. Biol. Med.* **145**, 67–69.

Lu, N. C., Hieb, W. F., and Stokstad, E. L. R. (1976). *Proc. Soc. Exp. Biol. Med.* **151**, 701–706.

Lu, N., Newton, C., and Stokstad, E. L. R. (1977). *Nematologica* **23**, 57–61.

Lu, N. C., Hugenberg, G., Jr., Briggs, G. M., and Stokstad, E. L. R. (1978). *Proc. Soc. Exp. Biol. Med.* **158**, 187–191.

Murfitt, R. R., Vogel, K. and Sanadi, D. R. (1976). *Comp. Biochem. Physiol. B.* **53**, 423–430.

Myers, R. F., and Balasubramanian, M. (1973). *Exp. Parasitol.* **34**, 123–131.

Myers, R. F., Buecher, E. J., and Hansen, E. (1971). *J. Nematol.* **3**, 197–198.

Nicholas, W. L. (1975). ''The Biology of Free-living Nematodes.'' Oxford Univ. Press (Clarendon), London and New York.

Nicholas, W. L., and Jantunen, R. (1963). *Nematologica* **9**, 332–336.

Nicholas, W. L., Dougherty, E. C., and Hansen, E. L. (1959). *Ann. N.Y. Acad. Sci.* **77**, 218–236.

Nicholas, W. L., Dougherty, E. C., Hansen, E. L., Holm-Hansen, G., and Moses, V. (1960). *J. Exp. Biol.* **37**, 435–443.

Nicholas, W. L., Hansen, E., and Dougherty, E. C. (1962). *Nematologica* **8**, 129–135.

Patel, T. R., and McFadden, B. A. (1978). *Nematologica* **24**, 51–62.

Petriello, R. P., and Myers, R. F. (1971). *Exp. Parasitol.* **29**, 423–432.

Pinnock, C. B., and Stokstad, E. L. R. (1975). *Nematologica* **21**, 258-261.
Pinnock, C. B., Shane, B., and Stokstad, E. L. R. (1975). *Proc. Soc. Exp. Biol. Med.* **148**, 710-713.
Rothstein, M. (1963). *Comp. Biochem. Physiol.* **9**, 51-59.
Rothstein, M. (1965). *Comp. Biochem. Physiol.* **14**, 541-552.
Rothstein, M. (1970). *Int J. Biochem.* **1**, 422-428.
Rothstein, M. (1974). *Comp. Biochem. Physiol. B.* **49** 669-678.
Rothstein, M., and Cook, E. (1966). *Comp. Biochem. Physiol.* **17**, 683-692.
Rothstein, M., and Coppens, M. (1978). *Comp. Biochem. Physiol. B.* **61**, 99-104.
Rothstein, M., and Gotz, P. (1968). *Arch. Biochem. Biophys.* **126**, 131-140.
Rothstein, M., and Mayoh, H. (1964a). *Biochem. Biophys. Res. Commun.* **14**, 43-47.
Rothstein, M., and Mayoh, H. (1964b). *Arch. Biochem. Biophys.* **108**, 134-142.
Rothstein, M., and Mayoh, H. (1965). *Comp. Biochem. Physiol.* **16**, 361-365.
Rothstein, M., and Mayoh, H. (1966). *Comp. Biochem. Physiol.* **17**, 1181-1188.
Rothstein, M., and Nicholas, W. (1969) *In* "Chemical Zoology" (M. Florkin and B. Scheer, eds.), Vol. 3, pp. 289-328. Academic Press, New York.
Rothstein, M., and Tomlinson, G. A. (1961). *Biochim. Biophys. Acta* **49**, 625-627.
Rothstein, M. and Tomlinson, G. A. (1962). *Biochim. Biophys. Acta* **63**, 471-480.
Roy, T. K. (1975). *Nematologica* **21**, 12-18.
Samoiloff, M. (1978). *Proc. Int. Congr. Plant Pathol., 3rd,* p. 145.
Sayre, F. W., Hansen, E., Starr, T. J., and Yarwood, E. A. (1961). *Nature (London)* **190**, 1116-1117.
Sayre, F. W., Hansen, E. L., and Yarwood, E. A. (1963). *Exp. Parasitol.* **13**, 98-107.
Sayre, F. W., Lee, R. T., Sandman, R. P., and Perez-Mendez, G. (1967). *Arch. Biochem. Biophys.* **118**, 58-72.
Smith, M. M., and Lee, D. L. (1963). *Proc. R. Soc. London Ser. B.* **157**, 234-257.
Taylor-Robinson, D. (1977). *In* "Mycoplasma Infection of Cell Cultures" (G. J. McGarrity, D. G. Murphy and W. W. Nichols, eds.), pp. 47-56. Plenum, New York.
Thirugnanam, M. (1976). *Exp. Parasitol* **40**, 149-157.
Tietjen, J. H., and Lee, J. J. (1975). *Cah. Biol. Mar.* **16**, 685-693.
Tilby, M. J., and Moses, V. (1975). *Exp. Gerontol.* **10**, 213-223.
Tomlinson, G. A. and Rothstein, M. (1962). *Biochim. Biophys. Acta* **63**, 465-470.
Tully, J. G. (1977). *In* "Mycoplasma Infection of Cell Cultures" (G. J. Mc Garrity, D. G. Murphy, and W. W. Nichols, eds.), pp. 1-34. Plenum, New York.
Vanfleteren, J. R. (1973). *Nematologica* **19**, 93-99.
Vanfleteren, J. R. (1974). *Nature (London),* **248**, 255-257.
Vanfleteren, J. R. (1975a). *Nematologica* **21**, 413-424.
Vanfleteren, J. R. (1975b). *Nematologica* **21**, 425-437.
Vanfleteren, J. R. (1976). *Nematologica* **22**, 103-112.
Vanfleteren, J. R. (1978). *Annu. Rev. Phytopathol.* **16**, 131-157.
Vanfleteren, J. R., and Avau, H. (1977). *Experientia* **33**, 902-904.
Vanfleteren, J. R., and Roets, D. E. (1972). *Nematologica* **18**, 325-338.
Yarwood, E. A., and Hansen, E. L. (1969). *J. Nematol.* **1**, 184-189.
Zuckerman, B. M., Himmelhoch, S., and Kisiel, M. (1973). *Nematologica* **19**, 117.

4

Action of Chemical and Physical Agents on Free-Living Nematodes

M. R. SAMOILOFF

Department of Zoology
University of Manitoba
Winnipeg, Manitoba R3T 2N2, Canada

I.	Introduction	81
II.	Methods of Assay and Probes	82
	A. Developmental Parameters	82
	B. Enzyme Activity	90
	C. Chemical Mutagens	91
	D. Radiation	93
	E. Carcinogenesis	96
III.	Prospects	96
	References	97

I. INTRODUCTION

During the postembryonic portion of their life cycle, free-living nematodes express a limited repertoire of developmental activities; these include limited cell growth, cell division, programmed cell death, differential gene expression, morphogenesis, and the establishment of positional information. These activities are manifested by the growth of the organism, molting, the development of the reproductive system, and the senescence of the organism.

Analyses of postembryonic development in free-living nematodes have used three types of experimental approach to disrupt these processes. Genetic dissection of postembryonic development has been mediated in *Caenorhabditis elegans* by the establishment of temperature-sensitive, postembryonic-development-arrest mutants (Hirsh and Vanderslice, 1976), temperature-sensitive gonad-development mutants (Abdulkader and Brun, 1978), and a post-embryonic cell division-deficient mutant (Albertson *et al.*, 1978). Laser-microbeam ablation experiments have demonstrated regions controlling

NEMATODES AS BIOLOGICAL MODELS
VOLUME 2

postembryonic growth and molting (Samoiloff, 1973) and the site of the receptors for mating attraction (Samoiloff *et al.*, 1973) in *Panagrellus redivivus*, and have been used as a probe for specific cellular function in *C. elegans* (Sulston and Horvitz, 1977). However, the most widely used method for specifically disrupting normal postembryonic development has been the use of chemical agents that interfere with normal biosynthetic activities.

Free-living nematodes can be synchronized at an early stage and grown in controlled axenic or monoxenic liquid medium containing a known concentration of a specific inhibitor, and the effects of that inhibitor on one or more of the parameters of postembryonic development can be determined. Utilization of inhibitors with known modes of action has provided a useful probe for the dissection of postembryonic development. Limitations and advantages of chemical inhibitors are discussed in Chapters 1 and 2 of this volume.

Because of the increasing knowledge of the events of postembryonic development of free-living nematodes and the effects of known inhibitors on these events, the organisms can be used as indicators in the detection of toxic effects of environmental agents. Such an indicator is needed since the number of potential environmental toxicants increases yearly, primarily because of an inability to predict toxicity. Environmental mutagens are readily detected by the rapid, inexpensive Ames test, using stocks of *Salmonella typhimurium* (Ames *et al.*, 1973a,b) either alone or in concert with a mammalian metabolizing system (Ames *et al.*, 1975). Few eukaryotic systems are suitable for routine rapid screening for toxic effects. Whole-mammal assays are expensive and time-consuming, and the use of mammalian cell cultures to detect specific effects, such as loss of contact inhibition, may be of limited applicability. A free-living nematode assay system for the detection of the effects of environmental toxicants on developmental processes provides a useful intermediate between the *S. typhimurium* system and mammalian whole-animal assays.

This chapter will examine the effects of various nongenetic methods of disrupting the postembryonic development of free-living nematodes and the methods used for the detection of environmental toxicants using *P. redivivus*. The effects of agents specifically produced to act on nematodes will not be discussed here. The mode of action of nematicides has been comprehensively reviewed (del Castillo, 1969; Castro and Thomason, 1971).

II. METHODS OF ASSAY AND PROBES

A. Developmental Parameters

1. Synchrony

To study the effects of inhibitors or toxicants on development, synchronous experimental populations can be used. Several methods of obtaining such synchrony are available. Alternatively, experiments can proceed with small numbers of individuals transferred to new media throughout the reproductive period, so that the test specimens are of a known age. The discussion that follows refers only to population synchrony.

Caenorhabditis deposits embryonated eggs that can be isolated to give rough age synchrony (Zuckerman *et al.,* 1971). Tighter age synchrony in *C. elegans* was obtained by permitting hermaphrodites to deposit eggs on the agar surface for a 4-hr period and then gently washing the hermaphrodites from the plate, leaving the eggs stuck to the agar, or by washing freshly hatched worms, which do not stick to agar, free from eggs stuck on agar (Byerly *et al.,* 1976). A degree of synchrony can also be obtained by collecting a population of animals of similar size, all in the size range of second-stage juveniles. This can be accomplished by filtration, either through a column of 300- to 500-μm diameter glass microbeads (Samoiloff and Pasternak, 1968; Gershon, 1970), or by filtration through fine mesh material (Myers *et al.,* 1971; Hieb and Rothstein, 1975). *Panagrellus redivivus* can be synchronized at the molt between the second and third larval stages by washing gravid females and placing them in salt solution. Larvae emerge from the female, but only grow to the L2–L3 molt, where they will remain until the addition of nutrient. Larvae may be held in this condition for several days with no effect on their subsequent growth. This method does not work with *C. elegans* larvae, which will complete postembryonic development through the fourth juvenile stage in the absence of nutrients.

2. Length

The most commonly used indicator for progress through postembryonic development is the length attained by individuals in a synchronously growing population. Each stage of a particular species has its characteristic length range, although this range will vary as a function of culture conditions. Growth occurs during the intervals between molts, with a cessation of growth during molts. *Panagrellus redivivus* undergoes a 5% decrease in length at each molt (M. R. Samoiloff, unpublished results). Growth of living animals can be determined by the video recording of these individuals and the measurement of stopped-frame tracings from the video screen (Samoiloff, 1973), or by microflash photography. Samples from a synchronously growing population can be taken and can be heat killed and stained by simple staining methods (Bird, 1971) such as 0.0025% cotton blue in lactophenol. Measurement can then be made through a compound microscope. An electronic device for measuring and counting individual nematodes has been reported (Byerly *et al.,* 1975). This device uses a constant-velocity flow of oriented nematodes through a transducer, detecting resistance changes in a 56-kHz detection current. This automated system is also sensitive to molting and reproductive activities.

The use of length attained by a synchronously growing population as an indicator of the toxic effects of environmental agents offers several advantages.

Quantitative data can be quickly obtained with a minimum of expense. Routine determination of length does not require highly trained personnel. Of the possible indicators, length attained represents the parameter most suitable for routine detection of sublethal toxic effects. However, several reservations concerning this method have been raised and must be considered. Slight differences in length may represent significant differences in volume (Kisiel *et al.*, 1972). There may be differential sensitivity to a particular agent at different times in the postembryonic period (Boroditsky and Samoiloff, 1973). The specific pattern of growth may vary from trial to trial (Pasternak and Samoiloff, 1970; Leushner and Pasternak, 1975). Any method of assay for toxic effects must be selective in terms of growth conditions, choice of experimental organism, and the actual numerical analysis used for the assay. The selection of an appropriate growth medium can be based on several different axenic media available for free-living nematodes (this volume, Chapter 3). One difficulty with axenic culture is that the addition of an agent of unknown action may act on the components of the medium, producing an effect on growth not related to a direct action on the biological system. Growth on an agar surface, such as NG agar (Brenner, 1974) or Czapek dox agar (Samoiloff and Pasternak, 1968), is unsatisfactory since the organisms must be immersed in the medium, and the agar may selectively bind the agent tested, removing it from solution. The best culture medium for toxicity assay is a minimal salt solution containing heat-killed food organisms and sterol. *Panagrellus redivivus* grows well in autoclaved M9 buffer (Brenner, 1974) containing 0.05 mg/ml of Baker's yeast and 5 mg/liter of cholesterol. *Caenorhabditis elegans* will also grow in this medium, although it will not grow on autoclaved *E. coli* (Hieb and Rothstein, 1968). A high frequency of dauer larvae is observed with *C. elegans* in this medium. *Panagrellus redivivus* is the indicator organism of choice because postembryonic development in this species is more readily entrained to external stimuli (Samoiloff, 1978).

Experiments in which starvation-synchronized *P. redivivus* second-stage juveniles were grown for 96 hr in an M9 autoclaved-yeast medium containing various concentrations of putative toxicants (M. R. Samoiloff, unpublished results) have demonstrated that the mean length attained by such populations is significantly different from the mean length of untreated controls only at extremely high concentrations of some putative toxicants (Table I). However, the standard deviation of most experimental and control populations was quite high. Although part of this variation is the result of sexual dimorphism that could be experimentally eliminated by the elimination of one sex by genetic manipulation or the recording of length and sex, this large variability represents a major drawback of mean population length as an indicator of toxicity.

My laboratory has been using an operational model for the postembryonic development of *P. redivivus* in which it is considered that each molt represents an interface between two developmental options—to continue development or to

TABLE I

**Mean Length Attained by L2 *P. redivivus* after 96 Hours of Growth in M9-Autoclaved Yeast[a]
Containing Known Concentrations of Toxicants[b]**

Toxicant	Concentration (M)					
	10^{-8}	10^{-7}	10^{-6}	10^{-5}	10^{-4}	10^{-3}
Cadmium	668[c]	642[c]	760[c]	666[d]	544[d]	412[e]
Hexachlorobenzene	713[c]	722[c]	563[d]	717[f]	818[f]	842[c]
Benzene	742[f]	754[d]	774[c]	715[f]	781[c]	719[f]
Ethyl methanesulfonate	808[c]	843[c]	925[c]	891[f]	908[f]	945[c]
Proflavine	534[f]	765[c]	623[c]	629[c]	414[d]	229[g]
5-Bromouracil	646[c]	610[c]	645[c]	662[c]	609[c]	649[d]

[a] 0.05 mg/ml.
[b] Control length = 830 ± 204 μm.
[c] Standard deviation of length between 150 and 200 μm.
[d] Standard deviation of length between 100 and 150 μm.
[e] Standard deviation of length between 50 and 100 μm.
[f] Standard deviation of length between 200 and 250 μm.
[g] Standard deviation of length less than 50 μm.

stop growth, pending an appropriate stimulus. This operational "interface" model has led M. R. Samoiloff (unpublished results) to propose that the frequency of completion of a particular molt is a good indicator of the degree of inhibition produced by a particular agent. These values are obtained by first obtaining the length distribution of a synchronous population following a fixed growth period in the agent tested. From the size distribution, the number of animals in the size range corresponding to each stage is determined, with the size at which a particular molt occurs included in the stage leading to that molt. The values p_1, p_2, and p_3 are calculated as follows:

$$p_1 = \frac{b + c + d}{a + b + c + d}$$

$$p_2 = \frac{c + d}{b + c + d}$$

$$p_3 = \frac{d}{c + d}$$

where a is the number of L2 animals, b is the number of L3 animals, c is the number of L4 animals, and d is the number of adults.

The value p_1 represents the frequency of the L2 animals in the initial synchronous population completing the L2–L3 molt. The value P_2 is the frequency of animals that have reached the L3 stage completing the L3–L4 molt, whereas p_3 represents the probability of L4 individuals completing the L4–adult molt. Standard growth conditions are established so that the value of p_3 of control populations is 0.5. This is accomplished in a 96-hr growth period in autoclaved M9-yeast with sterol. Table II presents the concentrations of several known toxicants at which a significant difference in the values of p_1, p_2, and p_3 from control populations is obtained. Each toxicant was tested at concentrations from 10^{-3} to 10^{-8} M, with each concentration tested on four replicates of 100 animals each. For each toxicant at least one of the three molts was sensitive to a particular toxicant, indicating that different mechanisms may be acting at each molt.

The use of these values for the detection of sublethal effects of toxicants requires only the determination of the number of L2, L3, L4, and adult animals in the experimentally grown population, rather than the exact length distribution in the population. Such determination based by stage is much more rapid than exact length measurement, and can be performed by an automated image-analysis system in which each nematode is fitted to an appropriate "window" corresponding to the size range of each stage. Such a system can be used to analyze 10 experimental populations in the time required for the size measurement of one experimental population.

The mechanisms by which the agents discussed here act to inhibit growth are obscure. The L2–L3 molt in *P. redivivus* appears to be regulated by the hindgut and cell bodies of the nerve ring (Samoiloff, 1973). Damage to the hindgut inhibits growth prior to cuticle formation, while damage to the cell bodies of the nerve ring block the molt at ecdysis.

Inhibition of RNA synthesis in *P. redivivus* does not necessarily block all subsequent protein synthetic activity. Both the cycles of collagen biosynthesis

TABLE II

Minimum Molar Concentration of Toxicants Causing Significant Reduction in Frequency of Completion of Second, Third and Final Molts in *P. redivivus*

Toxicant	p_1	p_2	p_3
Cadmium chloride	10^{-7}	10^{-8}	10^{-5}
Mercuric chloride	10^{-6}	10^{-7}	10^{-8}
Methyl mercury	10^{-7}	10^{-8}	10^{-8}
Selenium oxide	10^{-5}	10^{-5}	10^{-8}
Benzene	ND	10^{-8}	10^{-4}
Hexachlorobenzene	10^{-6}	10^{-7}	10^{-8}
Proflavine	10^{-8}	10^{-8}	10^{-4}
5-Bromouracil	10^{-8}	10^{-7}	10^{-8}

(Pasternak and Leushner, 1975) and the induction of synthesis of alcohol dehydrogenase (Kriger *et al.*, 1977) occur in the presence of actinomycin D, suggesting posttranscriptional control of these events separate from the events of growth. A mutant of *C. elegans* with abnormal postembryonic somatic blast cell division (Albertson *et al.*, 1978) normally grows to the adult stage, demonstrating that these divisions are not required for either growth or molting. This is consistent with the observation that inhibitors of DNA synthesis at least have no effect on the completion of the earlier molts. The blockage of protein synthesis or RNA synthesis inhibits growth, probably as the result of the inhibition of the synthesis of one or more essential precursors for growth. The hydroxylation blocking agent 1, 10-phenanthroline blocks growth completely (Noble *et al.*, 1978), demonstrating that, as well as a requirement for synthesis, growth requires the processing of products of biosynthesis.

Starvation blocks growth in *P. redivivus* early in postembryonic development, but does not stop larval growth of *C. elegans,* possibly indicating differences in pool size, pool utilization, or the nature of the stimuli for postembryonic growth. In summary, the inhibition of growth may act as an indicator for toxicity, but the nature of that toxicity and the target system cannot be inferred.

3. Reproduction

The ability to reproduce is the other obvious consequence of development. The development of the reproductive system from the four-celled gonad primordium found in the L1 or L2 stage has been described for *C. elegans* (Hirsh *et al.*, 1976) and *P. redivivus* (Hechler, 1970; Westgarth-Taylor and Pasternak, 1973; Boroditsky and Samoiloff, 1973). Mutants of *C. elegans* with altered development of the reproductive system or expression of secondary sex characteristics have been reported (Vanderslice and Hirsh, 1976; Hodgkin and Brenner, 1977; Nelson *et al.*, 1978; Ward and Miwa, 1978; Abdulkader and Brun, 1978), demonstrating the degree of susceptibility of gonad development to dissection by genetic probes. In dioecious species, such as *P. redivivus* or *T. aceti*, reproduction requires both normal development of the reproductive system and some mechanism to bring the two sexes together for copulatory activity to occur. In *P. redivivus* only adult males are competent to respond to a pheromone produced only by adult females (Cheng and Samoiloff, 1972). A mating attraction system has not been found in *C. elegans*.

Four types of assay for the development of the reproductive system have been used. Population growth, or population obtained from an initial inoculum in a fixed duration of time, can detect very slight differences in the reproductive ability of animals exposed to different treatments. The determination of generation time considers both the growth and the development of the reproductive system. Fecundity is the most rapid quantitative indicator of maturation of the reproductive system. Cytological observation of gonad morphology following a particular treatment may demonstrate the production of specific lesions in gonad development. Each of these assays has advantages for a particular type of study, although it is difficult to integrate the results obtained from the different types of assay.

Decreased population growth can be the result of lowered fecundity, increased generation time, or inefficient mate-finding. An increased generation time may be the consequence of retarded postembryonic development, delayed gonad maturation, abnormalities in the gametes, or inhibition of embryogenesis. Fecundity data are highly variable. A high population density, low levels of nutrients, or growth in a liquid medium causes an increase in the frequency of *endotokia matricida* in both *P. redivivus* and *C. elegans,* resulting in a significant decrease in mean fecundity under these conditions.

Reproductive activity will also have a marked influence on fecundity. A *P. redivivus* female will produce an average of approximately 100 progeny following a single mating, but will produce an average in excess of 300 progeny following matings with several males. A *P. redivivus* female will exhaust the sperm from a single mating in 3 days, but will be able to mate after this period and produce a second brood of offspring. The maximum number of offspring produced by a series of multiple matings of a *P. redivivus* female is 700 (S. L. Schulz, unpublished results). One *P. redivivus* male mated to many females can sire as many as 3000 progeny. Any expression of a change in reproductive capacity, whether detected by population growth, generation time, or fecundity, is usually presented as a relative value compared to control values for a particular set of conditions, rather than in terms of the production of a specific lesion in the reproductive system.

Inhibition of reproduction by the blocking of sexual attraction has been reported in *P. redivivus* (Cheng and Samoiloff, 1972). Adult females placed in 200 µg/ml actidione do not produce mating attractant, although this blockage is reversible. Males grown in 400 µg/ml hydroxyurea were not attracted to normal females, although these males had normal copulatory behavior. Females grown in hydroxyurea do not attract normal males, but will stimulate copulatory behavior. The treatment of adult animals with hydroxyurea had no effect on mating attraction or copulation. The requirement for protein synthesis for continued production of female pheromone has been reported in the animal parasitic nematode *Nippostrongylus brasiliensis.* Females of this species removed from actidione-treated mice do not produce sex pheromone (Bone and Shorey, 1978).

Agents other than those inhibiting DNA synthesis may selectively inhibit reproduction in free-living nematodes. It appears that all free-living nematodes have a requirement for steroid in order to reproduce (Hieb and Rothstein, 1968). Growth of *P. redivivus* in a steroid-free medium produces normal adult males and females blocked at the L4–adult ecdysis (Abdulrahman, 1975). Such females often have blistered cuticles and multiple vulvae. *Panagrellus redivivus* grown in

25 μg cordycepin or 50 μg/ml proflavine showed the same effect: normal males, but females blocked in the final ecdysis with cuticular blisters and multiple vulvae (M. R. Samoiloff and F. Kriger, unpublished results).

The similarity of the effects of steroid deprivation and low levels of cordycepin or proflavine are suggestive of a common mode of action. Cordycepin or proflavine will specifically block the hydrocortisone-induced synthesis of glutamine synthetase in the chick embryo neural retina (Sarkar *et al.*, 1973; Moscona and Wiens, 1975), probably by the selective inhibition of a polyadenylated transcript or by competitive binding to steroid receptors on chromosomes for cordycepin and proflavine, respectively. Similar cuticular blisters are observed in *P. redivivus* treated with 2000 μg/ml hydroxyurea (Westgarth-Taylor and Pasternak, 1973), in *C. briggsae* grown in 100 μg/ml aminopterin, and in *C. elegans* sterile adults grown from L2 animals irradiated with 20 krads of γ-radiation (M. R. Samoiloff, unpublished results).

Fecundity has been used in my laboratory as an indicator of the effects of several toxicants. To assay for fecundity, 100 L2 *P. redivivus* were grown individually in 0.5 ml of M9 steroid-autoclaved yeast medium with toxicant in 2.5-ml autoanalyzer cups for 120 hr. The virgin adults were washed free of toxicant and mating pairs were placed in fresh medium. The number of progeny produced by each pair was recorded. Counts were completed when the female entered *endotokia matricida*.

No progeny were obtained from crosses of animals grown in cadmium chloride at concentrations of 10^{-5} M or greater, and fecundity was significantly reduced at 10^{-7} and 10^{-6} M. No reproduction occurred in hexachlorobenzene in concentrations of 10^{-5} M or more, with significant reductions in fecundity at concentrations as low as 10^{-8} M. Selenium oxide stopped reproduction at 10^{-5} M, and mercuric chloride produced sterility at 10^{-6} M. Benzene had no effect on reproduction at 10^{-3} M. Proflavine blocked reproduction at 10^{-4} M, with significant decreases in fecundity at concentrations up to and at 10^{-8} M. At 10^{-8} and 10^{-3} M, 5-bromouracil produced a significant decrease in fecundity, but intermediate concentrations were normal. Two-hour exposures to either ethyl methanesulfonate or 5-bromouracil during postembryonic development produced marked reductions in the fecundity of *P. redivivus* (Samoiloff and Smith, 1971). The greatest sensitivity was during the molts, with the least sensitivity during the L4 stage.

4. Senescence

Although the use of free-living nematodes as models for the study of the mechanisms of aging as first proposed by Gershon (1970) has flourished (see Chapter 1, this volume), there have been few attempts to use these organisms for what has been termed "pharmacolongevity" (LaBella, 1972), investigations into the pharmacological prolongation of life. Free-living nematodes grown in axenic culture would seem to be ideal model organisms for such study. On the surface, any agent capable of increasing the mean life span of a synchronous population of nematodes would appear to be a likely anti-aging agent. However, slight differences in nutritional state can result in changes in the developmental period, with concomitant changes in the mean life span (Abdulrahman and Samoiloff, 1975; Croll *et al.*, 1977). Studies on anti-aging effects must be carried out with strict attention to other developmental effects of the agents tested.

Three classes of agents with possible anti-aging effects (LaBella, 1972) have been tested in free-living nematodes. α-Tocopherol (vitamin E), which acts as a general biological antioxidant, increases the mean life span of axenically grown *C. briggsae* (Epstein and Gershon, 1972) and *P. redivivus* (Buecher and Hansen, 1973), although further work with this agent has not been reported.

The drug Gerovitol H3 has been reported to increase longevity in rats (Aslan *et al.*, 1964). The active component of Gerovitol H3 is procaine hydrochloride. Procaine hydrochloride, used alone or as a component of Gerovitol H3, reversibly blocked the growth and reproduction of *C. briggsae* at a concentration of 18 mg/ml, and reversibly retarded growth and reproduction at concentrations of 15 and 9 mg/ml. However, these latter concentrations also significantly increased the mean life span and delayed the onset of the osmotic fragility characteristic of senescence (Zuckerman, 1974). At concentrations lower than 1 mg/ml (0.36 M*M*) procaine hydrochloride, no inhibition of growth was detected, and at concentrations less than 0.1 mg/ml, no effects on reproduction were observed, but the onset of osmotic fragility was delayed (Castillo *et al.*, 1975). Centrophenoxine, reported to decrease lipofuscin accumulation in guinea pigs (Nandy, 1968), did not increase longevity in *C. briggsae* in the concentration range 10–17 m*M*, but did block the age-related increase in specific gravity and osmotic fragility (Kisiel and Zuckerman, 1978). Growth in 6.8 m*M* centrophenoxine produced a 42% decrease in lipofuscin in 21-day-old nematodes. The effects of *p*-chlorophenoxyacetate (PCA) and dimethylaminoethanol (DMAE) on *C. briggsae* were also examined (Zuckerman and Barrett, 1978). The combination of these two agents at 6.8 m*M* each produced a significant decrease in both the number and area of lipofuscin granules in 21-day-old nematodes, whereas 6.8 m*M* of either product had no significant effect on lipofuscin.

Although the above reports comprise the only attempts at pharmacolongevity using free-living nematodes, the use of inhibitors in aging studies of free-living nematodes is common, resulting from Gershon's (1970) report that inhibition of reproduction with DNA-synthesis inhibitors has no effect on longevity.

B. Enzyme Activity

Inhibitors of the respiratory chain and uncouplers of oxidative phosphorylation have been tested in isolated mitochondria from free-living nematodes. Rothstein *et al.* (1970) reported that the cytochrome oxidase inhibitors, cyanide or azide, inhibited the respiration of isolated *T. aceti* mitochondria. The effect of inhibitors on isolated *C. elegans* mitochondria was reported by Murfitt *et al.* (1976). They reported that 30 μmole/mg of protein potassium cyanide or 5000 μmole sodium azide completely block state 3 respiration. Rotenone (35 pmole/mg of protein), which blocks NADH dehydrogenase, or antimycin A (60 pmole/mg of protein) acting between cytochromes *b* and *c,* completely blocks state 3 respiration in *C. elegans* mitochondria. Rutamycin (0.3 nmole/mg of protein), dicyclohexyl-carbodiimide (10 mmole/mg of protein), or tri-*n*-butyltin (1.5 mmole/mg of protein) completely stopped oxidative phosphorylation in isolated *C. elegans* mitochondria, and inhibited mitochondrial ATPase. Carbonyl cyanide *m*-chlorophenyl-hydrazone, pentachlorophenol, or 2,4-dinitrophenol at micromolar concentrations per milligram of protein uncoupled oxidative phosphorylation but had no stimulatory effect on mitochondrial ATPase. Treatment of whole animals with inhibitors of mitochondrial enzymes gives highly unsatisfactory results, either because of impermeability of the cuticle or because of some system of detoxification. For example, rotenone does not inhibit the in vivo metabolism of methanol in *P. redivivus,* but will inhibit the in vitro metabolism of alcohols (D. Burke and M. R. Samoiloff, unpublished results).

Panagrellus redivivus, when placed in 7% methanol or ethanol, will cease activity for a period of approximately 30 min, but will then resume normal activity. During this period the level of alcohol dehydrogenase increases threefold, and the level of aldehyde dehydrogenase increases sevenfold (Kriger *et al.,* 1977). Neither actinomycin D nor cordycepin inhibits this adaptation, but cycloheximide does block both the behavioral recovery and the increased enzyme activity. The adaptation does not occur in the presence of 100 mM pyrazole, an inhibitor of alcohol dehydrogenase.

A single electrophoretic species of alcohol dehydrogenase is found in extracts of mixed stages of *P. redivivus* (D. Burke, unpublished results). The alcohol and aldehyde dehydrogenase systems of *P. redivivus* represent the only free-living nematode enzyme systems examined with inhibitors both in vitro and in vivo, and may represent a model for differential protein synthesis paralleling cuticular collagen biosynthesis in *P. redivivus* as an experimental system. Blockage of hydroxylation of the nascent collagen with 1,10-phenanthroline stabilizes the polysome-nascent polypeptide complex, permitting the isolation of cuticular collagen mRNA, capable of translation using a wheat germ cell-free extract (Noble *et al.,* 1978).

C. Chemical Mutagens

After Brenner's (1974) genetic study, standard methods of mutagenesis with ethyl methanesulfonate (EMS) have been followed for the generation of the vast array of mutants in *C. elegans.* The methodologies for the generation of mutant lines are discussed by Herman and Horvitz (Volume 1, Chapter 6), and methods for the detection of agents causing mutagenesis will be discussed later in this section. However, mutagenesis per se can be used as a direct probe for the dissection of developmental or vital processes. The first attempt at such a probe was by Samoiloff and Smith (1971). These workers used 2-hr pulses of EMS or 5-bromouracil to detect abnormal development as measured by decreased fecundity subsequent to the pulse of mutagenesis. The authors hoped to be able to determine the time of the gene action events of postembryonic development that could be blocked by mutagenesis prior to these events but would not be blocked by mutagenesis after gene activity. However, these expectations were not met, since *P. redivivus* larvae showed differential sensitivity to these mutagens, with the greatest sensitivity at the time of the molts. Their results indicate that the ability of chemical mutagens to penetrate to the somatic or gonadal nuclei varies through postembryonic development. This approach was, however, used in conjunction with radiation-induced mutagenesis, as will be discussed in the next section.

Brenner (1974) estimated the number of essential genes on the X chromosome of *C. elegans* by first determining the frequency of lethal mutation on the X chromosome following standard EMS treatment and dividing this value by the calculated frequency of mutation per locus. Using experimentally determined values, Brenner (1974) calculated a lethal mutation frequency of 0.15 per mutagenized X chromosome, and a mutation rate of 5×10^{-4} per locus, resulting in an estimated 300 essential loci per X chromosome. Extending this value to the five autosomes, Brenner estimated 2000 essential loci per haploid genome of *C. elegans*. Similar calculations were performed using the *P. redivivus* X chromosome (Burke, 1978). The lethal mutation frequency per X chromosome of *P. redivivus* was found to be 0.135, with a mutation frequency per visible locus of 4.4×10^{-4}, resulting in an estimated 310 essential loci per X chromosome. These values are of importance for the detection of mutagenesis, using as an assay system the production of lethal mutations on the X chromosome, and in the target-theory analysis of radiation-induced blockage of postembryonic development.

With these numbers of essential loci on the X chromosome, methods for the detection of mutagenic properties of environmental agents become relatively simple. A mutation frequency of 10^{-5} mutations per locus should generate approximately 0.3% lethals, which could be detected by screening 500 gametes. Brenner (1974) detected lethal mutation by using an autosomal marker epistatic to an X chromosome marker to ascertain that mating had taken place, and then isolating the treated X chromosome in hemizygous males of *C. elegans*.

Caenorhabditis elegans can also be used to detect lethal mutations by exposing an individual heterozygous for two markers on any chromosome to a mutagen, selecting the heterozygous progeny, and scoring their progeny for the absence of individuals homozygous for one of the two markers. The dioecious *P. redivivus*, with XO males and XX females, has been used for the detection of mutagenic agents (Samoiloff, 1979). A mutation on the X chromosome designated b7 (Burke, 1978) causes an immobile, coiled posture of homozygous or hemizygous individuals in a liquid medium. Females homozygous for b7 are grown in putative mutagen-containing medium and are mated to wild-type males. At high mutation frequencies, greater than 5×10^{-5} per locus, significant differences in the sex ratio can be used as an indicator of mutagenesis. For lower mutation frequencies, the heterozygous females are isolated and crossed to wild-type males, and the progeny of these individuals are scored for the presence or absence of the b7 expression. The absence of b7 males in the progeny indicates a lethal mutation on the b7-marked X chromosome derived from the female grown in the putative mutagen. This method can be used to determine a spontaneous mutation frequency of 7×10^{-6} mutations per locus for the X chromosome of *P. redivivus*. Growth in 10^{-8} *M* proflavine produced approximately 10^{-4} mutations per locus, whereas 10^{-8} *M* 5-bromouracil produced 3×10^{-5} mutations per locus. Hexachlorobenzene produced 6×10^{-5} mutations per locus at a concentration of 10^{-7} *M*. Cadmium, selenium, or benzene had no detectable mutagenic action.

Several assumptions used in studies of the type described above must be pointed out. The notion of an essential locus refers to (1) a structural gene producing a product essential for embryogenesis or cellular survival, so that any amorphic or hypomorphic mutation results in lethality, or (2) a regulatory gene that functions in some essential process. It is assumed that the essential genes are of equal size, with equal rates of mutation. Mutagenesis of essential loci is assumed to occur at the same rate as for structural genes influencing visible characteristics. The proportion of essential genes to nonessential loci is not known.

D. Radiation

Nematodes are extremely resistant to ionizing radiation (Evans *et al.*, 1941; Thomas and Quastler, 1950; Wood and Goodey, 1957; Myers and Dropkin, 1959; Myers, 1960; Siddiqui and Viglierchio, 1970). This has been ascribed to the eutelic nature of nematodes, or to the syncytial nature of most nematode tissue. *Turbatrix aceti* and *P. redivivus* are sterilized at 40,000 rad doses of γ-radiation (Myers, 1960), whereas *C. elegans* is sterilized at 20,000 rads (M. R. Samoiloff, unpublished results). One percent of *P. redivivus* populations exposed to 1,500,000 rads survived after 24 hr, whereas 5% of the populations exposed to 960,000 rads survived for 24 hr (Myers, 1960), with survivors found in such populations weeks after irradiation. The behavior of *P. redivivus* is normal following radiation doses up to 200,000 rads, with uncoordinated behavior following doses between 400,000 and 500,000 rads, and a cessation of activity following doses in excess of 500,000 rads (Myers, 1960). Myers (1960) reported that pregastrula embryos of *P. redivivus* fail to develop following doses of 10,000 rads, although embryos irradiated after gastrulation were more radiation resistant. Second-stage juvenile females of *P. redivivus* complete postembryonic development following doses of γ-radiation of up to 70,000 rads, express delayed postembryonic development at doses up to 200,000 rads, and fail to develop at doses greater than 200,000 rads (M. R. Samoiloff, unpublished results). Some irradiated second-stage juvenile males fail to develop past the L2–L3 molt at doses as low as 10,000 rads, with no adult males found following irradiation at doses greater than 40,000 rads. This difference in radiation sensitivity may be due to a differential repair capacity of males and females, but is more likely due to the expression of somatic mutation on the X chromosome of essential cells in hemizygous males, resulting in developmental failure. This observation has permitted Samoiloff to determine the number of X-linked loci required for the postembryonic development of *P. redivivus* by the application of target theory. Target theory has been reviewed by Maynard Smith (1968), and can be expressed as

$$\ln S = N \ln R - Npk$$

where S is the proportion of survivors (or successes), N represents the number of targets, p is the dose-related probability of hitting a target, in this case the dose-related mutation frequency, k is the dose applied in rads, and r is the number of replicates of each target, for X-linked loci; $r = 1$ for males and $r = 2$ for females.

The determination of the distribution of length attained by males irradiated at the L2 stage under conditions in which similar doses permit all females irradiated at the L2 stage to reach the adult stage permits calculation of the percentage of animals successfully completing the second, third, and final molts. These values (p_1, p_2, and p_3—Section II, A,2) represent independent values of S. The dose-dependent mutation frequency of *P. redivivus* has been determined by the methods outlined in the previous section as 4×10^{-8} mutations per locus per rad. Calculations using these values permit the estimation of the number of radiation-sensitive targets required for postembryonic development, and completion of the second, third, and final molts. These values have been determined to be 2400, 180, 130, and 1800, respectively.

In this application of target theory, targets specifically refer to the radiation-sensitive X-linked loci per organism. This method cannot distinguish if the 180 targets required for completion of the L2–L3 molt represent 180 essential loci in one cell, 180 essential cells with a single sensitive locus each, or any other distribution of essential cells and loci. However, since it has been calculated that there are approximately 300 essential loci per X chromosome (Brenner, 1974; Burke, 1978) and slightly more than 500 somatic nuclei in second-stage juveniles (Sin and Pasternak, 1970; Sulston and Horvitz, 1977), there are more than 150,000 essential X-linked targets per organism. A dose of 10,000 rads should produce an average of 45 mutations on essential X-linked loci per organism, assuming a radiation-induced mutation rate in somatic cells similar to that determined for gonadal cells. Males treated with 10,000 rads as L2s showed no lethal effects, and less than 2% blockage at the L2–L3 molt, indicating that either extensive compensation for these lethals occurs, or the expression of these loci occurs prior to the second juvenile stage. Support for this latter view comes from the high radiosensitivity of early embryos (Myers, 1960), whereas support for the former view comes from laser-microbeam ablation studies (Samoiloff, 1973; White, cited in Sulston and Horvitz, 1977).

Laser irradiation of the region of L2 *P. redivivus* containing the gonad primordium failed to produce abnormalities of the adult gonad, suggesting that the gonad primordium has some capacity for regeneration or regulation (Samoiloff, 1973). Irradiation of the cell bodies of the nerve ring blocks ecdysis and causes approximately 20% lethality, and approximately 20% of the adults that developed have abnormal gonads. Irradiation of the nerve ring inhibits growth, with adult-length animals showing gonad abnormalities similar to those observed following growth in inhibitors of DNA synthesis. Irradiation of the region anterior to the gonad primordium produced a shift in the sex ratio, with an increased proportion of females, suggesting positional information anterior to the gonad primordium required for the development of the expression of the male phenotype. Laser-microbeam irradiation would appear to be the method of choice for the determination of the specific sites of the controlling regions and receptors and for the determination of specific cellular function.

Panagrellus redivivus, exposed to modulated low-energy microwave radiation at an athermal carrier frequency (2.45 GHz) showed behavioral changes characteristic for a particular modulation frequency (Samoiloff *et al.,* 1973). Exposure to low-energy microwave radiation modulated at 24 kHz will block the L2–L3 molt, although normal growth and development will resume following removal from the wave guide (M. R. Samoiloff, unpublished results). The effects of athermal microwave radiation on *P. redivivus* probably result from the resonance of some cellular dipole, with specific arrays of phospholipid on the cell membrane as the most logical site. Pulsed athermal microwave radiation at appropriate carrier frequencies may represent a highly useful physical probe for events that require cell/cell-surface interactions. Attempts to restore normal function in various uncoordinated mutants of *C. elegans* using pulsed athermal microwave irradiation have been unsuccessful to date (M. R. Samoiloff, unpublished results).

E. Carcinogenesis

Upon first consideration it would seem unlikely that eutelic nematode systems would be of any utility in the detection of carcinogenic agents. The only likely approach for the determination of carcinogenic activity would be to induce abnormalities in dividing postzygotic cells, which remain relatively undifferentiated during their dividing phase, or in developing gonadal tissue. Nematode embryos offer no significant biological advantage over the larger, more readily manipulated amphibian embryonic systems. Gonad development does not hold much promise as an indicator either. Abnormal gonad development has been reported following treatment with hydroxyurea (Boroditsky and Samoiloff, 1973) and nutritional stress (Abdulrahman, 1975). However, the genetic advantages of the free-living nematode systems may be used to detect mutagens as discussed previously, and evidence is accumulating that carcinogens are mutagens (Ames *et al.*, 1975; Nagao and Sugimura, 1978; Ames, 1978). Eukaryotic extracts added to the *Salmonella* test system and the development of new *Salmonella* tester strains have increased the utility of the "Ames test" for the detection of carcinogens, but have made this assay somewhat more laborious. Perhaps the nematode mutagenesis-detecting system will be more sensitive to mutagenesis induced by carcinogens. Mutagenesis testing using known carcinogens has not been performed to date using free-living nematodes, although such tests would demonstrate whether or not the system is useful.

III. PROSPECTS

The approach to the study of free-living nematodes using chemical or physical insults applied to developmental or adaptive processes has been widely used and has added to the overall understanding of these organisms. The use of this method has paralleled the use of genetic probes, primarily in *C. elegans*. While it may appear that the researchers using the three commonly studied free-living nematodes, *C. elegans, T. aceti* and *P. redivivus,* are following divergent pathways, this is in fact a demonstration of the August Krogh Principle: "For many problems there is an animal on which it can be most conveniently studied" (Krebs, 1975). Each of these species has its particular advantages for a particular approach to the study of general biological problems. *Caenorhabditis elegans* is a superb genetic system (Brenner, 1974; Edgar and Wood, 1977), *T. aceti* has major advantages in axenic culture (Hieb and Rothstein, 1975), and *P. redivivus* can be most readily entrained to environmental stimuli. The full potential of free-living nematodes as model systems will be completely realized only by the application of a range of experimental approaches and the integration of these into comprehensive models. The use of chemical and physical probes augments genetic dissection.

REFERENCES

Abdulkader, N., and Brun, J. L. (1978). *Rev. Nematol.* **1**, 27–37.

Abdulrahman, M. (1975). M.Sc. Thesis, University of Manitoba, Winnipeg, Manitoba, Canada.

Abdulrahman, M., and Samoiloff, M. R. (1975). *Can. J. Zool.* **53**, 651–656.

Albertson, D. G., Sulston, J. E., and White, J. G. (1978). *Dev. Biol.* **63**, 165–178.

Ames, B. N., McCann, J., Lee, F. D., and Durston, W. E. (1973a). *Proc. Natl. Acad. Sci. U.S.A.* **70**, 782–786.

Ames, B. N., McCann, J., Yamasaki, E. (1973b). *In* "Handbook of Mutagenicity Test Procedures" (B. J. Kilbey, M. Legator, W. Nichols, and C. Ramel, eds.), pp. 1–17. Elsevier, Amsterdam.

Ames, B. N., McCann, J., and Yamasaki, E. (1975). *Mutat. Res.* **31**, 347–363.

Ames, B. N. (1978). *Science* **204**, 587–593.

Aslan, A., Vrabiescu, A., and Domilescu, C. (1964). *Gerontologist* **4**, 33.

Bird, A. F. (1971). "The Structure of Nematodes." Academic Press, New York.

Bone, L. W., and Shorey, H. H. (1978). *Proc. Helminthol. Soc. Wash.* **45**, 264–266.

Boroditsky, J. M., and Samoiloff, M. R. (1973). *Can. J. Zool.* **51**, 483–492.

Brenner, S. (1974). *Genetics* **77**, 71–94.

Buecher, E. J., and Hansen, E. L. (1973). *Proc. Int. Congr. Plant Pathol., 2nd,* Abstr. 0210.

Burke, D. (1978). M.Sc. Thesis, University of Manitoba, Winnipeg, Manitoba, Canada.

Byerly, L., Cassada, R. C., and Russell, R. L. (1975). *Rev. Sci. Instrum.* **46**, 517–522.

Byerly, L., Cassada, R. C., and Russell, R. L. (1976). *Dev. Biol.* **51**, 23–33.

Castillo, J. M., Kisiel, M. J., and Zuckerman, B. M. (1975). *Nematologica* **21**, 401–407.

Castro, C. E., and Thomason, I. J. (1971). *In* "Plant Parasitic Nematodes" (B. M. Zuckerman, R. A. Rohde, and W. F. Mai, eds.), Vol. 2, pp. 289–296. Academic Press, New York.

Cheng, R., and Samoiloff, M. R. (1972). *Can. J. Zool.* **50**, 333–336.

Croll, N. A., Smith, J. M., and Zuckerman, B. M. (1977). *Exp. Aging Res.* **3**, 175–199.

del Castillo, J. (1969). *In* "Chemical Zoology" (M. Florkin and B. T. Scheer, eds.), Vol. 3, pp. 521–559. Academic Press, New York.

Edgar, R. S., and Wood, W. B. (1977). *Science* **198**, 1285–1286.

Epstein, J., and Gershon, D. (1972). *Mech. Ageing Dev.* **1**, 257–262.

Evans, T. C., Levin, A. J., and Sulkin, N. M. (1941). *Proc. Soc. Exp. Biol. N.Y.* **48**, 624–628.

Gershon, D. (1970). *Exp. Gerontol.* **5**, 7–12.

Hechler, H. C. (1970). *J. Nematol.* **2**, 355–361.

Hieb, W. F., and Rothstein, M. (1968). *Science* **160**, 778–779.

Hieb, W. F., and Rothstein, M. (1975). *Exp. Gerontol.* **10**, 145–153.

Hirsh, D., and Vanderslice, R. (1976). *Dev. Biol.* **49**, 220–235.

Hirsh, D., Oppenheim, D., and Klass, M. (1976). *Dev. Biol.* **49**, 200–219.

Hodgkin, J. A., and Brenner, S. (1977). *Genetics* **86**, 275–287.

Kisiel, M. J., and Zuckerman, B. M. (1978). *Age (Omaha, Nebr.)* **1**, 17–20.

Kisiel, M., Nelson, B., and Zuckerman, B. M. (1972). *Nematologica* **18**, 373–384.

Krebs, H. A. (1975). *J. Exp. Zool.* **194**, 221–226.

Kriger, F., Burke, D., and Samoiloff, M. R. (1977). *Biochem. Genet.* **15**, 1181–1191.

LaBella, F. (1972). *In* "Search for New Drugs" (A. Rubin, ed.), pp. 347–383. Dekker, New York.

Leushner, J., and Pasternak, J. (1975). *Dev. Biol.* **47**, 68–80.

Maynard Smith, J. (1968). "Mathematical Ideas in Biology." Cambridge Univ. Press, London and New York.

Moscona, A. A., and Wiens, A. W. (1975). *Dev. Biol.* **44**, 33–45.

Murfitt, R. R., Vogel, K., and Sanadi, D. R. (1976). *Comp. Biochem. Physiol. B.* **53**, 423–430.

Myers, R. F. (1960). *Nematologica* **5**, 56–63.

Myers, R. F., and Dropkin, V. H. (1959). *Plant Dis. Rep.* **43**, 311–313.

Myers, R. F., Chen, T. A., and Balasubramanian, M. (1971). *J. Nematol.* **3**, 85.

Nagao, N., and Sugimura, T. (1978). *Annu. Rev. Genet.* **12**, 117-159.

Nandy, K. (1968). *J. Gerontol.* **23**, 82-92.

Nelson, G. A., Lew, K. K., and Ward, S. (1978). *Dev. Biol.* **66**, 386-409.

Noble, S., Leushner, J., and Pasternak, J. (1978). *Biochim. Biophys. Acta.* **520**, 219-228.

Pasternak, J., and Leushner, J. R. A. (1975). *J. Exp. Zool.* **194**, 519-528.

Pasternak, J., and Samoiloff, M. R. (1970). *Comp. Biochem. Physiol.* **33**, 27-38.

Rothstein, M., Nicholls, F., and Nicholls, P. (1970). *Int. J. Biochem.* **1**, 695-705.

Samoiloff, M. R. (1973). *Science* **180**, 976-977.

Samoiloff, M. R. (1978). *Proc. Int. Congr. Plant Pathol., 3rd,* **4**, 145.

Samoiloff, M. R. (1979). "Environmental Contaminants Directorate (Canada) Report."

Samoiloff, M. R., and Pasternak, J. (1968). *Can. J. Zool.* **46**, 1019-1022.

Samoiloff, M. R., and Smith, A. C. (1971). *J. Nematol.* **3**, 299-300.

Samoiloff, M. R., McNicholl, P., Cheng, R., and Balakanich, S. (1973). *Exp. Parasitol.* **33**, 253-262.

Sarkar, P. K., Goldman, B., and Moscona, A. A. (1973). *Biochem. Biophys. Res. Commun.* **50**, 308-315.

Siddiqui, I. A. and Viglierchio, D. R. (1970). *Nematologica* **16**, 459-469.

Sin, W. C., and Pasternak, J. (1970). *Chromosoma* **32**, 191-204.

Sulston, J. E., and Horvitz, H. R. (1977). *Dev. Biol.* **56**, 110-156.

Thomas, L. J., and Quastler, H. (1950). *Science* **112**, 356-357.

Vanderslice, R., and Hirsh, D. (1976). *Dev. Biol.* **49**, 236-249.

Ward, S., and Miwa, J. (1978). *Genetics* **88**, 285-303.

Westgarth-Taylor, B., and Pasternak, J. (1973). *J. Exp. Zool.* **183**, 309-322.

Wood, F. C., and Goodey, J. B. (1957). *Nature (London)* **180**, 760-761.

Zuckerman, B. M. (1974). *In* "Theoretical Aspects of Aging" (M. Rockstein, ed.), pp. 177-186. Academic Press, New York.

Zuckerman, B. M., and Barrett, K. A. (1978). *Exp. Aging Res.* **4**, 133-139.

Zuckerman, B. M., Himmelhoch, S., Nelson, B., Epstein, J., and Kisiel, M. (1971). *Nematologica* **17**, 478-487.

part two

Physiology and Morphology

5

Respiration in Nematodes

H. J. ATKINSON

Department of Pure and Applied Zoology
The University of Leeds
Leeds LS2 9JT, England

I.	Introduction	101
II.	Diffusion of Oxygen into Nematodes	102
	A. Diffusion Equations	102
	B. Diffusion in Low Oxygen Regimes	106
III.	Oxygen and the Mitochondrion	109
	A. The Mitochondrion	109
	B. Electron Transport	110
	C. Oxidative Phosphorylation	114
	D. Adenylate Energy Charge	114
IV.	Oxygenases	115
V.	Factors Influencing Oxygen Demand	116
	A. Body Size and Interspecific Comparisons	116
	B. Body Size and Intraspecific Comparisons	118
	C. Trophic Forms	119
	D. Activity	120
	E. Dormancy	121
	F. Lipid Reserves and Starvation	123
	G. Temperature	126
	H. Other Factors	127
VI.	Oxygen Availability	127
VII.	Respiration in Low Oxygen Regimes	129
	A. General Effects of Suboptimal Oxygen Levels	129
	B. Effects on Oxygen Consumption	130
VIII.	Hemoglobin	132
IX.	Anaerobiosis	136
X.	Conclusions	137
	References	138

I. INTRODUCTION

The use of oxygen as a terminal oxidase can result in a far greater efficiency of energy metabolism than possible by anaerobiosis. For instance, aerobic

101

metabolism can yield 38 adenosine triphosphate (ATP) moles per glucose mole utilized, but the most efficient anaerobic pathway so far suggested for nematodes provides 5 ATP moles per mole of glucose (Barrett, 1976). There is such a large difference in metabolic efficiency between aerobic and anaerobic metabolism that most animals possess an array of adaptations which ensure that oxygen is usually available for oxidative phosphorylation. This chapter discusses the oxygen supply and demand in nematodes, the range of their respiratory adaptations, and the consequences of a limited oxygen availability.

II. DIFFUSION OF OXYGEN INTO NEMATODES

Oxidative phosphorylation occurs within the mitochondria of animal cells, and oxygen must be available at the site of utilization at a partial pressure of oxygen (P_{O_2}) of about 1 mm Hg (Chance *et al.*, 1964). The majority of available evidence suggests that thermal diffusion is solely responsible for the movement of oxygen through the cell cytoplasm to the mitochondria. The rate at which oxygen can pass through the respiring cell is such that diffusion from aerated water limits respiration in a sphere of protoplasm greater than about 1 mm in diameter (Krogh, 1941). Most higher animals possess respiratory organs to ensure that oxygen in their surroundings can pass into the animal through a large surface area. Circulatory organs also overcome the difficulties of oxygen supply by carrying oxygen in solution from peripheral to central tissues. The two systems together are so successful that oxygen supply hardly places any restriction on the size of higher animals. Furthermore, a combination of homeostatic controls normally accompanies these systems so that changes in oxygen demand have minimal effects on the P_{O_2} at the mitochondria of some tissues. Nematodes lack both specialized respiratory and complex circulatory organs and so their oxygen supply does not have this homeostatic control.

The sphere offers the lowest surface-to-volume ratio of any shape and so places the severest limitations on body size for a given level of aerobic metabolism. The cylinder offers a better surface-to-volume ratio, and improves contact with the substrate. Nematodes evolved a cylindrical shape and a form of sinusoidal movement, but the arrangement adopted enforces severe limitations on their organization (Lee and Atkinson, 1976). One factor limiting their body size is the supply of oxygen for an aerobic metabolism.

A. Diffusion Equations

The relationship between body size and oxygen availability in nematodes can be expressed using equations derived by Hill (1929)

$$P_c \geq mr^2/4k \qquad (1)$$

or, rearranging the terms:

$$r \leqslant 2\sqrt{kP_e/m} \tag{2}$$

where P_e is the oxygen partial pressure surrounding the tissue (atmospheres), r the radius of tissue (centimeters), k the permeability constant for oxygen diffusing into a tissue, usually taken as $8.4 \times 10^{-4} cm^2/atm/hr$ (Krogh, 1919) and m is the oxygen consumption of tissue (cubic centimeters of oxygen per cubic centimeter of tissue per hour).

Rogers (1962) applied these formulas to describe the diffusion of oxygen into nematodes. The equations emphasize the importance of body radius because its square is proportional to the P_{O_2} necessary to maintain a given level of oxygen consumption. Each of the parameters in the equation is subject to modifications by a variety of factors, which are given in Table I with some estimate of the likely variation.

The value for k may not be a constant for the whole nematode, but the range of variation may not be great. Chitin has a very low permeability, but its value is about 0.1 of that for muscle (Krogh, 1919). The basic component of the cuticle of nematodes is a form of collagen (Lee and Atkinson, 1976), but it is unlikely to have a much lower permeability than chitin, and the cuticle represents only a small proportion of the body radius. The passage of oxygen is due to thermal diffusion so that k is also dependent on temperature, but this factor is more than offset by the effect on oxygen consumption that such a change incurs. The apparent k may also alter if hemoglobin is present (Section VIII) or if the pseudocoelom occupies a large proportion of the body radius. A thorough mixing of the pseudocoelomic fluid during locomotion will result in the inner surface of the body wall and the outer surface of the intestine having a similar P_{O_2}. It is difficult to judge how frequently thorough mixing of the pseudocoelomic fluid does occur in nematodes.

The maximum environmental P_{O_2} normally experienced by a nematode will be in aerated water. Variations in different habitats may be important (Section VI). Two additional minor factors may arise during locomotion. A diffusion gradient may extend beyond the nematode into the surrounding fluid and this will be reduced by movement. There will also be a stationary layer of fluid at the surface of the nematode which will alter with the rate of movement of the animal.

These factors are important in aquatic insects that often show respiratory movements (Walshe, 1950), but such activity has not been detected even for larger nematodes. Some microenvironments such as soils are heterogeneous and this may influence the oxygen availability of those parts of the nematode in contact with a solid.

Body radius is a major factor influencing the oxygen availability to a nematode. The analysis becomes inaccurate if nematodes have elliptical cross sections or acquire a subspherical shape, as in the Heteroderidae. The fusiform

TABLE I

Some Modifers of Parameters in Equations (1) and (2)

Parameter	Modifying influences	Extent of variation[a]
k	Hemoglobin in tissues	$++$?
	Differences between tissues	$+$
	Pseudocoelomic mixing during locomotion	$+$
	Temperature (Q_{10}, 1.1)	$+$
m	Body size	$++$
	Contribution of anaerobiosis to metabolism	$++$
	Temperature	$++$
	Different oxygen requirements of tissues	$+$
	Changes in oxygen consumption by a tissue	$+$
P_e	Stationary layer around nematode	$+$
	Heterogeneous medium	$+$
r	Body size	$++$
	Cross section not circular	$+$
	Fusiform shape	$+$

[a] $++$, Major modifying factor; $+$, minor modifying factor.

shape of most species also ensures that oxygen is more readily supplied to the anterior and posterior regions than to the midregion. This may be of some benefit to the anterior and posterior neural ganglia because nerves are particularly susceptible to oxygen lack (Section VII,A). The oxygen consumption of a nematode is also related to body size (Section V,A) and a variety of other factors (Section V,B–H).

Rogers (1962) was aware of many of the modifying factors listed in Table I but attempted a comparison between oxygen consumption and availability for *Haemonchus contortus, Nematodirus* spp., and *Nippostrongylus brasiliensis* (*=muris*). He was able to show decreases in oxygen consumption at the level of oxygen availability predicted by the equation from their body radii. However, this may have been somewhat fortuitous because the oxygen levels fell rapidly over a short period, and it would have been preferable if the animals had been allowed to adapt to particular oxygen levels (Section VII,B). These nematodes also contain hemoglobin, which may profoundly influence their oxygen uptake at a low P_{O_2} (Section VIII).

The equation may be too imprecise to be useful in particular cases but it does have some value for its broad implications. The main point of interest is the nature of the relationship between oxygen availability, oxygen consumption, and body radius in nematodes across a whole array of sizes. This relationship is given in Fig. 1. A regression line for oxygen consumption against body radius has also been included. This relationship is similar to that obtained from the standard energy metabolism of nematodes (see Fig. 6) after assuming that body weight is proportional to the cube of body radius. This can be used to estimate the P_{O_2} necessary to support aerobic metabolism for the full size range of nematodes (Fig. 2). This emphasizes the adequacy of diffusion for many nematodes which have a body radius of less than 0.05 mm. A P_{O_2} of more than 15 mm Hg should provide an adequate oxygen supply to these animals within the likely limits of the parameters given in Table I. The two main exceptions are nematodes from low

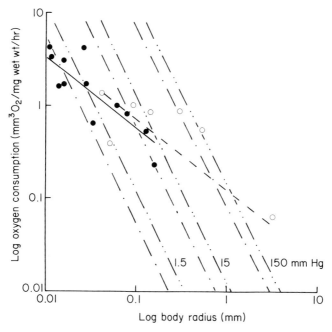

Fig. 1. The theoretical relationship between the maximum oxygen consumption limited by diffusion and body radius in nematodes. The relationship is given for partial pressures of oxygen of 1.5, 15, and 150 mm Hg at 20°C (—·—) and 37°C (—··—). Values for actual oxygen consumption are superimposed for the nematodes listed below. The regression lines for animal parasites (- - -) and other nematodes (—) were similar to those obtained from Eq. (4) in Fig. 6 after allowing for the relationship between body radius and wet weight ($P > .05$).

Animal parasites (○) listed in order of increasing body radius are *Nematospiroides dubius* (Bryant, 1973), *Nippostrongylus brasiliensis, Nematodirus filicolis, Haemonchus contortus, Litomosoides carinii, Ascardia galli,* and *Ascaris lumbricoides* (Rogers, 1962).

Other nematodes ● listed in order of increasing body radius are *Paratylenchus, Helicotylenchus* (Klekowski *et al.,* 1972); *Aphelenchus avenae, Ditylenchus dipsaci* (Rohde, 1960); *Panagrolaimus* (Klekowski *et al.,* 1972); *Panagrellus redivivus* (Santmeyer, 1956); *Caenorhabditis, Trichodorus christiei* (Rohde, 1960); *Aporcelaimus regius* (Nielsen, 1949); *Enoplus brevis, Enoplus communis* (Atkinson, 1973a) and *Pontonema vulgaris* (Nielsen, 1949).

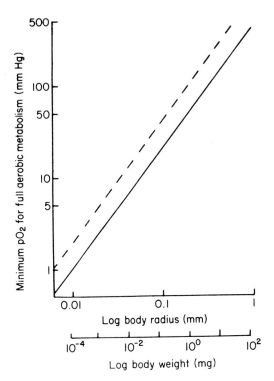

Fig. 2. Theoretical relationship between the minimum external partial pressure of oxygen required to support the aerobic metabolism of a nematode and body radius or body weight. Values at 20°C (—) and 37°C (---). The values are computed from Eq. (1) by substituting values for oxygen consumption (*m*) at a given weight and radius from the equation given in Fig. 6. The relationship between radius and body weight is estimated approximately from Andrassy (1956) using a ratio between length and maximum width of 30:1 (Wallace, 1971). The values at 37°C are estimated from the relationship at 20°C using a Q_{10} of 2.5 for oxygen consumption and a Q_{10} of 1.1 for *k*.

oxygen environments and the relatively few species that have a large body radius.

B. Diffusion in Low Oxygen Regimes

The model discussed above is probably of most value for determining if the environmental P_{O_2} is well above that required for the aerobic metabolism of a particular nematode. It becomes less reliable if the P_{O_2} is similar to this value, and it is an oversimplification to consider lower levels of oxygen supply as inadequate for an aerobic metabolism. The superficial layers of the nematode will be more readily supplied with oxygen than its central tissues, and an equation that estimates this effect is given by Hill (1929). This formula has been used to estimate the proportion of the cross section of a nematode that can be supplied by a given percentage of the P_{O_2} that is required to supply all its tissues (Fig. 3). This suggests that the cuticle, hypodermis, and musculature of the body wall will be supplied by 20% of the P_{O_2} required for all tissues, and that 4% is adequate for the cuticle and hypodermis only. It suggests that it is misleading to consider the oxygen supply to a nematode as either adequate or inadequate.

The analysis suggests that most adult trichostrongyle nematodes will be adequately supplied with oxygen if their environmental P_{O_2} is about 30 mm Hg (Section VI). A large species such as *H. contortus* with a body radius of about 0.1 mm can support a substantial aerobic metabolism at this level of oxygen availability. Even extremely large nematodes such as *Ascaridia galli* and *Ascaris lumbricoides* may receive sufficient oxygen in their environments to supply at least the hypodermis with oxygen.

Other factors also become important at a low environmental P_{O_2}. It is certain that hemoglobin has a respiratory function in at least one nematode, and this may influence the availability of oxygen to certain tissues at a low P_{O_2} (Section VIII). One adaptation that would favor the oxygen supply in a low oxygen regime is an increase in surface area without altering the volume of respiring tissue. *Nippostrongylus brasiliensis* has an enlarged, fluid-filled cuticle with 14 longitudinal

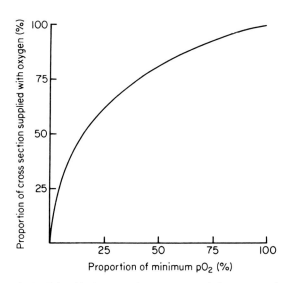

Fig. 3. Theoretical relationship between the percentage of the cross section of a nematode supplied with oxygen and values for P_{O_2} as a percentage of that required to supply all the nematode. Relationship derived from equations by Hill (1929). In a typical nematode, cuticle, hypodermis, and body wall musculature represent 50–60% of the cross-sectional area, and cuticle and hypodermis 10–30% of the cross-sectional area.

ridges (Lee, 1965). It seems likely that the fluid-filled cuticle consumes little oxygen relative to an equal volume of tissue. Perhaps this represents an increase in surface area that favors oxygen supply. The dimensions of the animal are given in Fig. 4 based on a drawing of the cross section of this animal (Lee, 1965). It can be calculated from these values that the circumference of a circle passing through the tips of each ridge is about 125% of that if the cuticle were closely applied to the hypodermis. Furthermore, the troughs between the ridges

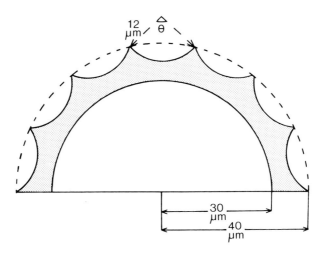

Fig. 4. Diagrammatic representation of half of a transverse section through *Nippostrongylus brasiliensis* to show the outer shape of the cuticle. Dimensions are based on those given by Lee (1965). It is suggested that the complex perimeter of the cross section may increase the surface area sufficiently to influence the oxygen relationships of the nematode. Assume that (1) the respiring cylinder of tissue is bounded by the hypodermis with a radius of 30 μm, and (2) the fluid-filled cuticle consumes little oxygen and has a permeability to oxygen similar to that of the tissues. It is possible to compare three possible arrangements: (i) Cuticle as in *N. brasiliensis* with 14 semicircular troughs. The angle θ subtended by each trough is $\simeq$ 93° and the sector has a radius of 12 μm.

$$\therefore \text{Perimeter of shape} = 14 \times \left(\frac{2\pi \times 12 \times 93}{360}\right) = 272.7 \ \mu m^2$$

(ii) Cuticle fluid filled as in *N. brasiliensis* but lacking semicircular troughs.

$$\text{Perimeter of shape} = 2 \pi \times 40 = 251.3 \ \mu m^2$$

(iii) Typical nematode cuticle with a ratio of its thickness to nematode diameter of 1:34 (Bird, 1971). In this case the ratio would be $\simeq$ 1.9 μm:63.8 μm.

$$2 \pi (30 + 1.9) = 200.4$$

Ratio of i:ii:iii = 1.36:1.25:1. This suggests that the cuticle of *N. brasiliensis* may increase the surface area of the nematode sufficiently to reduce the P_{O_2} required to supply its tissues to about 75% of that of a nematode with a more typical cuticle.

increase the perimeter of the shape, further giving a net gain for both factors of about 36% in surface area.

The head region of this animal is also bulblike because of a fluid-filled cuticle, but it lacks longitudinal ridges; perhaps they would limit the flexibility of the head. However, calculations from drawings given by Wright (1976) suggest that this arrangement also increases the surface area by about 25%. Clearly, this unusual cuticular arrangement may have other adaptive significances (Wright, 1976), but it does seem that the increase in surface area is sufficient to improve the oxygen supply to *N. brasiliensis* in the low oxygen environment of its habitat. The presence of hemoglobin in the cuticle may also favor the uptake of oxygen by this nematode (Section VIII).

There may be other factors that influence the availability of oxygen to a few nematodes. It can be calculated that food, particularly the blood ingested by some nematodes, may contain sufficient oxygen at least to prime an aerobic metabolism in the pharynx during the ingestion process. The behavior of the animal is one further factor that may influence the supply of oxygen in some habitats (Section VI).

III. OXYGEN AND THE MITOCHONDRION

Some phosphorylation of adenosine diphosphate (ADP) can occur in the cytosol and endoplasmic reticulum of cells, but this is not normally of quantitative significance in aerobic metabolism when compared with the rate of oxidative phosphorylation within their mitochondria.

A. The Mitochondrion

Mitochondria share a common basic structure in eukaryotic cells and have similarities to the organization of a bacterium. This has led to a suggestion that mitochondria evolved from symbiotic prokaryotes within an ancestral, anaerobic cell (Margulis, 1975). They consist of (1) an outer membrane that is freely permeable to many solutes, (2) an outer chamber, (3) an inner membrane that has a very selective permeability and is impermeable to H^+ and (4) a central matrix. Each of these four regions is associated with particular enzymes. The components of the electron-transport system, α-keto acid dehydrogenase and succinate dehydrogenase, all occur within the inner membrane, and the remainder of the tricarboxylic acid enzymes are within the matrix of the mitochondrion (Lehninger, 1975). Replenishment of these enzymes in the matrix and exchange diffusion of ADP for ATP are by specific carriers in the inner membrane (Williamson, 1976).

The assembly of the components of electron transport is believed to be arranged in a systematic manner and to occupy a definite area of the inner membrane. It follows that the maximum capacity of a cell for oxidative phosphorylation is proportional to the total area of inner mitochondrial membrane. This is a function of mitrochondrial number and also the complexity of the infoldings of the inner membrane that are termed cristae.

Two factors may influence the number of mitochondria and the extent of development of cristae in nematodes: (1) the extent of aerobic rather than anaerobic metabolism and (2) body size, which influences the oxygen consumption per unit volume of tissue (Section V,A). *Meloidogyne javanica* is an aerobe (Baxter and Blake, 1969), and the body wall musculature of the small, second-stage juvenile appears to have mitochondria with complex cristae, as do the cells of the very active median bulb of the female (Bird, 1971, Figs. 7A and 52, respectively). It is of interest to compare these mitochondria with those found in the tissues of *Ascaris*. Those in the noncontractile part (Reger, 1964) and the sarcoplasm (Rosenbluth, 1965) of the body wall musculature have few cristae, and those in the pharyngeal musculature are similar (Reger, 1966). This difference may be partly attributed to the difference in body weight between *Ascaris* and *M. javanica*. However, it is more likely that this difference is mainly due to changes in metabolism associated with the low P_{O_2} in the habitat of *Ascaris*. *Trichinella spiralis* also has mitochondria with few lamellar cristae (Boczon and Michejda, 1978).

Spermatozoa normally contain mitochondria with complex cristae (Threadgold, 1976), but those in the spermatozoa of *Ascaris* (Kaulenas and Fairbairn, 1968), *Aspiculuris tetraptera* (Lee and Anya, 1967), and *N. brasiliensis* (Jamuar, 1966) all appear to have few cristae. This may be related to a low P_{O_2} in the central tissues of these nematodes.

It is interesting that anaerobic strains of the yeast *Saccharomyces cerevisiae* contain few, large mitochondria-like structures, but those from aerobic conditions have mitochondria that have a form typical of aerobic yeasts (Lloyd, 1974). Clearly, a comparative study of the mitochondria of nematodes from aerobic, microaerobic, and anaerobic habitats would be of interest.

B. Electron Transport

Mitochondria are capable of passing electrons through a series of oxidation-reduction pairs, culminating in the reduction of oxygen to water. The three main divisions of the oxidation–reduction pairs are (1) NAD/NADH or NADP/NADPH, (2) FAD/FADH$_2$ or FMN/FMNH$_2$, and (3) a cytochrome chain. Unlike the other two pairs, the cytochrome chain is specific to oxidative phosphorylation. All cytochromes contain iron-porphyrin groups but, unlike that in hemoglobin, the iron undergoes valency changes when acting in electron transport. In a reduced ferrous state the cytochromes have spectra similar to that shown for oxyhemoglobin (see Fig. 10). Spectral differences are used to separate the cytochromes into the major classes a, b, and c. Unfortunately, many aspects of electron transport by cytochromes are not fully understood. For instance, cytochrome oxidase is often considered to be two distinct cytochromes, a and a_3, with only the latter reacting with oxygen. However, more recent work suggests that cytochrome oxidase is an oligomeric protein with two similar heme groups differing in their behavior toward ligands such as oxygen (Lehninger, 1975). An incomplete understanding of cytochromes in the classical electron transport system makes it more difficult to understand the unusual cytochrome systems of some nematodes.

Some emphasis has been placed on establishing the presence of cytochrome chains as an indicator of oxidative phosphorylation in nematodes. Cytochromes $a + a_3$, b, and c have been detected in *Turbatrix aceti* (Rothstein *et al.*, 1970) and cytochromes $a + a_3$ and c have also been found in *D. triformis* (Krusberg, 1960). Evidence for a cytochrome c has also been obtained for *Panagrellus silusiae* (Cooman, 1950), and *Pratylenchus scribneri* and *Caenorhabditis briggsae* (Deubert and Zuckerman, 1968). The juvenile of *T. spiralis* encysted in muscle contains all three cytochromes (Boczon and Michejda, 1978), and cytochrome oxidase has been found in adults (Goldberg, 1957). A complete, classical cytochrome chain occurs in adults of *H. contortus*, but this animal also possesses cytochromes b and c, which react with carbon monoxide (Pritchard in Barrett, 1976). There is also evidence, particularly from the work cited for *H. contortus* and *T. aceti*, for additional cytochromes to those of the classical cytochrome chain. *Ascaris* also contains at least one unusual cytochrome, but extrapolation of findings from this gigantic species to other nematodes requires caution. It is unfortunate, therefore, that the cytochromes of this species have been more fully studied than those of other nematodes. Much of past argument on the occurrence of cytochromes stemmed from the apparent conflict over the presence of cytochromes in an anaerobe. More recently, argument has centered more on the relative contribution of aerobic and anaerobic pathways in this animal, and it is certain that the animal is capable of some oxidative phosphorylation.

Keilin and Hartree (1949) demonstrated the presence of cytochromes in the muscles of *A. lumbricoides*, and there have been many subsequent investigations (Table II). Kikuchi *et al.* (1959) and Kikuchi and Ban (1961) seem in retrospect to have described much of the system in *Ascaris*. The more recent and fuller understanding is due to the elegant work of Cheah (1972, 1976), who used analytical techniques developed for cytochrome research by Chance (1957). A recently postulated scheme for the cytochrome chain in *Ascaris* is given in Fig. 5. The classical electron transport system is a minor pathway *in vitro* contribut-

TABLE II

Summary of Attempts to Detect Cytochromes in the Muscle of Ascaris[a]

	b-Type cytochrome	c₁-Type cytochrome	c-Type cytochrome	Cytochrome c oxidase			o-Type cytochrome
				a	$a + a_3$	a_3	
Detected	5, 6, 8, 9	9	5, 6, 8, 9	8, 9	2, 4, 5	8, 9	4?, 8, 9
Not detected		6, 7,	3		1, 3, 6, 7		

[a] Key to reference numbers: 1, Herrick and Thede (1945); 2, Van Grembergen et al. (1949); 3, Bueding and Charms (1952); 4, Kikuchi et al. (1959); 5, Kikuchi and Ban (1961); 6, Chance and Parsons (1963); 7, Lee and Chance (1968); 8, Cheah and Chance (1970); 9, Cheah (1976).

ing about 30% of succinate oxidation, and the majority of electron flow occurs down a branch from cytochrome b to cytochrome o, with oxygen being reduced to hydrogen peroxide.

A similar system also occurs in the large tapeworm *Moniezia* (Cheah, 1972), but the functional significance of branched cytochromes is unclear. The cytochrome o pathway has only one site coupled to ATP formation, it apparently has a lower oxygen affinity than cytochrome oxidase, and evidence for respiratory control has yet to be provided (Barrett, 1976). The classical electron transport system is fully active at a P_{O_2} of 1–5 mm Hg (Chance *et al.*, 1964), and it seems unlikely that modification of the cytochromes could improve the oxygen relationships of *Ascaris* as much as would occur by a reduction in body radius.

Comparison with microorganisms may provide useful indications of cytochrome function in nematodes. The yeast *S. cerevisiae* modifies the proportions of its cytochromes from a mole-to-mole ratio in aerobic conditions to relatively high levels of cytochromes b and c relative to a and a_3 in anaerobic conditions. In *Ascaris*, the cytochrome b, cytochrome c, and cytochrome $a + a_3$ are in ratios of about 6:4:1 [data from Cheah (1976)]. *Trichinella spiralis* also has more cytochrome b than c and the latter exceeds cytochrome $a + a_3$ levels. This may be a general feature of helminth cytochromes, but kinetic studies suggest that cytochrome $a + a_3$ is not rate-limiting, at least *in vitro*, for *T. spiralis* (Boczon and Michejda, 1978).

Lloyd (1974) lists four species of protozoa, three yeasts, and a fungus in which cytochrome o occurs. He considers that it and some other unusual cytochromes may act as a terminal oxidase in some organisms on the basis that a cytochrome that combines with carbon monoxide probably reacts with oxygen (Keilin, 1966). The yeast *S. cerevisiae*, like *Ascaris*, generates hydrogen peroxide in its respiratory chain and it may also form the super anion O_2^-. This yeast contains a cytochrome c peroxidase that is capable of oxidizing cytochrome c using hydrogen peroxide generated intramitochondrially by the respiratory chain (Lloyd, 1974). Possibly the cytochrome o system of *Ascaris* has a similar function, and this may play a protective role when the animal experiences oxygen levels at the upper end of its environmental range. Another possibility is that the cytochrome o of *Ascaris* occurs in different mitochondria from the classical cytochrome chain. This could arise if the contractile and noncontractile regions of the muscle cells have different mitochondria, or because the hypodermis is not excluded in the preparation of body wall musculature. The possibility that the hypodermis of

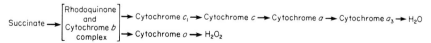

Fig. 5.   Postulated branched electron transport system of *Ascaris* muscle mitochondria. (After Cheah, 1976.)

Ascaris is aerobic is discussed elsewhere (Sections II,B, and VII,A). Overall, it seems that most nematodes contain a classical cytochrome chain, but in addition, some contain additional electron-transfer systems associated with low or fluctuating oxygen regimes.

C. Oxidative Phosphorylation

There are three main theories to explain the coupling of the energy released in electron transport to the formation of ATP from ADP in the process of oxidative phosphorylation (Lehninger, 1975). A discussion of these theories is beyond the scope of this chapter, but the functional significance of oxidative phosphorylation is to conserve the energy released when electrons move from NADH to oxygen. This energy is approximately 7.2 times that required to form ATP from ADP. The process is not 100% efficient, but on average each of the oxidation steps of the tricarboxylic acid cycle results in the formation of three ATP molecules for each oxygen atom reduced. This gives an average P:O ratio of 3:1. This represents a highly efficient process conserving about 40% of the energy released when a pair of electrons pass from NADH to oxygen. Each of the three phosphorylations occurs at a specific stage of the electron-transport system. Failure to obtain a P:O ratio of 3:1 and an insensitivity of the system to specific inhibitors that uncouple the sites of phosphorylation from electron transport are both evidence for systems other than oxidative phosphorylation via the classical electron-transport chain. A value of P:O well below 3 has been reported for *Ancylostoma caninum* (Warren, 1970). This may also occur in *Ascaris* but the situation is complex because the cytochrome *o* system has a P:O ratio of 1:1 rather than 3:1 as in the classical electron-transport chain. The net result is that the value will be a hybrid dependent on the relative activities of these two chains and will be depressed further by any phosphorylation that does not depend upon oxygen. An additional difficulty that arises from investigating P:O ratios is that oxidative phosphorylation is notoriously labile, and it is difficult to isolate mitochondria in which electron transport and phosphorylation remain closely coupled (Lehninger, 1975).

D. Adenylate Energy Charge

Atkinson and Walton (1967) proposed a measure of the proportion of high energy phosphate bonds in the adenylate pool of a cell and called it the adenylate energy charge. It indicates the balance between energy-forming and energy-utilizing processes in the cell, and a modification of its value can indicate a change in metabolism or a loss of metabolic integrity. The main justification for discussing the adenylate energy charge in this chapter is that it can offer advantages over measurements of oxygen consumption as an indication of metabolic change.

Respirometry is normally carried out over an extended period and is unsuitable for detecting initial or transient changes in metabolism. A second difficulty arises with nematodes such as endoparasites of animals, for experimental conditions cannot approximate those of their normal habitats. The adenylate energy charge overcomes both difficulties by giving an indication of the metabolic state at a particular instant *in vivo*.

Stimulation to hatch is followed by an increase in oxygen consumption in the second-stage juvenile of *G. rostochiensis* (Atkinson and Ballantyne, 1977a), but a change in the adenylate energy charge gives an earlier indication of the termination of dormancy (Atkinson and Ballantyne, 1977b). In *Ascaris,* stimulation to hatch also influences the ATP/ADP ratio (Beis and Barrett, 1975), and this probably alters the adenylate energy charge. Its value also changes in *Nematodirus battus* and *N. brasiliensis* from hosts developing immunity to these nematodes (Ballantyne *et al.,* 1978). The measurement of adenine nucleotides has also been proposed as an indicator of the efficacy of antihelmintics (Bryant *et al.,* 1976). It must be emphasized that the adenylate energy charge does not measure the same parameter as oxygen uptake, but it can be useful for investigating physiological stresses when technical difficulties prevent the use of respirometry.

IV. OXYGENASES

A small part of the oxygen consumed by nematodes is due to oxygenases and is not associated with oxidative phosphorylation. Oxygenases are enzymes that presumably catalyze the activation of oxygen and the subsequent incorporation of either one or two atoms of oxygen into one molecule of various substrates (Hayaishi, 1974). They play a central role in the detoxification of drugs or pesticides within the animal. *Panagrellus redivivus* oxidizes the organophosphate phorate to its sulfoxide and sulfone, and this was inhibited by low oxygen regimes but not by carbon monoxide. It seems that the metabolism of this pesticide is dependent on a mixed-function oxidase, and comparison with other animals suggests that this is located in the microsomal fraction of the cells (Le Patourel and Wright, 1974).

Oxygenases also play an important role in the biosynthesis, transformation, and degradation of amino acids, lipids, sugars, porphyrins, vitamins, and hormones (Hayaishi, 1974). They may be required for cuticle synthesis in *Ascaris* (Fairbairn, 1970), and a polyphenol oxidase is involved in the tanning of the cyst of *Globodera (Heterodera) rostochiensis* (Ellenby and Smith, 1967). Nematodes also contain peroxidases (Messner and Kerstan, 1978), although the level of catalase in *Ascaris* is low (Laser, 1944). Possibly the perienteric hemoglobin of *Ascaris* decomposes hydrogen peroxide, since this appears to be the function of a high-affinity hemoglobin of the yeast *S. cerevisiae* (Lloyd, 1974).

V. FACTORS INFLUENCING OXYGEN DEMAND

It is convenient to separate factors influencing the oxygen demand of an animal into endogenous factors and environmental influences on oxygen uptake. In general this is useful, although it can be difficult to partition the importance of endogenous and exogenous components when considering the influence of a factor such as activity on oxygen demand.

A. Body Size and Interspecific Comparisons

The oxygen consumption per unit weight decreases with increasing body size for both homeotherms and most poikilotherms. This relationship between the oxygen consumed by an individual (Y) and its body weight (X) can be expressed by

$$Y = aX^b \tag{3}$$

or if oxygen consumption is expressed as a unit weight basis

$$Y/X = aX^{(b-1)} \tag{4}$$

If the relationship holds, logarithmic transformation of Y or Y/X and of X should give a linear plot whose gradient is b or $(b-1)$, depending on which form of the equation is used (Zeuthen, 1953; Hemmingsen, 1960).

Zeuthen (1947) considered that nematodes formed a limited array of poikilotherms in which oxygen consumption was weight-independent ($b = 1.0$). He considered that there were two further arrays, one for acellular animals ($b = 0.67$) and a second for larger poikilotherms ($b = 0.67$). Essentially, acellular forms had a lower rate of metabolism at a given size than the larger poikilotherms, with the nematodes and certain other small poikilotherms forming an intermediate link. Hemmingsen (1960) suggested a value of $b = 0.75$ for all poikilotherms with body weights greater than 0.1 g of wet weight. The factors determining this value are poorly understood. It may represent an evolutionary compromise between the advantages of maintaining a high level of metabolism with increasing size and the limitations imposed by the reduced effectiveness of surface-dependent processes in larger animals.

A weight-dependent oxygen consumption was not established in nematodes by

Hemmingsen (1960), but von Brand (1962) considered that body size may be important from the data published for nematodes at that time. More values are now available by single-animal techniques for marine nematodes (Wieser and Kanwisher, 1961) and nematodes from soil (Nielsen, 1949; Klekowski et al., 1972). These values have been pooled by covariance analysis to show a weight-dependent metabolism with $b = 0.79$ (Atkinson, 1976). This value differs significantly from that expected for a weight-independent metabolism ($b = 1.0$), or from the value generally considered to indicate a surface-dependent metabolism ($b = 0.67$). The value for nematodes does not differ significantly from $b = 0.75$, relating body size and metabolism in most poikilotherms (Hemmingsen, 1960). There seems no reason to accept any value for nematodes other than that typical for poikilotherms as a whole (Klekowski et al., 1972; Atkinson, 1976). The equations given in Fig. 6 also suggest a standard energy metabolism from the oxygen consumption at a given weight which is intermediate between that for acellular organisms and most larger poikilotherms (Hemmingsen, 1960).

The relationship between oxygen consumption and body size has broad implications. It is essential that this factor be taken into account when comparing any physiological process that may be influenced by the rate of metabolism. This extends far beyond comparisons of the rates of respiration of nematodes. For instance, Wright and Newall (1976) suggest that nitrogen excretion in nematodes may be dependent on body size. The data available for total nitrogen excreted per gram of wet weight per 24 hr (Table 1 in Wright and Newall, 1976) can be adjusted for the animal parasites and free-living nematodes to a standard temperature of 20°C using a Q_{10} of 2.0. The data are too imprecise to show a significant regression coefficient, and further work is necessary to establish this relationship. However, nitrogen metabolism may remain a constant proportion of total metabolism for nematodes of all sizes. In this case the power relationship to body size would be $b = 0.75$ (Fig. 6), and the available data would suggest that a nematode of 1.0 μg wet weight would excrete 1.92 μmoles N/g wet weight/hr. This catabolism of nitrogen provides some of the energy requirements of an animal. It is difficult to be confident of the relationship between nitrogen excreted and oxygen consumed in nematodes, but in mammals it is assumed that 1 g of nitrogen excreted signifies the utilization of 5.94 liters of oxygen (Hardy, 1972). If this is also true for nematodes, it can be shown that the rate of nitrogen excretion for a wet weight of 1 μg contributes about 10 ± 7% to the standard energy metabolism given in Fig. 6. This value is approximate because of insufficient data, but it is of the same order as that of about 20% in mammals (Hardy, 1972). The actual value for an individual nematode will depend upon its physiological condition and diet, but the relationship does show how body size influences interpretations of physiological and biochemical processes. A second example can also be given for a comparison of lipid reserves (Section V,F).

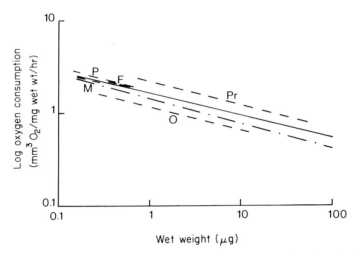

Fig. 6. Oxygen consumption and body weight of nematodes. Data based on single animal techniques. (— · —) Regression line obtained by Klakowski *et al.* (1972); (—) regression line obtained by covariance analysis by Atkinson (1976). Data from Nielsen (1949), Wieser and Kanwisher (1961), and Klekowski *et al.* (1972). Regression line for data at 20°C from Atkinson (1976) can be expressed as follows [see Eq. (4), Section V,A]

$$Y/X = 1.6X^{-0.216}$$

where Y = cubic millimeters of oxygen per individual per hour and X = wet weight in milligrams. This is termed the standard energy metabolism of nematodes in this chapter. Klekowski *et al.* (1972) divided soil nematodes into trophic groups. Regression lines based on the data compiled by them for single-animal techniques have been superimposed onto the figure. The trophic groups are given below with their estimated energy metabolism as a percentage $\pm$ SE of the standard energy metabolism at their grand mean dry weights; the abbreviation for each regression line in the figure and the number of species in each analysis are given in parentheses: predators, 132 $\pm$ 24% (Pr; 8); omnivores, 70 $\pm$ 12% (O; 7); fungal feeders, 105 $\pm$ 30% (F; 3); microbivores, 99 $\pm$ 18% (M; 18); and plant parasites, 119 $\pm$ 18% (P; 9). Covariance analysis indicates that all the regression lines have similar slopes of $(b-1) = -.216$ ($p > .1$) and that there are differences in levels of metabolism between the groups ($p < .05$). However, further analysis suggests that no group is clearly different from the overall value for the standard energy metabolism.

B. Body Size and Intraspecific Comparisons

The relationship between oxygen consumption and body weight may depart from $b = 0.75$ in some cases (Hemmingsen, 1960). This can occur because the size range between individuals is small in comparison to the array of sizes for all nematodes. This influences the accuracy of determinations of b and also allows other factors to influence the value of b in a manner that is not sustained for all species.

A weight-dependent metabolism was shown for *Ascaris* by Krüger (1940). Zeuthen (1955) believed that *Pontonema vulgaris* (data from Nielsen, 1949) showed a triphasic growth curve with $b = 0.4$ in juveniles and $b = 1.0$ in mature adults. A similar change in exponent was also recorded for *Enoplus communis* (Wieser and Kanwisher, 1960). However, in neither case was the analysis proved by statistics, and it is difficult to establish such a relationship with imprecise data in any other way. Hemmingsen (1960) considered that the data given by Nielsen (1949) showed an intermediate value. More recently, *Panagrolaimus rigidus* has been considered not to show a triphasic growth curve but to have an intraspecific value for b of 0.64 ± 0.12 (Klekowski *et al.*, 1974). A similar value was also obtained for *Nematospiroides dubius* (Bryant, 1974), but these values are not clearly different from $b = 0.75$.

Further work may suggest that such departures do occur commonly during the development of nematodes, but this does not question the value of the interspecific value for b (Hemmingsen, 1960). An exceptional value for b of greater than 1.0 does occur in *N. dubius* during the transition from a noninfective to an infective juvenile (Bryant, 1973). A further factor may be the inclusion of gravid females in any comparison. The eggs they contain will influence the body weight of the female, but the eggs and the female tissues may have different oxygen consumptions on a unit weight basis.

Low oxygen availability also results in an exceptional value of $b > 1.0$ in males of the marine nematode *Enoplus brevis* but not in males of the related species *E. communis* (Atkinson, 1973a). This would arise if all males of *E. communis* and larger individuals of *E. brevis* lapsed more readily into a quiescent state at a low P_{O_2} than do smaller males of *E. brevis*. The intraspecific value for b in males of both these species, $b = 0.33$, is a good example of an unusual value for b within a restricted size range of nematodes. This value suggests a correlation between oxygen consumption and body radius that estimates the length of the diffusion pathway in nematodes. This ensures that increasing body size does not require a higher environmental level of oxygen to support an aerobic metabolism. This is clearly of adaptative significance to the enoplids over a limited size range in a low oxygen environment and its value can be appreciated without implying a causal relationship (Atkinson, 1973a).

C. Trophic Forms

Ecologists may wish to calculate the energy required for the maintenance of nematode communities. After allowance for body size, it is of value to determine if the level of metabolism is dependent on their way of life. Teal and Wieser (1966) deserve credit for attempting such an analysis for marine nematodes based on the structure and size of the buccal cavity. Unfortunately they did not allow for differences in body size, and too little is known of marine nematodes to be sure of the validity of the recognition of trophic groups from buccal cavities (Klekowski *et al.*, 1972). An analysis of trophic forms has been carried out using free-living species and marine nematodes (Klekowski *et al.*, 1972). Some of these data are best omitted because they are not based on single-animal techniques and so may misrepresent the relationship between metabolism and body size (Ellenby, 1953). It also seemed inappropriate to include marine species as one trophic group (Klekowski *et al.*, 1972) and it is better to exclude them from the analysis. The remaining data are separated into (1) parasites of plants, (2) fungal feeders, (3) microbivorous nematodes, (4) omnivorous, and (5) predators. A summary of the analysis is given in Fig. 6, but the data are somewhat inadequate because of the small number of species in some trophic groups. For instance, five of the eight species of omnivores belong to the genus *Eudorylaimus,* and this may misrepresent that trophic group. The validity of the groups may also be questioned, and a somewhat different classification is given by Yeates (1971). It seems advisable at present to conclude that a general equation such as that in Fig. 6 is the best available approximation for estimating the requirements for the maintenance of nematode communities after the assessment of body size and abundance (Klekowski *et al.*, 1972).

D. Activity

Some nematodes may possess musculature with a marked anaerobic capacity (Section IX), but it can be assumed that most species receive sufficient oxygen to allow the energy for contraction to be supplied by oxidative phosphorylation. Changes in locomotor activity will modify the oxygen consumption of aerobic muscle and so influence the oxygen uptake of the individual. Difficulties arise in estimating the contribution of activity to the total oxygen consumption because activity in water varies between species (Wallace and Doncaster, 1964) and is also influenced by the medium for movement. Nielsen (1949) found that urethane inhibited locomotion but reduced oxygen consumption by about 5% for most soil nematodes. Two species that were active in water showed higher contributions of locomotion to oxygen uptake of 15% for *Plectus cirratus* and 20% for *Mononchus papillatus*. The oxygen uptake of *N. dubius* was not influenced by activity in water (Bryant, 1973). *Caenorhabditis elegans* also shows changes in locomotor activity that do not have a detectable influence on oxygen uptake (Dusenbery *et al.*, 1978). However, the bacterial feeder *Pelodera chitwoodi* apparently shows a change in activity in the presence of bacteria, and the increase in locomotion and presumably pharyngeal activity during feeding is sufficient to influence its oxygen consumption (Mercer and Cairns, 1973).

It seems that measurements of oxygen uptake are often relatively insensitive to changes in the activity of nematodes. Measurements of lipid utilization can be made over prolonged periods and may provide a better basis for estimating the energy cost of locomotion in different media. *Meloidogyne javanica* utilizes lipid reserves more readily in soils in the presence of host roots than in water, but conserves energy reserves longest in soils without plant roots (Van Gundy *et al.,* 1967). Many infective-stage juveniles of animal-parasitic nematodes are unable to feed, and they show increased activity when stimulated by conditions indicative of the presence of the host. At other times they are less active and conserve lipid reserves (Barrett, 1969a; Croll, 1972). Recently, an attempt has been made to estimate the energy cost of locomotion in *Ancylostoma tubaeforme*. The induction of continuous hyperactivity for 8 hr with neostigmine bromide or by frequent stimulation caused a 10% loss of lipid reserves (Nwosu, 1978). The animals normally remain inactive for long periods but the results suggest that prolonged hyperactivity does have a large effect on the energy use of a nematode of this body size. This may be consistent with a low energy requirement for the less extreme activity of many soil nematodes in water. Except during prolonged hyperactivity, it may be that nematodes, like other microanimals, require less than 50% of the energy consumed by the individual for locomotion, and in some cases activity may contribute only a small part of this value (Zeuthen, 1947).

E. Dormancy

Keilin (1959) considered dormancy a condition in which there was a lowered metabolism, and considered latent life as a state in which there was no detectable metabolism. Evans and Perry (1976) correctly asserted that the primacy of metabolic activity was not supported by work in other fields and preferred to use dormancy as a general term for all resting states. The extent of a reduced metabolism is not a basis on which to define dormancy but it often accompanies such states. It has adaptative significance in prolonging survival and it is therefore of interest to estimate the size of this effect.

Dormant states of various types occur in cyst nematodes (Evans and Perry, 1976; Lee and Atkinson, 1976). Changes in oxygen consumption have been detected during the termination of dormancy prior to hatching (Atkinson and Ballantyne, 1977a). The level of oxygen consumption of cysts in water was 0.48 mm^3 O$_2$/mg dry weight/hr, which represents about 10% of the value expected for second-stage juveniles with a wet weight of $\simeq 0.3$ μg. This is calculated using the data in Fig. 6 after due allowance for the weight of the cyst wall. This reduced oxygen consumption may be partly due to the enclosure of the eggs within the cyst (Onions, 1955), but it is probably also due to a lowered metabolism. The rate of utilization of lipid reserves of the unhatched juvenile is approximately that expected from measurements of oxygen consumption (A. J. Ballantyne, unpublished data). Measurements of the rate of depletion of lipid reserves do give a good indication of the level of metabolism of these animals during dormancy. In the highly persistent species *Globodera pallida,* lipid reserves have been measured in eggs from cysts that have remained in soil for periods from 1 to 14 years (Fig. 7). These measurements show a 50% lipid decline after 8.1 ± 0.5 years in soil, whereas hatched juveniles have been shown to use half their lipid reserves within 34 ± 2 days at 20°C (R. M. J. Storey, in preparation). This represents a difference of nearly 100-fold between hatched juveniles at 20°C and those unhatched in field soils. Some of this difference is due to differences in the temperatures of field soils and the laboratory, but even so, the unhatched juveniles clearly have a reduced, dormant metabolism. Dormancy also influences the oxygen consumption of *C. briggsae.* The oxygen consumption of dauer larvae is somewhat less than that of adults, shows a lower value for Q_{10}, and the dormant stage has a greater tolerance of anoxia (Anderson, 1978).

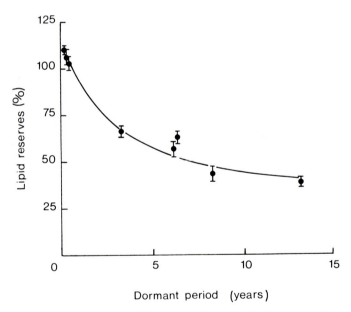

Fig. 7. Changes in relative neutral lipid content of juveniles of *Globodera pallida* during dormancy within cysts in field soils. Each mean ± SE is based on at least 100 individuals hatched from about 100 cysts. Neutral lipid was measured using microdensitometry [see Croll (1972)]. All results are expressed as a percentage of a standard population of *G. rostochiensis* as used in other work (Atkinson and Ballantyne 1977a,b). Field populations were collected from Humberside, England. (Unpublished data from R. M. J. Storey.)

F. Lipid Reserves and Starvation

There may be an interrelationship between starvation and dormancy in some nematodes but this is inadequately studied at present. Some nematodes show a decrease in respiration after isolation from their habitats, e.g., *P. redivivus* (Santmeyer, 1956), *M. javanica* (van Gundy *et al.*, 1967), and *Pelodera chitwoodi* (Mercer and Cairns, 1973). Cooper and van Gundy (1970) found no such change for *Caenorhabditis* sp. or *Aphelenchus avenae* and suggested that some reported decreases in oxygen consumption after separation of the nematodes from their habitats were artifacts. It is true that *Pontonema vulgaris, Dorylaimus obtusicaudatus, M. papillatus,* and *Plectus granulosus* showed no such effect over 4 days (Nielsen, 1949), and *E. brevis* and *E. communis* maintained a constant oxygen uptake in seawater for 7 days (Atkinson, 1972).

Additional data are now available from measurements of lipid reserves and this allows further analysis. This is an example of how changes in lipid reserves can supplement results from measurements of oxygen consumption. Many nematodes are aerobic and use their lipid reserves when starved. The percentage loss of lipid and oxygen consumption are interrelated and so both are a function of body size. It is assumed from calculations based on tripalmitin that 1 g of lipid requires about 2.0 liters of oxygen for its complete oxidation and yields 39.8 kJ (Hardy, 1972). By using the standard energy metabolism of nematodes (Fig. 6) it is possible to convert the oxygen consumption of a nematode into the quantity of lipid required if it alone fuels the energy metabolism. From this it is possible to plot the length of time that this metabolism can be supported by lipid reserves of 10 and 40% of the dry weight (Fig. 8). This may underestimate the actual survival of nematodes at 20°C during starvation. For instance, substantial mortality of infective juveniles of *G. rostochiensis* occurs at 20°C from 4 to 6 weeks after hatching, and not as soon as implied in Fig. 8 (R. M. J. Storey, in preparation). Third-stage juveniles of *A. tubaeforme* can also modify their level of metabolism in response to environmental stimuli (Croll, 1972). Both these nonfeeding, infective stages of very different nematodes seem able to remain inactive with a level of metabolism well below that indicated by Eq. (4) in Fig. 6. It may be that other species may also show a much reduced metabolism during starvation and this would prolong the survival of nematodes beyond the range suggested by Fig. 8.

However, this relationship does permit some conclusions, whatever the change in metabolism with starvation. Large soil nematodes may survive starvation for about 10 times as long as a very small species with the same original percentage lipid content. It is perhaps to be expected that any decrease in oxygen consumption during starvation would occur more readily in smaller than larger nematodes, and this may explain the difference between species considered

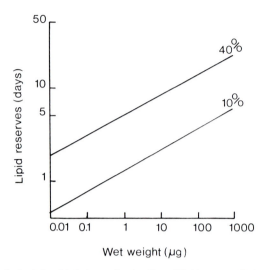

Fig. 8. Theoretical relationship between the duration of lipid reserves during starvation at 20°C and body size in nematodes. Values for neutral lipid reserves of 10 and 40% of dry weight are taken as extreme values for nematodes. One microgram of lipid is assumed to require 2.015 mm³ of O_2 for its oxidation (Hardy, 1972). Oxygen consumption for a given wet weight is calculated from the equation in Fig. 6. The relationship emphasizes the influence of body size on the ability of nematodes to withstand starvation. However, some species may show more prolonged survival than suggested by this relationship and this must involve a lower level of metabolism during starvation than that given in this figure.

earlier. Large size in soil nematodes might be particularly associated with intermittent feeding. This may occur in the large plant ectoparasite *Longidorus macrosoma,* which takes over a year to complete its life cycle and apparently develops and multiplies only when growing hosts are available (Cotten, 1976). The combination of a low temperature and a slowing of metabolism with starvation may explain the prolonged survival of this species at low temperatures in soil without roots (Yassin, 1969).

Adults of animal parasitic nematodes have a relatively low lipid content. The highest value in a table collated from the literature by Barrett (1976) is for adult *N. brasiliensis* with a value of 11.9% of the dry weight. This nematode has a wet weight of about 4–10 μg (Lee 1971) and so its lipid content is equivalent as an energy reserve at a given temperature to the 20–30% found in many small soil nematodes (Fig. 8). This is broadly correct even if about 5% of the lipid content of nematodes is not available for energy metabolism. This suggests another example of the need to take account of body size when comparing nematodes.

A better way of indicating the relative importance of lipid and carbohydrate to a nematode is to plot their relative contribution to its energy reserves. Glycogen has an energy value upon oxidation of about 16.8 kJ/g, but a neutral fat has a corresponding value of about 37.7 kJ/g (Lehninger, 1975). The relative energy value of glycogen and lipid reserves, after some allowance for structural lipids, is plotted in Fig. 9 for a range of nematodes. The data suggest that the metabolism of adult *N. brasiliensis,* and presumably some other trichostrongyles, is substantially aerobic, and this agrees with evidence from their oxygen relationships (Sections II,B and VI). It may be that even the low energy reserve stored as lipid in other animal parasites has some quantitative significance in energy metabolism. Many of these parasites are large and any lipid utilization during starvation in aerobic conditions *in vitro* may have little effect on their lipid reserves for this reason (Fig. 8) rather than because of the nature of their metabolism (von Brand, 1973). Most species are believed to be unable to catabolize their stored lipid (Fairbairn, 1970), and it has been suggested that the lipid could be incorporated into eggs, or be a stored excretory product (Barrett,

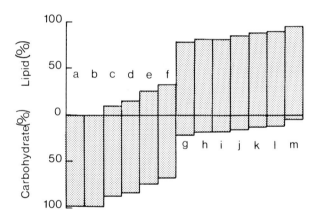

Fig. 9. Relative calorific value of lipid and carbohydrate in nematodes, assuming utilization is by an aerobic metabolism. The relative contribution of the dry weight of lipid and carbohydrate has been calculated after allowing for the ratio of about 2.3:1 for the calorific value for lipid and carbohydrate, respectively (Hardy, 1972). Allowance is made for lipid that is not available for energy metabolism by assuming none of the lipid in *Eustrongylides* (4.4% dry wt) and *Porrocaecum decipiens* (3.7% dry wt) represents an energy reserve. The following are species with over 50% of their energy reserves as carbohydrate: a, *Eustrongylides ignotus* (juveniles); b, *Porrocaecum decipiens* (juvenile); c, *Ascardia galli* (adult); d, *Heterakis gallinae* (adult); e, *Trichinella spiralis* (muscle juvenile); f, *Ascaris lumbricoides* (adult).

The following are species with over 50% of their energy reserves as lipid: g, *Ascaris lumbricoides* (eggs); h, *Nippostrongylus brasiliensis* (adult); i, *Aphelenchoides sp.* (adult); j, *N. brasiliensis* (third-stage juvenile); k, *Ancylostoma caninum* (third-stage juvenile); l, *Cooperia punctata* (third-stage juvenile); m, *Strongyloides ratti* (third-stage juvenile). Values for lipid and carbohydrate reserves from data compiled by Barrett (1976).

1976). The second explanation is unconvincing, and it may be that lipid utilization occurs in these animals during periods of greater oxygen availability. Alternatively, some tissues may have a lesser anaerobic capacity than others (Section IX).

G. Temperature

Nematodes are poikilotherms and are subject to the effects of temperature on their metabolism. The relationship has been expressed using the Arrhenius equation for *A. caninum* and *Eustrongylides ignotus* (von Brand, 1973). This approach may have its value, but the uptake of oxygen of a whole animal is due to separate oxygen-utilizing processes and they may not all be equally affected by a change in temperature. Changes in oxygen consumption with temperature can also be expressed by Q_{10}:

$$Q_{10} = \left(\frac{k_1}{k_2} \right)^{10/(t_1 - t_2)} \tag{5}$$

where k_1 and k_2 are oxygen consumptions at temperatures t_1 and t_2, respectively. Frequently Q_{10} has a high value at temperatures near 0°C and low values near the upper range of that tolerated by the animal (Hardy, 1972; Jones, 1972).

Caenorhabditis elegans has been reported as having a constant Q_{10} of 2.5 between 15° and 30°C (Dusenbery *et al.*, 1978). However, the data seem to suggest a higher Q_{10} of about 4 at 10°–15°C and a value of 1.2–1.7 at 25°–30°C. This would be the type of change frequently found in animals (Hardy, 1972; Jones, 1972). A low Q_{10} of 1.63 is also shown by *Heterodera oryzae* between 20° and 36°C (Reversat, 1977). Complex changes in Q_{10} may occur for *Ditylenchus dipsaci* within its normal environmental range (Bhatt and Rohde, 1970). Large fluctuations in Q_{10} may be correlated with changes in behavior with temperature. The oxygen uptake of infective juveniles of *N. brasiliensis* is enhanced by the high temperatures typical of mammalian skin (Wilson, 1965). This thermal response occurs in many skin-penetrating juveniles of animal-parasitic nematodes and it favors their invasion of homeotherms. Changes in chemotactic behavior and locomotion also occur with temperature in *C. elegans,* but in this case this is not associated with unusual values of Q_{10} (Dusenbery *et al.*, 1978). Some animals show acclimation of their metabolism to temperature in a complex manner favoring homeostatic regulation of their metabolism (Prosser, 1964). *Heterodera oryzae* may show some rudimentary acclimation to increasing temperature (Reversat, 1977). It has also been suggested that *P. redivivus* and *T. aceti* show adaptations to cold (Cooper and Ferguson, 1973), but the experimental basis for this has been questioned (Dusenbery *et al.*, 1978). More work on this is required using species from habitats such as the marine intertidal zones, which have large diurnal and seasonal changes in temperature.

H. Other Factors

The oxygen consumption of a nematode may be influenced by a variety of exogenous factors, but difficulties sometimes arise when interpretating the significance of such effects. Ionic changes in the incubation medium influence N. brasiliensis (Schwarbe, 1957) and Necator americanus (Fernando, 1963), and phosphate buffers suppress the oxygen uptake of H. oryzae (Reversat, 1977). However, urea at an osmotic pressure of 2 atm increases the oxygen uptake of H. oryzae (Reversat, 1977); the osmotic pressure of non-ionic solutions also influences Pratylenchus penetrans, Aphelenchoides ritzemabosi, Anguina spp., and D. dipsaci (Bhatt and Rohde, 1970). Such work attempts to understand the effects of ionic changes and matrix potentials on the metabolism of nematodes in soil. However, the effects recorded could also be due to (1) hyperactivity induced by physiological stress, (2) altering tissue hydration, or (3) changes in muscle tone required to maintain body turgor (Beadle, 1937). This means that it is not often possible to relate the changes in energy requirements directly to the increased work necessary to withstand osmotic or ionic stress (Potts and Parry, 1964).

Respirometry has also been used to study the effects of nematicides on nematodes (Ritzrow and Kampfe, 1971), but difficulties arise if changes are ascribed to a specific effect. For instance, oxygenases may be involved in the detoxification process (Section IV). Respirometry is inherently insensitive for detecting initial effects on metabolism. The adenylate energy charge may be a more sensitive indicator of such effects (Section III,D) and the measurement of adenine nucleotides has been proposed as a suitable indicator of anthelmintic efficacy (Bryant et al., 1976). Such an approach would be particularly appropriate if the nematode is suspected of having a considerable anaerobic capacity.

VI. OXYGEN AVAILABILITY

Nematodes are aquatic animals and an oxygen lack is more likely to occur in water than air for two main reasons. The oxygen content of fresh water at 20°C is about 0.03% of that in air and the diffusion rate through water is about 10^{-5} of that in air. Nematodes will be deprived of an adequate oxygen supply in water when the diffusion pathway from air to the microenvironment does not exceed the utilization of oxygen by microorganisms or other processes more able than nematodes to use a low P_{O_2}.

The oxygen regimes of the microhabitats of nematodes can be divided into three main categories: (1) normally aerobic, with a P_{O_2} usually above the minimum required to maintain approximately the level of oxygen consumption recorded in aerated water; (2) microaerobic, in which the oxygen availability is variable and may be less than required for the oxygen uptake in aerated water; and (3) virtually anoxic, in which oxygen is not available at levels of quantitative significance to the energy metabolism of the nematode. Von Brand (1946) considered freshwater and marine muds below superficial layers and the central lumen of the alimentary tract of vertebrates as examples of anaerobic habitats.

Microaerobic habitats are important for determining the limitations of an aerobic existence in nematodes. Examples of such habitats are the superficial layers of marine muds (Wieser and Kanwisher, 1961; Teal and Wieser, 1966; Atkinson, 1977), the upper layers of sewage sludge (Abrams and Mitchell, 1978), and the lumen of parts of the vertebrate gut close to its lining (Rogers, 1962). The P_{O_2} close to the mucosa of the small intestine of the duck is about 25 mm Hg (Crompton *et al.*, 1965), and the values for the mucosa of the rat are probably similar (Rogers, 1949a). Recently it has been suggested that the aqueous content of the rat intestinal lumen may have a P_{O_2} of 40–50 mm Hg, and this may increase when this host is parasitized with the cestode *Hymenolepis diminuta* (Podesta and Mettrick, 1974). It may be difficult to measure P_{O_2} with a polarographic oxygen electrode because of calibration difficulties in some media; the oxygen consumption of the electrode may influence the P_{O_2}, and the geometry of the electrode in the habitat is not similar to that of the nematode. Oxidation–reduction potentials may be an alternative approach for distinguishing an aerobic from an anaerobic habitat (Atkinson, 1977), but this does not indicate the actual P_{O_2} (Beadle, 1958).

Microaerobic habitats frequently involve a marked oxygen gradient. Oxygen pentrates about 2 cm into the mud inhabited by *E. brevis*. This represents a distance of about four adult body lengths, and clearly the level of oxygen availability will depend on the precise position of the nematode in the mud (Atkinson, 1977). *N. brasiliensis* also shows a marked behavioral response that ensures its precarious aerobic position in the oxygen gradient of the intestinal lumen (Alphey, 1972). Perhaps nematodes can detect oxidation–reduction potentials through their chemosensory responses to ions (Ward, 1973; Dusenbery, 1974). *Trichostrongylus colubriformis* also maintains an aerobic position by burrowing beneath the mucus and superficial layers of the mucosa (Barker, 1973). Intestinal lumen dwellers such as *Ascaris* may occupy a different microhabitat. The aerobic nature of *Ascaris* is in doubt because of its large body radius (Section II,B), which also prevents it from maintaining close contact with the mucosa. Oxygen derived from intermittent contact with the mucosa may support some aerobic metabolism (Smith, 1969a), and the intestine of man contains an appreciable volume of flatus often with over 5% oxygen (Calloway, 1968). Disagreement on *Ascaris* is centered on the relative contributions of efficient aerobic pathways in an animal with an inadequate oxygen supply and a well-established but energetically inefficient anaerobic metabolism. This dispute cannot be solved by the biochemical approach alone, and the physiology of this animal *in vivo* requires much greater study.

VII. RESPIRATION IN LOW OXYGEN REGIMES

A. General Effects of Suboptimal Oxygen Levels

Nematodes are not equally susceptible to the effects of a lack of oxygen and some processes also seem more susceptible to anoxia than others. Oxygen is required for nuclear division in most animals (James, 1971), and nematodes may not be an exception. Eggs of *A. lumbricoides* can survive anoxia but require oxygen for development (Brown, 1928). Oxygen also influences egg production in *H. contortus* (Le Jambre and Whitelock, 1967). In contrast, hatching of juveniles of *A. lumbricoides, Ascardia galli,* and *Toxocara mystax* can occur in virtually anoxic conditions (Rogers, 1960); this probably represents an adaptation to animal parasitism. Hatching in many other nematodes may be dependent on oxygen, as in *M. javanica* (Baxter and Blake, 1969). There is also an increase in oxygen consumption by *G. rostochiensis* prior to hatching (Atkinson and Ballantyne, 1977a).

Locomotion is dependent on some oxygen in many nematodes. Second-stage juveniles of *M. javanica* show reduced activity at low oxygen concentrations (Baxter and Blake, 1969). *Pelodera punctata* shows a loss of mobility at a P_{O_2} below 15 mm Hg, and is inactive at less than 5 mm Hg (Abrams and Mitchell, 1978). Adults of *N. brasiliensis* also become inactive in anaerobic conditions; this also occurs in *E. communis*. This species survives an oxygen lack for about 24 hr, but many marine nematodes survive in a quiescent state for several days (Wieser and Kanwisher, 1959). Quiescence may also be induced by anoxia in the infective stage of the mermithid *Romanomermis culicivorax*, and a low P_{O_2} for 36 hr does not suppress its subsequent infectivity (Brown and Platzer, 1978).

Acrobeloides buetschlii (Nicholas, 1975) and *P. punctata* (Abrams and Mitchell, 1978) can also survive anoxia for several days. *Caenorhabditis briggsae* shows impaired movement after several hours, followed by a loss of movement and then death after 24 hr (Nicholas and Jantunen, 1964). Pharyngeal activity and feeding are also influenced by low oxygen availability in *E. brevis* and *E. communis* (Atkinson, 1977).

Loss of activity may indicate an inadequate anaerobic capacity in the muscles of certain nematodes. However, nerves are particularly sensitive to anoxia and it may be coordination that is impaired rather than muscle contraction. Most nerve cells of nematodes are within the hypodermis which may remain aerobic at a low P_{O_2} (Section II,B). This, together with at least a limited capacity for contraction in anoxic conditions, probably explains the ability of most nematodes to remain active except in very low oxygen regimes. It may be that not all nematode tissues are equally susceptible to an oxygen lack, and central tissues will experience anoxia most frequently. The growth and reproduction of many nematodes may be particularly susceptible to an oxygen lack. The growth of *C. briggsae* is reduced at a P_{O_2} of 12 mm Hg, which is probably a higher level of oxygen availability than that required to prevent locomotion. The rate of the buildup of populations of *Caenorhabditis* sp. and *Aphelenchus avenae* was also reduced by even a short period of anoxia (Cooper *et al.*, 1970).

B. Effects on Oxygen Consumption

Body radius is the key factor in influencing the minimum level of oxygen required by a nematode (Section II,A). The relationship between oxygen consumption and P_{O_2} has been studied frequently in work on the oxygen requirements of nematodes. Animals are sometimes divided into (1) respiratory conformers, whose rate of respiration drops with oxygen throughout the P_{O_2} range 0–155 mm Hg, and (2) regulators, which maintain a more or less constant oxygen uptake over at least a part of this range (Prosser, 1961). *Ascaris lumbricoides* (Krüger, 1936; Laser, 1944) and *Litomosoides carinii* are respiratory conformers, but many other nematodes are regulators.

It has been argued that regulators such as *C. briggsae* possess a respiratory pigment (Bryant *et al.*, 1967) or more specifically, a cytochrome system such as that in *N. brasiliensis* (Roberts and Fairbairn, 1965). Dependency may indicate a diffusion limitation (Krogh, 1916), but fundamental interpretations may be misleading because oxygen availability may not influence all tissues of a nematode equally.

Bair (1955) demonstrated that oxygen consumption was regulated to a lower P_{O_2} by *Pelodera strongyloides* than by the soil nematode *C. elegans*. However, more recent work has suggested that these results were affected by the experimental technique and that *C. elegans* regulates its oxygen consumption above a P_{O_2} of 27 mm Hg (Anderson and Dusenbery, 1977). This is a value similar to that determined for *C. briggsae* (Bryant *et al.*, 1967). It has been argued that such a relationship may be expected for nematodes from environments where oxygen is freely available (Atkinson, 1976). It cannot be an adaptation to microaerobic habitats because there must be a critical P_{O_2} for all animals below which any regulatory mechanisms fail (Newell, 1970).

The relationship between oxygen consumption and oxygen availability may be more complex for nematodes from microaerobic habitats. It is implied in much work that the oxygen consumption in aerated water indicates an optimal level of metabolism. This level of oxygen availability is outside the normal environmental range for a microaerobic nematode and may induce an abnormal level of metabolism. Atmospheric oxygen levels may only occur for *E. brevis* at the surface of its mud habitat, and exposure of the animal would represent an abnormal and potentially hazardous situation. Hyperactivity at such times may enable the nematode to regain its normal position within the mud. This suggests that the relationship between oxygen consumption and oxygen partial pressure for animals in a respirometer may provide a misleading impression of the oxygen requirements of a microaerobic nematode. Perhaps the minimum P_{O_2} required by the animal can be estimated from the standard energy metabolism for nematodes. The oxygen consumption of *E. brevis* and *E. communis* is calculated as a percentage of the standard energy metabolism in Table III. The comparison may be too approximate to be valuable, but it does suggest that *E. brevis* can maintain 50% of the standard energy metabolism at a P_{O_2} of less than 35 mm Hg.

There is further evidence to suggest that the relationship between oxygen consumption and oxygen availability is complex in some nematodes. The oxygen consumption of *E. brevis* and *E. communis* on transfer from a P_{O_2} of 135 to 35 mm Hg is only 67% of that after prolonged exposure to these conditions (Atkinson, 1973b). It may be that the oxygen consumption of these nematodes at a low P_{O_2} is not directly limited by the diffusion of oxygen but is less than the maximum imposed by the supply of oxygen. *Enoplus brevis* must cope with changes in oxygen availability in its microenvironment (Atkinson, 1977) and diurnal changes in temperature of up to 10°C with the ebb and flow of the sea onto the intertidal salt marsh in which it lives (Atkinson, 1972). Both factors will modify the balance between the oxygen supply and demand of the nematode. An oxygen consumption below the maximum dictated by diffusion would enable the nematode to maintain its aerobic metabolism in spite of fluctuations in oxygen

TABLE III

Relationship between Oxygen Consumption as a Percentage of Standard Energy Metabolism and the Partial Pressure of Oxygen in Seawater Bathing the Nematodes[a]

Partial pressure of oxygen (mm Hg)	Oxygen consumption (mm^3O_2/mg dry wt/hr at 15°C)		Percentage of standard energy metabolism at 15°C	
	E. brevis	*E. communis*	*E. brevis*	*E. communis*
135	2.91 ± 0.08	1.72 ± 0.07	146	101
75	1.93 ± 0.09	1.34 ± 0.05	97	79
35	1.55 ± 0.06	0.86 ± 0.04	78	51
12	0.35 ± 0.06	0.2 ± 0.04	18	12

[a] Data are based on a mean dry weight for *Enoplus brevis* and *E. communis* of 32.4 and 110.5 μg, respectively. The ratio of weight to dry weight is 3.41:1. The standard energy metabolism for wet weight at 20°C is calculated from the equation in Fig. 6. The standard energy metabolism is adjusted to 15°C using a Q_{10} of 2.0. (Data based on Atkinson 1972, 1973a.)

supply or temperature. This may represent rudimentary homeostasis in an animal that lacks the flexibility of specialized respiratory or circulatory organs but inhabits a microaerobic habitat.

Some nematodes may possess an interesting way of supplementing their oxygen supply. *Mesodiplogaster lheritieri* apparently swallows air at an air–agar interface at a frequency that is influenced by low oxygen availability and high carbon dioxide levels. It is suggested that this is not fully explained by accidental ingestion during feeding (Klingler and Kunz, 1974). The physiological significance of this has not been quantified, but air swallowing has also been reported for other Rhabdita (Klingler and Kunz, 1974), for a marine nematode (Hopper and Meyer, 1966), and even for *A. lumbricoides* (Harpur, 1964). More work is needed to determine what contribution this behavior has to the supply of oxygen to these nematodes.

VIII. HEMOGLOBIN

Hemoglobins undoubtedly have respiratory functions (Jones, 1972) in invertebrates, and nematodes are unlikely to be an exception to this generalization. At least 18 species of animal parasitic nematodes contain hemoglobin, and the coloration of other species suggests that many more contain the pigment (Lee and Smith, 1965). The free-living adult of *Mermis subnigrescens* contains hemoglobin in the hypodermal cords, particularly in the anterior region near the so-called "chromotrope" (Ellenby and Smith, 1966). The free-living marine nematode *E. brevis* contains substantial quantities of hemoglobin in the pharynx, anterior body wall, hypodermal cords of both sexes, and in the copulatory muscles of the male. A second species, *E. communis,* contains much less of the pigment and its occurrence is restricted to isolated areas of the hypodermal cords in the tail region of males (Ellenby and Smith, 1966).

Hemoglobin has been found in the cuticle, body wall, and perienteric fluid of several species of animal parasitic nematodes (Lee and Smith, 1965) and is present in the hypodermis of *E. brevis* and *E. communis, M. subnigrescens* (Ellenby and Smith, 1966), and *A. lumbricoides* (Smith and Lee, 1963). The discontinuous distribution of the hemoglobin suggests that it serves particular tissues rather than increasing the oxygen supply to all the nematode. Its presence in the hypodermis may be correlated with the susceptibility to anoxia of the neural tissue that it encloses. Hemoglobin also occurs in the fluid-filled layer of the cuticle of *N. brasiliensis* (Lee, 1965). In this position it is well placed to extract oxygen from the surroundings and to serve the adjacent tissue, which again is the hypodermis.

The molecular weight of the body wall hemoglobin of *Ascaris* is approximately 37,000, but it has a single heme per molecule (Smith, 1969b). The perienteric fluid hemoglobin of *Ascaris* has a much higher molecular weight of 280,000 (Smith, 1969b), but this seems an unusual pigment in many ways. Hemoglobins of *A. lumbricoides* and *Strongylus* (Davenport, 1949a,b) and *H. contortus* (Rogers, 1949b) have a greater relative absorption of the β peak than the α peak, and this is considered to be a characteristic of hemoglobins with a high oxygen affinity. The absorption spectrum of the pharynx of *E. brevis* is given with that for *E. communis* in Fig. 10. The difference between the species is due to the presence of hemoglobin in the pharynx of *E. brevis*. At least some nematode hemoglobins have the additional features of an unusually low affinity for carbon monoxide and a slow rate of dissociation from oxygen (Lee and Smith, 1965). The half-time of dissociation of the perienteric-fluid hemoglobins of *Strongylus* and *Ascaris* are exceptionally long (Davenport, 1949a,b), but the body wall hemoglobin of *Ascaris* also dissociates much more slowly than mam-

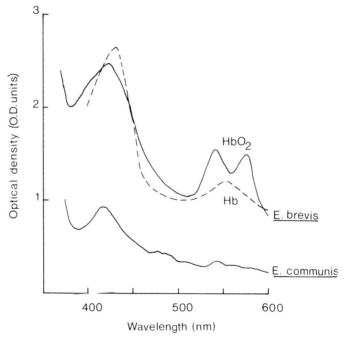

Fig. 10. A typical absorption spectrum for oxyhemoglobin (HbO_2) and deoxyhemoglobin (Hb) of *Enoplus brevis* measured in the pharyngeal region of the animal *in vivo*. A similar spectrum is given for *Enoplus communis*, which lacks hemoglobin in its pharynx [see Atkinson (1975) for experimental details]. Mean values $\pm$ SE for the α, β, and soret absorption peaks of the oxyhemoglobin are 577.6 ± 0.6, 543.6 ± 0.5, and 421.7 ± 1.9 nm, respectively. The corresponding values for the visible and soret peaks of deoxyhemoglobin are 555.2 ± 0.9 and 432.2 ± 1.3 nm. Mean values for visible absorption peaks for species of animal-parasitic nematodes are 577.6 and 547.7 nm for oxyhemoglobin and 555 nm for the deoxygenated pigment. (Atkinson, 1975.)

malian hemoglobin or myoglobin (Gibson and Smith, 1965). The oxygen affinities of the perienteric-fluid hemoglobin of *Strongylus* and *Ascaris* (Davenport, 1949a,b) are very high, and even the body wall hemoglobins of some species are half oxygenated at low values of P_{O_2}—for example, 0.1–0.2 mm Hg for *N. brasiliensis,* 0.05 mm Hg for *H. contortus* and less than 0.01 mm Hg for *Ascaris* (Rogers, 1962). This is apparently correlated with their function at a low tissue P_{O_2}. Evidence for this is provided by comparison with *Syngamus trachea,* which is a large nematode from the high-oxygen environment of the trachea (Rose and Kaplan, 1972). This nematode contains a hemoglobin that is 50% oxygenated at a P_{O_2} of 9.00 mm Hg. The situation is less clear for *E. brevis. In vivo,* the hemoglobin of this nematode is half oxygenated by an external P_{O_2} of about 12 mm Hg (Atkinson, 1975), but oxygen demand by the tissues will ensure that this underestimates its oxygen affinity.

The perienteric fluid hemoglobins may not have a respiratory function. That in *A. lumbricoides* may act as a hematin source which is incorporated into the gametes (Smith and Lee, 1963), and it is interesting to note that the heme content of the perienteric fluid of *Ascaris* does decrease if eggs are formed *in vitro* (Smith, 1969b). The tissue hemoglobins may either facilitate oxygen diffusion or act as short term oxygen stores. Vertebrate hemoglobins (Wittenberg, 1966) and myoglobin (Wittenberg, 1970; Wittenberg *et al.*, 1975) are capable of considerably enhancing the diffusion of oxygen through solution. The effect is probably greatest *in vivo* when the oxygen gradient ensures that the pigment is fully unloaded at the site of utilization but leaves the pigment fully oxygenated at the surface of the tissue (Wittenberg, 1970). This effect is likely to be of greater physiological significance than an oxygen store in hemoglobins capable of facilitated diffusion. The body wall hemoglobins of animal parasitic nematodes, such as *Ascaris,* do deoxygenate *in vivo* (Davenport, 1949a), but they have too low a rate of dissociation from oxygen in vitro to facilitate diffusion (Wittenberg, 1966). Possibly, the pigments are affected by extraction. It is known that the oxygen affinity of mammalian myoglobin is subject to control by cellular phosphates (Benesch and Benesch, 1969). There is some evidence for a similar control process in invertebrates (Wittenberg, 1970), but this may not occur in some annelids (Mangum, 1976).

The respiratory function of the hemoglobin of *E. brevis* has been investigated. The oxygen consumption of this species has been compared with that of *E. communis,* which contains very little hemoglobin (Fig. 10). The only difference between the oxygen consumption of the two nematodes was an indication that smaller *E. brevis* were less affected by a low P_{O_2} of 12 mm Hg than were larger individuals of either species (Atkinson, 1973a). The main conclusion from this work was that hemoglobin may serve particular tissues, but it did not cause a large increase in the overall oxygen consumption of *E. brevis* at a low P_{O_2}. Further work indicated that the pharyngeal hemoglobin showed a progressive deoxygenation for animals maintained at a P_{O_2} between 20 and 5 mm Hg (Atkinson, 1975). A partial deoxygenation is a prerequisite for an appreciable faciliated diffusion of oxygen to occur. Furthermore, stimulation of the animal by electric shock treatment resulted in a partial deoxygenation of its hemoglobin at a higher P_{O_2} than required at rest. This would again be necessary for facilitation to occur, although it may merely indicate an oxygen storage function. The hemoglobin of the pharynx has been shown to enable *E. brevis* to feed more successfully than *E. communis* at values of P_{O_2} between 40 and 15 mm Hg (Fig. 11). The feeding of *E. brevis* is also reduced by P_{O_2} levels at which the hemoglobin is at least 50% deoxygenated at rest. These results show that the hemoglobin in the pharynx of *E. brevis* has a respiratory function during feeding. The changes in oxygenation satisfy the requirements for facilitated diffusion but do not prove that this occurs. It may be that the hemoglobin acts merely as an oxygen store in this nematode and in other species (Atkinson, 1977).

A pigment with a slow dissociation constant and a high oxygen affinity may act as a short-term oxygen store for aerobic activity in muscles at a low P_{O_2}. In other tissues, hemoglobin may protect the cells from an oxygen lack caused by either an increased oxygen demand by adjacent tissues or by a short-term fluctuation in external oxygen availability. It may be that some nematode hemoglobins

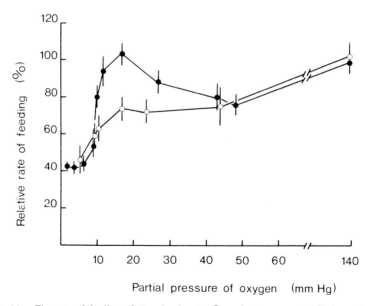

Fig. 11. The rate of feeding of *Enoplus brevis* (●) and *E. communis* (○) in mud and the ambient partial pressure of oxygen. Values are plotted as running grand means for triads of means ± SE (Atkinson, 1977). *Enoplus brevis* contains hemoglobin in its pharyngeal region but *E. communis* lacks the pigment (see Fig. 10).

are also able to facilitate oxygen diffusion. This would mean that not all nematode hemoglobins share a low rate of oxygen dissociation or that extraction of these pigments results in different properties from those found *in vivo*. Work with *E. brevis* does suggest that hemoglobin is a respiratory adaptation in nematodes to microaerobic habitats, irrespective of whether the pigment acts primarily to facilitate oxygen diffusion or as an oxygen store.

IX. ANAEROBIOSIS

A detailed study of anaerobic metabolism is beyond the scope of this chapter, but it is relevant to consider if nematodes have an ability to offset any shortfall in aerobic metabolism by increasing anaerobic pathways, as may occur in other animals (Beadle, 1961). This may take place in *Caenorhabditis* sp. and *Aphelenchus avenae,* which utilize lipid reserves in aerobic conditions but depend upon carbohydrate when the availability of oxygen is low (Cooper and van Gundy, 1970). The rate of utilization of carbohydrate is very low and this suggests that these nematodes experience a severe metabolic reduction during anoxia.

Anaerobic metabolisms can be separated into those based on (1) glycolysis and (2) additional metabolic pathways (Hochachka and Somero, 1976). Glycolysis is normally a homolactic fermentation in animals and produces 2 moles of ATP/mole of glucose. Its main feature is that a high rate of ATP production can be achieved in the pathway. The potential energy of the lactate can also be conserved if pyruvate can be reformed when oxygen is again available, or by passage of lactate to tissues not experiencing anoxia. It is ideal, therefore, as a short-term measure for a high rate of ATP production during an oxygen lack (Hochachka and Somero, 1976). Tissues that are homolactate fermenters frequently show an increase in oxygen consumption after anoxia during repayment of an oxygen debt. Such an effect has been recorded for *A. lumbricoides* (Laser, 1944), *E. ignotus* (von Brand, 1947), *Strongyloides ratti* (Barrett, 1969b), and *Heterodera oryzae* (Reversat, 1977). However, it may be an oversimplification to attribute such changes in a whole animal to an oxygen debt because the transfer of the nematode from a low to a high P_{O_2} may also stimulate activity or other oxygen-utilizing processes. It can also be of little adaptative significance to a nematode that does not normally experience a fluctuation in oxygen availability.

Nonglycolytic pathways have been most fully studied in *Ascaris* (Barrett, 1976). These systems may allow the formation of 5 moles of ATP/mole of glucose in *Ascaris* and may be as much as 7 moles of ATP/mole of glucose in the muscle of the oyster (Hochachka and Somero, 1976). This is an improvement over the efficiency of a homolactate fermentation in the absence of a subsequent aerobic metabolism. The nonglycolytic pathways are particularly suited to a prolonged oxygen lack, but in the muscle of the oyster they are capable of a lower rate of ATP production than homolactate fermentation and may limit the power output during contraction (Hochachka and Somero, 1976). The low levels of cytochromes and hemoglobin in the body wall muscles of *Ascaris* and the dependence of the organism on glycolysis are features shared with the inactive, white skeletal muscles of vertebrates (Lehninger, 1975). The absence of phosphagens in the muscle of *Ascaris* (Barrett, 1973) also suggests that this muscle has a low power output.

For a few nematodes, anaerobiotic pathways may enable the level of energy metabolism to remain unaffected by anoxia, but there is much evidence to show that the range of metabolic activities of many species is curtailed by a lack of oxygen (Section VII,A).

X. CONCLUSIONS

The high efficiency of an energy metabolism that uses oxygen as a terminal oxidase is such that nematodes normally possess adaptations to maintain an aerobic existence whenever possible. The level of organization limits the range of respiratory adaptations available to nematodes, and they have homeostatic capabilities different from those of higher animals.

The rate of metabolism is inversely related to body size, as it is in many other invertebrates. Account must be taken of this when nematodes are compared in any way that is influenced by metabolic rate. A wide range of other factors also influences the oxygen uptake of a nematode, and in some rudimentary homeostasis may occur in response to fluctuations in oxygen supply and perhaps temperature.

A small body radius and a high environmental partial pressure of oxygen may ensure that many species rarely experience anoxia. Other species inhabit environments in which the oxygen supply may be limited, and they show adaptations, such as hemoglobin, to maintain an aerobic metabolism. The life processes of many of these species are severely curtailed by a lack of oxygen, and prolonged anaerobic survival is normally due to the induction of quiescence. The remaining species inhabit anoxic environments or possess a body radius that limits the oxygen supply to some tissues in a microaerobic habitat. These animals may rely on anaerobic metabolism to maintain an appreciable part of their energy metabolism. They may utilize different anaerobic pathways from those used by nematode species that normally experience only short periods of oxygen lack.

ACKNOWLEDGMENTS

I thank Professor D. L. Lee for helpful criticism of this chapter and Mr. R. M. J. Storey for providing unpublished data. My work on cyst nematodes is supported by the Agricultural Research Council, and that on marine nematodes was financed by the Science Research Council.

REFERENCES

Abrams, B. I., and Mitchell, M. J. (1978). *Nematologica* **24**, 456-462.
Alphey, T. J. W. (1972). *Parasitology* **64**, 449-460.
Anderson, G. L. (1978). *Can. J. Zool.* **56**, 1786-1791.
Anderson, G. L., and Dusenbery, D. B. (1977). *J. Nematol.* **9**, 253-254.
Andrassey, I. (1956). *In* "English translation of Selected East European Papers in Nematology" (B. M. Zuckerman, M. W. Brzeski, and K. H. Deubert, eds.). Univ. of Massachusetts, East Wareham, Massachusetts (1967).
Atkinson, D. E., and Walton, G. M. (1967). *J. Biol. Chem.* **242**, 3239-3241.
Atkinson, H. J. (1972). Ph.D. Thesis, University of Newcastle-upon-Tyne, England.
Atkinson, H. J. (1973a). *J. Exp. Biol.* **59**, 255-266.
Atkinson, H. J. (1973b). *J. Exp. Biol.* **59**, 267-274.
Atkinson, H. J. (1975). *J. Exp. Biol.* **62**, 1-9.
Atkinson, H. J. (1976). *In* "The Organization of Nematodes" (N. A. Croll, ed.), pp. 243-272. Academic Press, New York.
Atkinson, H. J. (1977). *J. Zool.* **183**, 465-471.
Atkinson, H. J., and Ballantyne, A. J. (1977a). *Ann. Appl. Biol.* **87**, 159-166.
Atkinson, H. J., and Ballantyne, A. J. (1977b). *Ann. Appl. Biol.* **87**, 167-174.
Bair, T. D. (1955). *J. Parasitol.* **41**, 191-203.
Ballantyne, A. J., Sharpe, M. J., and Lee, D. L. (1978). *Parasitology* **76**, 211-220.
Barker, I. K. (1973). *Parasitology* **67**, 307-314.
Barrett, J. (1969a). *Parasitology* **59**, 3-17.
Barrett, J. (1969b). *Parasitology* **59**, 859-875.
Barrett, J. (1973). *Int. J. Parasitol.* **3**, 393-400.
Barrett, J. (1976). *In* "The Organization of Nematodes" (N. A. Croll, ed.), pp. 11-70. Academic Press, New York.
Baxter, R. I., and Blake, C. D. (1969). *Ann. Appl. Biol.* **63**, 191-203.
Beadle, L. C. (1937). *J. Exp. Biol.* **14**, 56-70.
Beadle, L. C. (1958). *J. Exp. Biol.* **35**, 556-566.
Beadle, L. C. (1961). *Symp. Soc. Exp. Biol.* **15**, 120-131.
Beis, I., and Barrett, J. (1975). *Dev. Biol.* **42**, 188-195.
Benesch, R., and Benesch, R. B. (1969). *Nature (London)* **221**, 618-622.
Bhatt, B. D., and Rohde, R. A. (1970). *J. Nematol.* **2**, 277-285.
Bird, A. F. (1971). "The Structure of Nematodes." Academic Press, New York.
Boczon, K., and Michejda, J. W. (1978). *Int. J. Parasitol.* **8**, 507-513.
Brown, H. W. (1928). *J. Parasitol.* **14**, 141-160.
Brown, B. J., and Platzer, E. G. (1978). *J. Nematol.* **10**, 110-113.
Bryant, C., Nicholas, W. L., and Jantunen, R. (1967). *Nematologica* **13**, 197-209.
Bryant, C., Cornish, R. A., and Rahman, M. S. (1976). *In* "Biochemistry of Parasites and Host-

Parasite Relationships'' (H. Van den Bossche, ed.), pp. 599–604. North-Holland Publ., Amsterdam.

Bryant, V. (1973). *Parasitology* **67**, 245–251.

Bryant, V. (1974). *Parasitology* **69**, 97–106.

Bueding, E., and Charms, B. (1952). *J. Biol. Chem.* **196**, 615–627.

Calloway, D. H. (1968). *In* ''Handbook of Physiology'' Section 6 (C. F. Code, ed.), Vol. V, pp. 2839–2859. Am. Physiol. Soc., Waverley Press, Baltimore, Maryland.

Chance, B. (1957). *In* ''Methods in Enzymology'' (S. P. Colowick and N. O. Kaplan, eds.), Vol. IV, pp. 273–329. Academic Press, New York.

Chance, B., and Parsons, D. F. (1963). *Science* **142**, 1176–1180.

Chance, B., Schoener, B., and Schindler, F. (1964). *Oxygen Anim. Org. Proc. Symp. 1963*, pp. 367–388.

Cheah, K. S. (1972), *In* ''Comparative Biochemistry of Parasites'' (H. Van den Bossche, ed.), pp. 417–432. Academic Press, New York.

Cheah, K. S. (1976). *In* ''Biochemistry of Parasites and Host-Parasite Relationships'' (H. Van den Bossche, ed.), pp. 133–143. Academic Press, New York.

Cheah, K. S., and Chance, B. (1970). *Biochim. Biophys. Acta.* **223**, 55–60.

Cooman, E. P. (1950). *Ann. Soc. R. Zool. Belgium* **81**, 5–13.

Cooper, S. C., and Ferguson, J. H. (1973). *J. Nematol.* **5**, 241–245.

Cooper, A. F., and Van Gundy, S. D. (1970). *J. Nematol.* **2**, 305–315.

Cooper, A. F., Van Gundy, S. D., and Stolzy, L. H. (1970). *J. Nematol.* **2**, 182–188.

Cotten. J. (1976). *Ann. Appl. Biol.* **83**, 407–412.

Croll, N. A. (1972). *Parasitology* **64**, 355–368.

Crompton, D. W. T., Shrimpton, D. H., and Silver, I. A. (1965). *J. Exp. Biol.* **43**, 473–478.

Davenport, H. E. (1949a). *Proc. R. Soc. London Ser. B* **136**, 255–270.

Davenport, H. E. (1949b). *Proc. R. Soc. London Ser. B* **136**, 271–280.

Deubert, K. H., and Zuckerman, B. M. (1968). *Nematologica* **14**, 453–455.

Dusenbery, D. B. (1974). *J. Exp. Zool.* **188**, 41–47.

Dusenbery, D. B., Anderson, G. L., and Anderson, E. A. (1978). *J. Exp. Zool.* **206**, 191–198.

Ellenby, C. (1953). *J. Exp. Biol.* **30**, 475–491.

Ellenby, C., and Smith, L. (1966). *Comp. Biochem. Physiol.* **19**, 871–877.

Ellenby, C., and Smith, L. (1967). *Comp. Biochem. Physiol.* **21**, 51–57.

Evans, A. A. F., and Perry, R. N. (1976). *In* ''The Organization of Nematodes'' (N. A. Croll, ed.), pp. 383–424. Academic Press, New York.

Fairbairn, D. (1970). *Biol. Rev.* **46**, 29–72.

Fernando, M. A. (1963). *Exp. Parasitol.* **13**, 90–97.

Gibson, Q. H., and Smith, M. H. (1965). *Proc. R. Soc. London, Ser. B.* **157**, 234–257.

Goldberg, E. (1957). *Exp. Parasitol.* **5**, 367–382.

Hardy, R. N. (1972). ''Temperature and Animal Life.'' Arnold, London.

Harpur, R. P. (1964). *Comp. Biochem. Physiol.* **13**, 71–85.

Hayaishi, O. (1974). *In* ''Molecular Mechanisms of Oxygen Activation'' (O. Hayaishi, ed.), pp. 1–28. Academic Press, New York.

Hemmingsen, A. M. (1960). *Rep. Steno. Mem. Hosp. Nord. Insulinlab.* **9**, 1–110.

Herrick, C. A., and Thede, M. (1945). *J. Parasitol.* **31**, 18–19.

Hill, A. V. (1929). *Proc. R. Soc. London Ser. B* **104**, 39–96.

Hochachka, P. W., and Somero, G. N. (1976). *In* ''Adaptation to Environment: Essays on the Physiology of Marine Animals'' (R. C. Newell, ed.), pp. 279–314. Butterworth, London.

Hopper, B. E., and Meyer, S. P. (1966). *Nature (London)* **209**, 899–900.

James, W. O. (1971). ''Cell Respiration.'' English Univ. Press, London.

Jamuar, M. P. (1966). *J. Cell. Biol.* **31**, 381–396.

Jones, J. D. (1972). "Comparative Physiology of Respiration." Arnold, London.

Kaulenas, M. S., and Fairbairn, D. (1968). *Exp. Cell. Res.* **52**, 233–251.

Keilin, D. (1959). *Proc. R. Soc. London Ser. B* **150**, 149–191.

Keilin, D. (1966). "The History of Cell Respiration and Cytochrome." Cambridge Univ. Press, London and New York.

Keilin, D., and Hartree, E. F. (1949). *Nature (London)* **164**, 254–259.

Kikuchi, G., and Ban, S. (1961). *Biochim. Biophys. Acta.* **51**, 387–389.

Kikuchi, G., Ramirez, J., and Barron, E. S. G. (1959). *Biochim. Biophys. Acta.* **36**, 335–342.

Klekowski, R. Z., Wasilewska, L., and Paplinska, E. (1972). *Nematologica* **18**, 391–403.

Klekowski, R. Z., Wasilewska, L., and Paplinska, E. (1974). *Nematologica* **20**, 61–68.

Klingler, J., and Kunz, P. (1974). *Nematologica* **20**, 52–60.

Krogh, A. (1916). "The Respiratory Exchange of Animals and Man." Longmans, Green, New York.

Krogh, A. (1919). *J. Physiol. (London)* **52**, 391–408.

Krogh, A. (1941). "The Comparative Physiology of Respiratory Mechanisms." Univ. of Pennsylvania Press, Philadelphia, Pennsylvania.

Krüger, F. (1936). *Zool. Jahrb. Abt. Allg. Zool. Physiol.* **57**, 1–56.

Krüger, F. (1940). *Z. Wiss. Zool.* **152**, 547–570.

Krusberg, L. R. (1960). *Phytopathology* **50**, 9–22.

Laser, H. (1944). *Biochem. J.* **38**, 333–338.

Lee, D. L. (1965). *Parasitology* **55**, 173–181.

Lee, D. L. (1971). *Parasitology* **63**, 271–274.

Lee, D. L., and Anya, A. D. (1967). *J. Cell Sci.* **2**, 537–544.

Lee, D. L., and Atkinson, H. J. (1976). "Physiology of Nematodes," 2nd ed., Macmillan, New York.

Lee, I., and Chance, B. (1968). *Biochem. Biophys. Res. Commun.* **32**, 547–553.

Lee, D. L., and Smith, M. H. (1965). *Exp. Parasitol.* **16**, 392–424.

Le Jambre, L. F., and Whitelock, J. H. (1967). *J. Parasitol.* **53**, 887.

Le Patourel, G. N. J., and Wright, D. J. (1974). *Pestic. Biochem. Physiol.* **4**, 144–52.

Lehninger, A. L. (1975). "Biochemistry. The Molecular Basis of Cell Structure and Function," 2nd ed. Worth, New York.

Lloyd, D. (1974). "The Mitochondria of Microorganisms." Academic Press, New York.

Mangum, C. P. (1976). *In* "Adaptation to Environment: Essays on the Physiology of Marine Animals" (R. C. Newell, ed.), pp. 191–278. Butterworth, London.

Margulis, L. (1975). *Symp. Soc. Exp. Biol.* **29**, 21–38.

Mercer, E. K., and Cairns, E. J. (1973). *J. Nematol.* **3**, 201–208.

Messner, B., and Kerstan, U. (1978). *Acta Histochem.* **62**, 244–253.

Newell, R. C. (1970). "The Biology of Intertidal Animals." Logus Press, London.

Nicholas, W. L. (1975). "The Biology of Free-living Nematodes." Oxford Univ. Press (Clarendon), London and New York.

Nicholas, W. L., and Jantunen, R. (1964). *Nematologica,* **10**, 409–418.

Nielsen, C. O. (1949). *Nat. Jutl.* **2**, 1–131.

Nwosu, A. B. C. (1978). *Int. J. Parasitol.* **8**, 85–88.

Onions, T. G. (1955). *Q. J. Microsc. Sci.* **96**, 495–513.

Podesta, R. B. and Mettrick, D. F. (1974). *Int. J. Parasitol.* **4**, 277–292.

Potts, W. T. W., and Parry, G. (1964). "Osmotic and Ionic Regulation in Animals." Pergamon, Oxford.

Prosser, C. L. (1961). *In* "Comparative Animal Physiology" (C. L. Prosser and F. A. Brown, Jr., eds.), 2nd ed., pp. 153–197. Saunders, Philadelphia, Pennsylvania.

Prosser, C. L. (1964). *In* "Handbook of Physiology," Section 4 (D. B. Dill, ed.), pp. 11–25. Am. Physiol. Soc., Waverley Press, Baltimore.

Reger, J. F. (1964). *J. Ultrastruct. Res.* **10**, 48–57.

Reger, J. F. (1966). *J. Ultrastruct. Res.* **14**, 602–617.

Reversat, G. (1977). *Nematologica* **23**, 369–381.

Rohde, R. A. (1960). *Proc. Helminthol. Soc. Wash.* **27**, 160–164.

Ritzrow, H., and Kampfe, L. (1971). *Nematologica* **17**, 325–335.

Roberts, L. W., and Fairbairn, D. (1965). *J. Parasitol.* **51**, 129–138.

Rogers, W. P. (1949a). *Aust. J. Sci. Res. Ser. B* **2**, 157–165.

Rogers, W. P. (1949b). *Aust. J. Sci. Res. Ser. B* **2**, 399–407.

Rogers, W. P. (1960). *Proc. R. Soc. London Ser. B* **152**, 367–386.

Rogers, W. P. (1962). "The Nature of Parasitism." Academic Press, New York.

Rose, J. E., and Kaplan, K. L. (1972). *J. Parasitol.* **58**, 903–906.

Rosenbluth, J. (1965). *J. Cell. Biol.* **26**, 579–591.

Rothstein, M., Nicholls, F., and Nicholls, P. (1970). *Int. J. Biochem.* **1**, 695–705.

Santmeyer, P. H. (1956). *Proc. Helminthol. Soc. Wash.* **23**, 30–36.

Schwarbe, C. W. (1957). *Am. J. Hyg.* **65**, 325–337.

Smith, M. H. (1969a). *Nature (London)* **223**, 1129–1132.

Smith, M. H. (1969b). *In* "Chemical Zoology" (M. Florkin and B. T. Scheer, eds.), Vol. 3, pp. 501–520. Academic Press, New York.

Smith, M. H., and Lee, D. L. (1963). *Proc. R. Soc. London Ser. B.* **157**, 234–257.

Teal, J. M., and Wieser, W. (1966). *Limnol. Oceanogr.* **11**, 217–222.

Threadgold, L. T. (1976). "The Ultrastructure of the Animal Cell," 2nd edition. Pergamon, Oxford.

Van Grembergen, G., Van Damme, R., and Vercruysse, R. (1949). *Enzymologica,* **13**, 325–342.

Van Gundy, S. D., Bird, A. F., and Wallace, H. R. (1967). *Phytopathology,* **57**, 559–571.

von Brand, T. (1946). *Biodynamica Monogr.* **4**, 137–278.

von Brand, T. (1947). *Biol. Bull.* **92**, 162–166.

von Brand, T. (1962). *In* "Nematology" (J. N. Sasser and W. R. Jenkins, eds.), pp. 233–241. Univ. North Carolina Press, Chapel Hill, North Carolina.

von Brand, T. (1973). "Biochemistry of Parasites." Academic Press, New York.

Wallace, H. R. (1971). *In* "Ecology and Physiology of Parasites" (A. H. Fallis, ed.) Univ. of Toronto Press, Toronto.

Wallace, H. R. and Doncaster, C. C. (1964). *Parasitology,* **54**, 313–326.

Walshe, B. M. (1950). *J. Exp. Biol.* **27**, 73–95.

Ward, S. (1973). *Natl. Acad. Sci. U.S.A.* **70**, 817–821.

Warren, L. G. (1970). *Exp. Parasitol.* **27**, 417–423.

Wieser, W., and Kanwisher, J. (1959). *Biol. Bull. Mar. Biol. Lab. Woods. Hole* **117**, 594–600.

Wieser, W., and Kanwisher, J. (1960). *Z. Vergl. Physiol.* **43**, 19–36.

Wieser, W., and Kanwisher, J. (1961). *Limnol. Oceanogr.* **6**, 262–270.

Williamson, J. R. (1976). *In* "Mitochondria: Bioenergetics, Biogenesis and Membrane Structure." (L. Packer and A. Gómez-Puyou, eds.), pp. 79–107. Academic Press, New York.

Wilson, P. A. G. (1965). *Exp. Parasitol.* **17**, 318–325.

Wittenberg, J. B. (1966). *J. Biol. Chem.* **241**, 104–114.

Wittenberg, J. B. (1970). *Physiol. Rev.* **50**, 559–636.

Wittenberg, B. A., Wittenberg, J. B., and Caldwell, P. R. B. (1975). *J. Biol. Chem.* **250**, 9038–9043.

Wright, D. J., and Newall, D. R. (1976). *In* "The Organization of Nematodes" (N. A. Croll, ed.), pp. 163–210. Academic Press, New York.

Wright, K. A. (1976). *In* "The Organization of Nematodes" (N. A. Croll, ed.), pp. 71–105. Academic Press, New York.

Yassin, A. M. (1969). *Nematologica* **15,** 169–178.

Yeates, G. W. (1971). *Pedobiologica* **11,** 173–179.

Zeuthen, E. (1947). *C. R. Lab. Calsb. Ser. Chim.* **26,** 17–161.

Zeuthen, E. (1953). *Q. Rev. Biol.* **28,** 1–12.

Zeuthen, E. (1955). *Annu. Rev. Physiol.* **17,** 459–482.

6

Osmotic and Ionic Regulation in Nematodes

D. J. WRIGHT

Department of Zoology and Applied Entomology
Imperial College
London SW7 2BB, England

D. R. NEWALL

Department of Plastic and Reconstructive Surgery
The Royal Victoria Infirmary
Newcastle-upon-Tyne NE1 4LP, England

I. Introduction	143
II. Techniques	144
A. Estimation of Volume Changes in Nematodes	144
B. Estimation of Inorganic Ions and Osmotic Pressure in Nematodes	146
C. Incubation Media for Studies on Osmotic and Ionic Regulation in Nematodes	147
III. Mechanisms in Nematode Osmoregulation and Possible Structures Involved in Transport Processes in Nematodes	149
A. Evidence for Active Regulation of Ions by Nematodes	152
B. Evidence for the Production of a Urine by Nematodes	152
C. Permeability of the Body Wall of Nematodes to Salts and Water .	155
IV. Specialized Aspects of Osmotic and Ionic Regulation in Nematodes .	156
A. Hatching: Osmotic and Ionic Influences	156
B. Locomotion	160
C. Desiccation Survival (Anhydrobiosis) and Salt Loss	160
V. Conclusions	161
References	162

I. INTRODUCTION

In a recent review on osmotic and ionic regulation in nematodes we considered the osmoregulatory ability of nematodes largely from a comparative ecological

143

viewpoint (Wright and Newall, 1976). It is our intention here to deal principally with the techniques available for such studies on nematodes, the possible mechanisms and structures involved in regulation, and specialized aspects of nematode biology where osmotic and/or ionic stress may be important.

Osmotic concentrations expressed originally as freezing-point depression values have been converted to the molarity of a NaCl solution with an equivalent osmotic pressure by the relationship $-1.0°C = 293$ mM NaCl (Krogh, 1939).

II. TECHNIQUES

A. Estimation of Volume Changes in Nematodes

1. Length Measurement Method

Nematodes possess an anisometric cuticle with the result than an increase in length leads to an increase in volume and vice versa (see Lee and Atkinson, 1976), and length measurements have been used in several studies on volume regulation (Stephenson, 1942; Lee, 1960; Wright and Newall, 1976). However, an investigation into the relationship between change in length and change in volume in three species of nematode, *Globodera (Heterodera) rostochiensis*, *Enoplus brevis,* and *Enoplus communis,* revealed interspecific differences and significant nonlinearities in this relationship (Fig. 1) that may be related to differences in the structure of the outer layers of the cuticle (L. Smith, personal communication).

Thus interspecific comparisons based on length measurements alone are not completely valid, and the measurement of length does not give an absolute value for a change in volume. If this method must be used, account should be taken of the change in the diameter of the nematode. The relationship $V_2/V_1 = l_2 \, d_2^2/l_1 \, d_1^2$, where l is the length of the nematode and d the maximum diameter (Andrássey, 1956), gives a better estimate of volume change (Wright and Newall, 1976).

Accurate measurement of changes in the length and diameter of nematodes is difficult, especially with the more active species. Various methods have been used. Lee (1960) measured the relative change in length from camera lucida drawings of nematodes. This method can be useful, but only with relatively slow-moving species, and the method is also time-consuming if done accurately. A better technique is to photograph the nematode and then project and measure the negative at a fixed magnification (Newall, 1980).

The accuracy of the latter method may be illustrated by the following example. Ellenby (1974) observed a 5% increase in water content of *G. rostochiensis* juveniles 30 min after liberation from the egg in tap water. It can be calculated that such a change in water content would be accompanied by an increase in

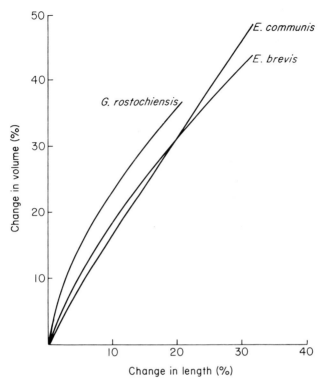

Fig. 1. Relationship between change in length and change in volume in *Globodera rostochiensis* J₂, *Enoplus communis,* and *Enoplus brevis.* The volumes were calculated from enlarged photographs, assuming each species to be circular in cross section. (After Wright and Newall, 1976.)

volume of about 18%. The average increase in length observed under these conditions was also about 5% (Ellenby, 1974) and, referring to data based on the photographic technique (Fig. 1), it can be seen that this increase in length would be associated with an increase in volume of about 16%. In view of the method used by Ellenby to measure the length of the liberated juveniles (a traversing hairline micrometer eyepiece), the two results are in close agreement.*

In any of the above methods it is essential that the nematode is confined to a small area (but without direct pressure on the nematode) so that measurements are not distorted by movements in the vertical plane or by excessive curvature of the body. Furthermore, the use of narcotics to reduce the movement of the nematodes is inadvisable because of probable effects on muscle tone. It is known, for example, that propylene phenoxitol (Ellenby and Smith, 1964), a

---

*An alternative to photographing nematodes may be to use a suitable high-resolution video system.

commonly used narcotic for nematodes, affects the length-to-volume relationship of several species (Wright and Newall, 1976).

2. *Other Methods*

Two other methods used with nematodes are (a) the direct measurement of water content using tritiated water (Marks *et al.,* 1968; Castro and Thomason, 1973) and (b) the indirect measurement of water content by comparison of wet and dry weights (Myers, 1966). Both methods are preferable to the measurement of length, but they are unsatisfactory for most studies with nematodes since, unlike the last method, they are discontinuous (i.e., destructive) and relatively large numbers of animals are required. In the wet-/dry-weight method there is also the problem of accurately determining the wet-weight point.

B. Estimation of Inorganic Ions and Osmotic Pressure in Nematodes

In large nematodes such as *Ascaris lumbricoides,* where it is possible to sample pseudocoelomic fluid, the measurement of ionic concentration and osmotic pressure may be carried out by conventional means (Hobson *et al.*, 1952a,b). Even in nematodes having a volume as low as 0.1 μl it has recently been found possible, using a micropipet with a tip diameter of $1-2$ μm, to remove sufficient pseudocoelomic fluid to determine the depression of freezing point and hence the osmotic pressure (L. Smith, personal communication). However, for the many species of nematode below this size range, the body fluids of which cannot at present be sampled, the direct measurement of osmotic pressure or ionic concentration is not possible.

An alternative has been to measure the total ionic concentration of an individual or small group of nematodes using flame spectrophotometry (Myers, 1966; Croll and Viglierchio, 1969; Wright and Newall, 1976), carbon rod atomizer atomic absorption spectrophotometry, which is capable of detecting concentrations of heavy metals in individual marine nematodes as low as 0.01 ppm (R. Howell, personal communication), or by exchanging the ion to be measured for a radioactive isotope of known specific activity (Wright and Newall, 1976). The drawback of these techniques is that they do not distinguish between free and bound or inter- and intracellular concentrations of ions, and it is these values that are of interest to the research worker. However, two recently introduced methods offer some hope in this direction.

First, energy-dispersive X-ray analysis (Chandler, 1977) has been used to measure changes in ionic concentration in nematodes (Atkinson and Ballantyne, 1979; Womersley, 1978). This technique is, for example, capable of detecting a net uptake or loss of calcium from an individual juvenile of *G. rostochiensis* of only about 1×10^{-7} g dry weight (Atkinson and Ballantyne, 1979) (see Section IV,A). Second, ion-specific liquid ion-exchanger microelectrodes (Walker,

1971) have been developed to measure intracellular activities of ions in living cells. The application of this technique to nematodes is only in its early stages, but its future looks very promising (L. Smith, personal communication).

C. Incubation Media for Studies on Osmotic and Ionic Regulation in Nematodes

1. Criticism of Single-Salt Solutions

Many of the earlier studies on nematodes used various concentrations of single-salt solutions, usually NaCl. The use of such solutions does not, however, give a valid impression of regulatory ability under natural conditions (see Lee, 1960; Wright and Newall, 1976), and is not to be recommended. Newall (1980) found that *E. brevis* and *E. communis* failed to regulate their volume when placed in 0.55 *M* NaCl (which is approximately isosmotic with 100% seawater), but in an osmotically equivalent medium that also contained Ca and K ions, volume regulation occurred (Fig. 2).

The importance of Ca and K in the external medium for normal volume regulation has been demonstrated in other invertebrate groups (Pantin, 1931; Ellis, 1937), and the above findings are consistent with them and with the principles of antagonistic phenomena (Lockwood, 1960). An increase in the permeability of the body wall of nematodes could result from a loss of Ca to a low-Ca medium (see Rojas and Tobias, 1965; Smith, 1971), or changes in the external concentration of either Ca or K could well result in an influx of Na, or of

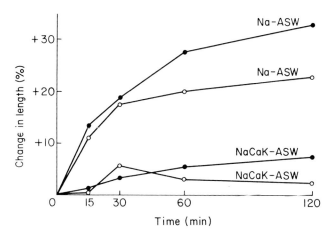

Fig. 2. Change in length of *E. communis* (○) and *E. brevis* (●) in artificial seawater (ASW), in which Na is the only cation (Na–ASW), or in which Na, Ca, and K are the only cations (NaCaK–ASW). Na–ASW contains 550 m*M* Na. (After Wright and Newall, 1976.)

any other cation down its concentration gradient. This would lead to an increase in the internal osmotic pressure and an increase in water uptake.

These effects may have contributed to the recovery of length in hypertonic single-salt solutions observed in several species of marine nematodes (Croll and Viglierchio, 1969; Viglierchio, 1974), and would explain why the rate of recovery was dependent upon the passive permeability of the nematode to the external cations and anions. Indeed, Viglierchio (1974) stated that *Monhystera disjuncta* was nonregulating on the basis of its failure to control body volume in 1.0 *M* NaCl. For further discussion of work on nematodes in "unbalanced" salt solutions, see Wright and Newall (1976).

2. Criticism of Distilled Water, Urea and Other Nonelectrolytes

Apart from the use of "unbalanced" salt solutions, there has also been an unfortunate tendency to incubate freshwater and soil species of nematodes in distilled water. Although nematodes in very dilute media may appear, at least initially, to be reasonably "normal," their activity and survival can be reduced (Wright, 1973; A. A. F. Evans, personal communication), and although some species are able to regulate their volume in distilled water (Stephenson, 1942; Wright and Newall, 1976), others are not (Myers, 1966). It is also known that second-stage juveniles of *G. rostochiensis* rapidly lose Na in media below 16 mM NaCl (Fig. 3), which is the approximate mean concentration of soil water (Wallace, 1971). Similar results have been obtained with *Panagrellus redivivus* and *Aphelenchus avenae* (Myers, 1966).

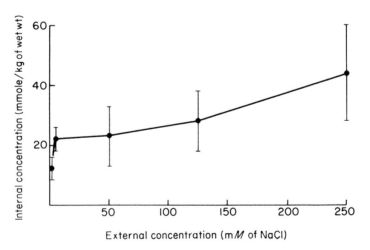

Fig. 3. Internal Na concentration of *G. rostochiensis* J_2 after 24 hr in NaCl-supplemented artificial tap water. The vertical lines represent ± the sample standard deviation. (After Wright and Newall, 1976.)

It would seem advisable therefore to maintain such species in artificial media containing at least 16 mM NaCl in addition to any other ions (see Section II,C,3).

Some studies on osmoregulation in nematodes have been made using urea (Blake, 1961; Viglierchio et al., 1969), glucose, or other sugar solutions (Myers, 1966). The use of such media is open to the above-mentioned criticisms, and, in addition, with urea it is difficult to distinguish between osmotic and toxic effects. Because of the permeability of nematodes to urea, it is probable that the toxic effects are the more important.

3. Balanced Salt Solutions

Specific Ringer solutions for nematodes await the accurate measurement of extracellular ionic concentrations. However, some existing media are reasonably suitable.

For marine and brackish-water species of nematodes, media based on a simplified, artificial seawater (Pantin, 1964) are appropriate (Wright and Newall, 1976). For freshwater and soil species, various Ringer or artificial tap water solutions can be used. For example, 35% (v/v) frog Ringer (Pantin, 1964) has been found to be approximately isotonic with P. redivivus (A. Stephenson, personal communication), and 33% (v/v) earthworm Ringer is suitable for the in vitro survival of the sperm of P. redivivus (Duggal, 1978). Newall (1980) has used an artificial Newcastle tap water (Greenaway, 1971) supplemented with NaCl for juveniles of G. rostochiensis. Various other media have been used by Myers (1966), Brenner (1974), and Vanfleteren (1975).

With some animal-parasitic species of nematodes there is the problem that many other factors can limit survival. For example, in an apparently isosmotic saline solution A. lumbricoides appears to undergo necrosis rapidly and is thus slowly dying, at best (see Wright and Newall, 1976).

III. MECHANISMS IN NEMATODE OSMOREGULATION AND POSSIBLE STRUCTURES INVOLVED IN TRANSPORT PROCESSES IN NEMATODES

There is little direct evidence for osmotic or ionic regulation in nematodes, and present knowledge is based largely on the ability of species to survive, or regulate their volume, in hyposmotic or hyperosmotic media. Present data on the regulatory ability of different species of nematodes are summarized in Table I.

As in the majority of marine invertebrates, marine species of nematodes are isomotic with seawater. The species examined showed little or no ability to regulate their volume in hyposmotic solutions, but possessed in most cases some regulatory ability in hyperosmotic conditions. As might be expected, the

TABLE I

Summary of Osmotic Relations in Nematodes

Species	Environment	Relation to environment	Volume regulation	Reference
Deontostoma californicum	Marine	Isosmotic	0/ – –	Croll and Viglierchio (1969)
D. timmerchioi	Marine	Isosmotic	0/ – –	Viglierchio (1974)
D. antarcticum	Marine	Isosmotic	0/ – –	Viglierchio (1974)
Monhystera disjuncta	Marine	Isosmotic	0/0	Viglierchio (1974)
Enoplus communis	Marine	Isosmotic	0/ –	Wright and Newall (1976)
E. brevis	30% Seawater +	Hyper/isosmotic	+ +/–	Wright and Newall (1976)
Rhabditis terrestris	Soil water	Hyperosomotic	+ + +/–	Stephenson (1942)
Panagrellus redivivus	Soil water	Hyperosmotic		Myers (1966)
Aphelenchus avenae	Soil water	Hyperosmotic		Myers (1966)
Panagrolaimus davidi	Freshwater	Hyperosmotic	/0	Viglierchio (1974)
Globodera rostochiensis 12	Soil water/plants	Hyperosmotic	+ + +/ – –	Wright and Newall (1976)
Hammerschmidtiella diesingi	Cockroach hindgut	Isosmotic	0/ –	Lee (1960)
Aspicularis tetraptera	Mouse intestine	Isosmotic?	+ +/ – –	Anya (1966)
Angusticaecum sp.	Tortoise colon	Hyperosmotic?		Pannikar and Sproston (1941)
Ascaris megalocephala	Horse intestine	Hyposmotic	0/0	Schepfer (1932)
A. lumbricoides	Pig intestine	Hypo/hyperosmotic	0/0	(See Wright and Newall, 1976)

[a] Degree of volume regulation in hypotonic solutions (+) and in hypertonic solutions (–).

brackish-water species *E. brevis* shows a greater degree of volume regulation, particularly in hyposmotic media.

Soil and freshwater nematodes are hyperosmotic with their normal environment and in at least two species, *Rhabditis terrestris* and *G. rostochiensis,* volume regulation is good in hyposmotic solutions but poor or absent in hyperosmotic ones.

With animal-parasitic nematodes the picture is less clear, and while some species appear to be isosmotic with the body fluid of the host, others may be hypo- or hyperosmotic to their environment. For example, whether *A. lumbricoides* is hyposmotic to pig intestinal fluid *in vivo* remains an open question (Wright and Newall, 1976). In some species volume regulation appears to be very limited. However, volume regulation has only been studied in adult parasites, and an examination of other stages in the life cycle, where drastic changes in the nematode's environment often occur, could reveal considerable regulatory abilities.

Thus most species of nematode are only capable of regulating their body volume over a limited range of external concentrations.

The mechanisms by which animals reduce changes in their volume and maintain greater internal ionic stability are twofold: (1) isosmotic intracellular regulation (Schoffeniels, 1967; Lang and Gainer, 1969), where there is an active alteration of the cellular osmotic pressure to conform with that of the extracellular fluid. This is usually achieved by changes in the free-amino acid pool of the cell; (2) anisosmotic extracellular regulation (Schoffeniels, 1960), which complements the first mechanism and involves the maintenance of extracellular body fluids either hypo- or hyperosmotic to the external environment. A non-ionic osmotically active component may also be involved in this process.

Although there is no direct evidence for the first mechanism in nematodes, it is almost certainly present since it is generally thought to be the more primitive mechanism and is found in euryhaline invertebrates that show no anisosmotic extracellular regulation (Potts, 1958; Duchâteau-Bosson *et al.,* 1961). The latter mechanism has been demonstrated in *A. lumbricoides* (Hobson *et al.,* 1952a; Harpur and Popkin, 1965; Kümmel *et al.,* 1969) and more tentatively in an *Angusticaecum* species [see Wright and Newall (1976)]. Evidence for this process occurring in other species of nematodes is based on the indirect evidence of the active regulation of ions, the production of a hypo- or isotonic urine, and the possession of low permeability to salts and water. The relatively high hydrostatic pressure of most nematodes may be involved in this type of regulation.

Non-ionic substances may be important in either of these mechanisms in nematodes, for example, in *R. terrestris* (Stephenson, 1942) and *A. lumbricoides* (Hobson *et al.,* 1952a), and it would be advantageous under hyperosmotic conditions for an animal to remain hypoionic by increasing its non-ionic component.

A. Evidence for Active Regulation of Ions by Nematodes

No active uptake or elimination of ions has been demonstrated in nematodes, although it is apparent from the ionic composition of the body fluid of some larger species of nematodes that regulation of ions does occur. For example, the relatively constant concentration of Na, K, Ca, and Mg in the body fluid of *A. lumbricoides,* despite considerable fluctuations in their external concentrations, suggests that all of these ions are actively regulated (Hobson *et al.,* 1952b; Kümmel *et al.,* 1969).

With the smaller species of nematodes, only total body ionic concentrations can be determined at present, and this information is of little use without a knowledge of their distribution between the intra- and extracellular compartments. A comparative study of changes in Na content of the marine nematode *E. communis* and the closely related estuarine species *E. brevis* showed that after 24 hr of incubation in various concentrations of seawater, *E. brevis* loses relatively less Na than does *E. communis* in hypotonic conditions (Fig. 4). This difference may indicate some degree of Na regulation by *E. brevis* in dilute seawater. There is also some evidence that *P. redivivus* and *A. avenae* can regulate their internal K content (but not their Na content) in hypertonic media (Myers, 1966), and that juveniles of *G. rostochiensis* can regulate Na in a range of artificial tap water solutions supplemented with various concentrations of NaCl (Fig. 3).

Within nematode tissues it is known, for example, that the muscle cell membrane of *A. lumbricoides* apparently has the ability to regulate K and Cl levels within very narrow limits, regardless of changes in the extracellular fluid (Arthur and Sanborn, 1969).

B. Evidence for the Production of a Urine by Nematodes

The intestine, various hypodermal gland cells, and the "excretory system" have all been suggested as sites of urine production in nematodes capable of volume regulation in hypotonic media (Wright and Newall, 1976).

(1) The nematode intestine consists of a single layer of cells with a luminal border of microvilli indicative of a secretory and/or absorptive function. Evidence for the importance of the gut in the removal of excess fluid has been found in three species of nematode. In *R. terrestris* rapid pumping of the intestine was observed in distilled water (Stephenson, 1942), and the gut was considered to be the major route for the removal of water. In *A. lumbricoides* the feces are hypotonic to the body fluid of the nematode, both *in vitro* and *in vivo* (Harpur and Popkin, 1965), and the intestine has been shown to be the most important route for the removal of soluble, nitrogenous waste products (Paltridge and Janssens, 1971). The transport of water across the intestine of the latter species has been demonstrated, and during this process the intercellular spaces of the

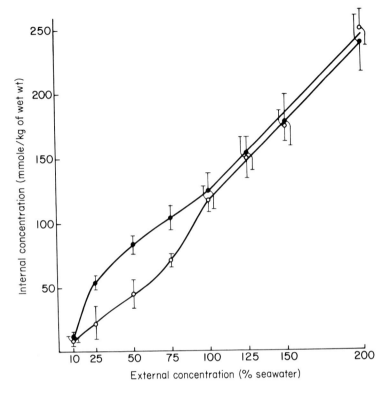

Fig. 4. Internal Na concentration of *E. communis* (○) and *E. brevis* (●) after 24 hr in various dilutions and concentrations of seawater. The vertical lines represent ± the sample standard deviation. (After Wright and Newall, 1976.)

intestine dilate (Harpur and Popkin, 1973), as in other epithelial cells (Diamond, 1971). Finally, when the anus of *E. brevis* is ligatured, the nematode fails to regulate in hypotonic dilutions of seawater, suggesting that the intestine is the major route for the removal of water.

(2) Hobson *et al.* (1952a) suggested that regulation of ions in *A. lumbricoides* involved structures in the body wall, either in the hypodermis or the somatic musculature. More recently, the ultrastructural examination of several species of nematodes has shown structures within the hypodermis that may have a transport function. In three genera of animal-parasitic nematodes belonging to the family Trichuroidea, various hypodermal gland cells are found that form structures called "bacillar bands" (Wright and Chan, 1973), and these cells closely resemble cells involved in osmotic and ionic regulation in arthropods and vertebrates (Wright, 1963; Oschman and Berridge, 1971). Similar cells have been observed in the free-living nematode *Acanthonchus duplicatus* (Wright and Hope, 1968).

In the plant-parasitic nematode *Xiphinema index,* there are invaginations of the outer hypodermal membrane which have continuity with the endoplasmic reticulum. This, together with the presence of microtubules, has suggested a transport function (Roggen *et al.,* 1967).

(3) Structures have been described as "excretory" organs in nematodes on morphological rather than physiological evidence (Wright and Newall, 1976). There appear to be two basic types of "excretory system" in nematodes, glandular and tubular. The former is probably more primitive (Bird, 1971) and is found in many freshwater and marine species; it consists of a single "renette" or ventral gland cell, usually within the pseudocoelom, connected to an exterior pore by a duct region. The latter system characteristically has a long "canal" in each lateral hypodermal cord; these are connected by a transverse canal to an excretory duct. This tubular system may be reduced in some genera and in others is sometimes associated with a pair of ventral glands (Bird, 1971). In both systems there is often an ampulla in the duct region near the excretory pore.

Recent ultrastructural and histochemical evidence strongly suggests that the nematode "excretory gland" has in fact a predominantly secretory rather than excretory function (Wright and Newall, 1976), although it has been proposed on morphological grounds that the gland in an *Anisakis* sp. is also associated with osmoregulation and excretion (Lee *et al.,* 1973).

The tubular excretory system of nematodes is, however, thought to have an osmoregulatory and/or excretory function in at least some species, and ultrastructural investigations have shown the presence of fine, branching tubules, or canaliculi, surrounding the lumen of the lateral canals in all species examined (Wright and Newall, 1976). How fluid flow is produced in the canal system is not clearly understood. It is clear, however, that the system is not "protonephridial," as described by Chitwood and Chitwood (1950), Prosser (1973), and Wilson and Webster (1974), since no protonephridial, current-producing structures, cilia, or flagella, have been found in the excretory system of nematodes (Bird, 1971; Narang, 1972). Harris and Crofton (1957) proposed that fluid is passed into the lumen of the canal system in *A. lumbricoides* by a pressure filtration mechanism analogous to the vertebrate kidney. However, subsequent physiological and ultrastructural work has failed to support this idea (Kümmel *et al.,* 1969; Dankwarth, 1971). Cytopempsis is not thought to be a likely mechanism for fluid transport in *A. lumbricoides* (Dankwarth, 1971) or in the free-living and plant-parasitic species investigated (Narang, 1972).

Kümmel *et al.* (1969) and Dankwarth (1971) have proposed that urine formation in *A. lumbrocoides* is a secretory process, as it is in the Malpighian tubules of insects. The former authors found that the "excretory fluid" of *A. lumbricoides* was approximately isosmotic with its body fluid, but when the Na and K levels of the external medium were changed, there were corresponding changes in Na and K concentrations in the excretory fluid, although the level of these ions in the body fluid remained very constant. On the basis of these results it was suggested that the excretory system in *A. lumbricoides* was involved in ionic regulation. Dankwarth (1971) has suggested that the outer region of the excretory canal, which contains numerous microtubules and mitochondria and an outer membrane with many infoldings, is structurally capable of transporting water and solute. This was based on the "standing gradient" theory of water transport (Diamond and Bossert, 1967; Oschman and Berridge, 1971), although it applies equally to the more recent "electro-osmotic" theory (Hill, 1975a,b). The canaliculi or vesicles in the inner region of the canal may similarly be important in fluid transport into the canal lumen. Other regions of the nematode, notably the intestine, possess membrane convolutions and/or local (intercellular) spaces. Much more experimental evidence is needed before it can be concluded if this resemblance to systems in other animals is anything more than structural.

In nematodes with a functional alimentary canal and an excretory system, the flow of fluid through the former is probably much greater, and its role in excretion and volume regulation is probably the more important (Wright and Newall, 1976). However, under certain circumstances—for example, in nonfeeding infective juveniles—the excretory system may be capable of removing most if not all of the water gained by the nematode, even when it is immersed in distilled water (Wright and Newall, 1976). This appears to be the case in three species of animal parasites belonging to the family Strongyloidea: *Nippostrongylus muris, Ancylostoma caninum* (Weinstein, 1952), and *Ancylostoma tubaeforme* (Croll *et al.*, 1972). When the third-stage infective juveniles of these species were placed in various concentrations of NaCl solutions, an inverse relationship was observed between the rate of pulsation of the excretory ampulla and the external osmotic pressure. Weinstein (1952) calculated that the amount of fluid expelled by the excretory system of *N. muris* and *A. caninum* when they were placed in distilled water was equivalent to their body volume in 11 and 75 hr, respectively. The observations of Croll *et al.* (1972) suggest that *A. tubaeforme* is intermediate to the former species in its rate of fluid expulsion via the ampulla. It is interesting to note that in the preinfective second-stage juvenile of *A. tubaeforme* the ampulla also pulsates in dilute media, although less regularly than in the infective stage. Croll *et al.* (1972) conclude that this may be because of the gross changes in fluid levels that occur in the intestine during feeding.

C. Permeability of the Body Wall of Nematodes to Salts and Water

The body wall of nematodes is differentially permeable to water and to various nonelectrolytes and ions. The outer cortical layer of the cuticle with its lipid "membrane" (Bird, 1971) is probably the major permeability barrier (Arthur and Sanborn, 1969).

The permeability of *A. avenae* and *Caenorhabditis briggsae* and of *E. brevis* and *E. communis* to water has been measured using tritium oxide (Castro and Thomason, 1973; Searcy *et al.*, 1976) and deuterium oxide (Newall, 1980), respectively. The estimated permeability constants for these species are as follows: *E. brevis* and *E. communis*, 3.4×10^{-4} cm/sec; *A. avenae*, 1.3×10^{-6} cm/sec (calculated from Castro and Thomason, 1973; and *C. briggsae*, $\sim 2.0 \times 10^{-6}$ cm/sec (calculated from Searcy *et al.*, 1976). Because of poor stirring within the nematodes, it is likely that these are underestimates (Wright and Newall, 1976).

These permeability constants are within the range found for other aquatic organisms and, as would be expected, the marine species are much more permeable to water than the freshwater *A. avenae* and *C. briggsae*. Cyanide treatment increases water permeability in *A. avenae* (Marks *et al.*, 1968).

Estimates for the permeability to water of the infective juveniles of *N. muris* and *A. caninum* give values of 6.1×10^{-6} cm/sec and 0.8×10^{-6} cm/sec, respectively (both values calculated from Weinstein, 1952).

These are of the same order of magnitude as *A. avenae* and *C. briggsae*. It can be calculated, on the assumption that the internal osmotic pressure of *A. avenae* is equivalent to 50 mM NaCl, that in tap water the rate of urine production would have to be equivalent to 30% of the body weight of *A. avenae* per day to maintain a constant body volume. Similar calculations for *N. muris* and *A. caninum* confirm the potential of the excretory system in these species for osmoregulation.

The permeability of animal- and plant-parasitic nematodes to non-ionic, osmotically active substances present in their hosts may be an important mechanism in the toleration of hyperosmotic conditions (i.e., allowing them to remain hypoionic as discussed earlier). This has been shown to be the case in some other parasites (Read *et al.*, 1959), and it is known for example that the anionic discrepancy in the body fluid of *A. lumbricoides* is made up mainly by fatty acids (Bueding and Farrow, 1956).

Finally, a decrease in the ability of nematodes to withstand osmotic stress with increasing age has been reported for *C. briggsae* (Zuckerman *et al.*, 1971), *A. avenae*, and *Bursaphelehcnus lignicola* (N. Ishibashi, personal communication). In the first species this change appears to be related to changes in the external cuticular surface, which deteriorates with age and probably results in the observed increased permeability to water found in older nematodes (Searcy *et al.*, 1976).

IV. SPECIALIZED ASPECTS OF OSMOTIC AND IONIC REGULATION IN NEMATODES

A. Hatching: Osmotic and Ionic Influences

Changes in the permeability of the egg shell of nematodes to water and/or solutes have been shown to be associated with the hatching process of several species. Permeability changes in the egg shell are induced by enzymes released by the juvenile within the egg, as for example in *Trichostrongylus retortaeformis*, or, possibly, directly by an exogenous hatching stimulus (see Arthur and Sanborn, 1969; Lee and Atkinson, 1976).

Most species of nematodes hatch spontaneously when the juvenile has reached a suitable stage of development and the physical conditions of the environment are favorable. However, some *Heterodera* spp., *G. rostochiensis,* and *Globodera pallida* only show a substantial rate of hatch in the presence of substances released by certain plants (Lee and Atkinson, 1976). The hatching process of *G. rostochiensis* is the best documented (Clarke and Perry, 1977), and recent work has shown that the metabolism of the unhatched juvenile is stimulated within 24 hr of exposure to potato root diffusate (Atkinson and Ballantyne, 1977a,b), and that water is taken up by the juveniles before and immediately after hatching (Ellenby and Perry, 1976).

Two theories have been proposed for the mode of action of potato root diffusate.

Clarke *et al.* (1978) found that when hatched juveniles of *G. rostochiensis* were transferred from distilled water to 0.4 *M* sucrose or trehalose, which is equivalent to an estimate for the osmotic pressure of the egg shell fluid (Clarke and Hennessy, 1976), the water content of the juveniles fell from 72% (wt/wt) to 67% (wt/wt). The latter value is the one found in unhatched juveniles in eggs equilibrated with distilled water (Ellenby and Perry, 1976). The movement of hatched second-stage juveniles of *G. rostochiensis* was also drastically reduced in 0.4 *M* sugar solutions, but on dilution to below 0.1 *M,* active movement was restored.

It was suggested, therefore, that the action of potato root diffusate was to increase the permeability of the egg shell to solutes, in particular to trehalose, the major osmotically active constituent of the egg fluid. This would lead to a drop in osmotic pressure, hydration, and increased activity of the juveniles, and, eventually, to hatching (Clarke *et al.,* 1978). In support of this hypothesis it is known that the osmotic pressure of the external medium influences the hatching of several species of nematodes including *G. rostochiensis* (Wallace, 1956, 1957, 1966; Dropkin *et al.,* 1958; Clarke and Perry, 1977). For example, concentrations greater than 0.01 *M* NaCl significantly reduce hatching in *Heterodera schactii*. It is also known that juveniles and egg shells of *G. rostochiensis* untreated with "diffusate" are not permeable to sugar solutions (Ellenby, 1968; Kämpfe, 1962). Clarke *et al.* (1978) proposed that the egg shell of *G. rostochiensis* provided an upper limit to the water content of unhatched juveniles and that the constituents of the egg fluid (mainly trehalose) limit the water content within the egg. Their work suggests that a water content of 69% (wt/wt) is required for the hatched juveniles of *G. rostochiensis* to become fully mobile.

Evidence for a similar mechanism has been reported for eggs of *Ascaris suum* (*lumbricoides*) (Clarke and Perry, 1980), a species which, like *G. rostochiensis,* requires an external stimulus to hatch. Barrett (1976) studied the ascaroside membrane of the egg of *A. suum*, which is the major permeability barrier in the egg shell, but failed to detect any chemical or conformational changes in the membrane during the onset of increased permeability following the stimulus to hatch. He suggested that either the permeability change in the egg shell was due to localized chemical or conformational changes not detectable by present techniques, or the change was due to mechanical damage to the membrane by the juvenile as it became active.

The fact that the unhatched juveniles of some other species, for example, *Meloidogyne javanica* and *H. schactii*, initially have a greater water content than *G. rostochiensis* has been suggested as a reason for their hatching spontaneously (Clarke *et al.*, 1978). This theory has recently been examined for *H. schactii* (Perry *et al.*, 1980). It was found that the movement of hatched juveniles of *H. schactii* was greater at osmotic pressures of between 0.3 and 0.4 *M* trehalose than had been previously found for *G. rostochiensis* juveniles, and that the osmotic pressure of the egg shell fluid of the former species was apparently less. Perry *et al.* (1980) have proposed that these two factors may explain why *H. schactii* hatches so readily in water.

In considering the theory of Clarke *et al.* (1978), several points must be borne in mind: (1) Leakage of trehalose from eggs of *G. rostochiensis* treated with potato root diffusate has yet to be demonstrated, although leakage of trehalose from eggs of *A. lumbricoides* that have been stimulated to hatch has been observed (Fairbairn, 1961). (2) It is difficult to determine accurately the osmotic pressure of the egg shell fluid ("periovic" fluid). (3) It is possible that the small degree of hydration associated with the increased activity of the unhatched juveniles may not be the cause of activation, but rather that hydration is the result of increased activity leading to changes in muscle tone and thus of water content. (4) The above experiments were performed in distilled water or pure sugar solutions.

An alternative primary action for potato root diffusate has been suggested by the work of Atkinson and co-workers. They have shown that there is a significant increase in the Ca content of both the unhatched juvenile and the egg shell of *G. rostochiensis* after soaking in "diffusate" for 48 hr. This is not due merely to an increase in the external Ca concentration, since no such change took place in autoclaved diffusate (Atkinson *et al.*, 1980). Analysis of cysts of *G. rostochiensis* showed that the Ca content of the periovic fluid started to rise after incubation in the diffusate for 24–36 hr. After 48 hr, however, the Ca content of the periovic fluid began to decrease, probably because the rate of uptake of Ca by the eggs exceeded the rate of diffusion of Ca through the cyst wall. It was thought possible, therefore, that the uptake of Ca by the juvenile may have been initiated by the time metabolic changes are established (Atkinson and Ballantyne, 1977a,b).

This uptake of Ca by the juvenile of *G. rostochiensis* prior to hatching is of particular interest. Ellenby and Gilbert (1957, 1958) had shown earlier that Ca can synergize suboptimal concentrations of potato root diffusate, and recent work suggests that hatching in *G. rostochiensis* is a Ca-mediated process (Atkinson and Ballantyne, 1979). The latter authors found that low concentrations of ruthenium red and lanthanum chloride (compounds thought to be specific inhibitors of Ca-mediated processes) (Berridge, 1975; Mikkelson, 1976) inhibited the hatching of juveniles. This inhibition did not appear to be due to any toxic action on the juvenile, since both compounds failed to prevent hatching when applied to eggs that had been soaked in diffusate alone for 1 week. Also, the action of these compounds may be relatively specific since at concentrations causing 90% inhibition of hatching in *G. rostochiensis* they had no effect on hatching in *H. schactii*.

Atkinson and Ballantyne (1979) also found that two Ca ionophores (A 23187, Eli Lilly; and X Br 537A, Roche) synergized the hatching of a 1971 population of eggs of *G. rostochiensis* in dilute diffusate and had a significant effect on hatching by themselves in a 1975 population that had a greater rate of spontaneous hatching in water. These Ca ionophores are lipophilic compounds that are able to sequester Ca and then pass freely through membranes (Truter, 1976; Simon *et al.*, 1977), thus allowing Ca to bypass normal control mechanisms in the membrane and to initiate a number of Ca-mediated systems. Thus, it was suggested that the ionophores interfered with the normal control of hatching in *G. rostochiensis* by allowing Ca to pass through the egg shell or through the body wall of the juvenile. However, because of the hydrophilic nature of the "hatching factor" in potato root diffusate, it was thought unlikely that this itself was an ionophore.

The above experiments with ionophores do not provide proof of the site of action on the egg of *G. rostochiensis* since it is known, for example, that X Br 537A can release sequestered Ca from cell organelles, etc., and therefore can work even in a Ca-free medium.

Whether the "hatching factor" acts on the egg shell or the juvenile is as yet unconfirmed, but the former site of action is perhaps the more likely considering the hydrophilic nature of the hatching factor. The hatching factor may in this case bind to a receptor on the egg shell, which thus activates the binding and transport of Ca at a second site (perhaps analogous to an ionophore system), thereby increasing the Ca concentration in the egg fluid and juvenile (H. J. Atkinson, personal communication).

By analogy with other Ca-mediated processes (Berridge, 1975; Forman *et al.*, 1976; Duncan, 1976), Atkinson *et al.* (1980) have proposed that the hatching factor could act in a number of ways on *G. rostochiensis*, for example, on cyclic nucleotides, hormonal systems, or on the musculature. They also point out that the changes in permeability of the egg shell associated with hatching (Clarke and Perry, 1977) can also be a Ca-mediated effect in other animals.

Thus, it seems likely that the results of Clarke *et al.* (1978) and of Atkinson's group are not mutually exclusive in that changes in the concentration of Ca could lead to permeability changes in the egg shell and subsequent loss of trehalose and other solutes from the egg shell fluid, an increase in the internal Ca concentration of the juveniles and a decrease in the osmotic pressure of their immediate surroundings being required for the activation of hatching. However, the initial action would seem to be on Ca.

B. Locomotion

Movement of nematodes in the soil has been shown to depend upon its moisture content (Wallace, 1971), and there has been some debate about the relative importance of osmotic pressure and the matric potential on the water balance of soil species.

The matric potential is the force that resists the loss of water from the soil and is made up of two components: (1) adhesive forces binding water molecules to solid surfaces, and (2) the surface tension at air–water interfaces. The matric potential is distinct from the osmotic pressure exerted by solutes dissolved in soil water, although their free-energy measurements are the same.

Until there is evidence of the direct effect of the matric potential rather than osmotic pressure on the water balance of soil-inhabiting nematodes, we are in broad agreement with Lee and Atkinson (1976). That is, although volume regulation may ensure that osmotic stress does not seriously reduce activity where there is sufficient free water for locomotion to occur, in soils of high matric potential it is survival mechanisms such as cryptobiosis that predominate.

C. Desiccation Survival (Anhydrobiosis) and Salt Loss

Cycles of desiccation and rehydration have been shown to be more stressful to nematodes than desiccation alone (Todd *et al.*, 1970; Schmidt *et al.*, 1974; Perry, 1977). This does not appear to be due to a loss in the ability to control the rate of dehydration in fourth-stage juveniles of *Ditylenchus dipsaci* (Perry, 1977) or in second-stage juveniles of *Anguina tritici* (Womersley, 1978). Perry (1977) proposed that loss of salts could reduce the ability of nematodes to survive repeated desiccation. Womersley (1978) monitored the loss of Na, K, Mg, and Ca ions in second-stage juveniles of *A. tritici* exposed to various periods and cycles of desiccation and rehydration. He found that salts were lost through the cuticle both during desiccation and, to a greater extent, during rehydration, and that *A. tritici* appeared to be largely incapable of controlling salt loss during rehydration. In particular, no ability was found to regulate the loss of Na. However, as Womersley points out, this may in part be an artifact due to the use of distilled water as the revival medium (see Section II,C,2).

Under natural conditions the major osmotic/ionic problem encountered by nematodes during a desiccation/rehydration cycle is likely to be the increase in internal osmotic pressure due to the loss of water. While this may be counteracted to some extent by the loss of salts, other mechanisms must be involved. Womersley (1978) has suggested that some nematodes (those capable of anhydrobiosis) may be able to reduce osmotically active carbohydrates such as glucose to relatively osmotically inactive compounds such as sorbitol (by hexose interconversion) or alditols and cyclitols (via a pentose phosphate pathway). In addition, these nematodes may be able to store salts in an inactive state. Mechanisms of this kind are found in the cockroach *Periplaneta americana,* where a reduction in the hemolymph volume by 50% causes only a slight increase in the osmotic pressure (Wall, 1970). Tucker (1977a,b,c) has shown that dehydration results in an increase in the Na:K ratio in the fat body tissue of *P. americana* because of the deposition of sodium urate; this is reversed on rehydration. In nematodes, fat body tissue does not occur, and Womersley (1978) has suggested the gut as a possible deposition site.

V. CONCLUSIONS

The first observation that must be made is that this field of study has not advanced significantly since we last reviewed it (Wright and Newall, 1976). There is, however, some cause for optimism in the application of energy-dispersive X-ray analysis and ion-specific liquid ion-exchanger microelectrodes to the problems of measuring the activities of inorganic ions in nematode body fluids.*

We are also rather concerned to find that many workers have still not heeded our plea for the use of more physiological media in their experiments. The natural environment of a nematode cannot be glass-distilled water or single-salt solutions, and no serious conclusions about the behavior of nematodes can be drawn from results obtained under such conditions, especially when they involve their long-term survival.

It is encouraging to see an increase in interest in the role of osmotic stress in such diverse subjects as hatching, cryptobiosis, aging, and movement, but here again it must be stressed that these investigations should be carried out under conditions that closely resemble those in the nematode's environment.

---

*Future studies might most usefully be applied to euryhaline species (Lee and Atkinson, 1976) since these might provide a real insight into the mechanisms used by nematodes to withstand osmotic stress.

162 D. J. Wright and D. R. Newall

ACKNOWLEDGMENTS

We are grateful to Dr. H. J. Atkinson, Mr. R. Howell, Professor N. Ishibashi, Dr. R. N. Perry, Mr. L. Smith, and Dr. C. Womersley for permission to mention unpublished work and for useful discussions, and to Miss Tricia Wood for reading and correcting the manuscript for this chapter.

REFERENCES

Andrássey, I. (1956). In "English translation of Selected East European Papers in Nematology" (B. M. Zuckerman, M. W. Brzeski, and K. H. Deubert, eds.), pp. 1-15. Univ. of Massachusetts, East Wareham, Massachusetts (1967).
Anya, A. O. (1966). Parasitology 56, 583-588.
Arthur, E. J., and Sanborn, R. C. (1969). In "Chemical Zoology" (M. Florkin and B. T. Scheer, eds.), Vol. 3, pp. 429-464. Academic Press, New York.
Atkinson, H. J., and Ballantyne, A. J. (1977a). Ann. Appl. Biol. 87, 159-166.
Atkinson, H. J., and Ballantyne, A. J. (1977b). Ann. Appl. Biol. 87, 167-174.
Atkinson, H. J., and Ballantyne, A. J. (1979). Ann. Appl. Biol. 93, 191-198.
Atkinson, H. J., Taylor, J. D., and Ballantyne, A. J. (1980). Ann. Appl. Biol. 94, 103-109.
Barrett, J. (1976). Parasitology 73, 109-121.
Berridge, J. M. (1975). Adv. Cyclic Nucleotide Res. 6, 1-98.
Bird, A. F. (1971). "The Structure of Nematodes." Academic Press, New York.
Blake, C. D. (1961). Nature (London) 192, 144-145.
Brenner, S. (1974). Genetics 77, 71-94.
Bueding, E., and Farrow, G. W. (1956). Exp. Parisitol. 5, 345-349.
Castro, C. E., and Thomason, I. J. (1973). Nematologica 19, 100-108.
Chandler, J. A. (1977). "X-Ray Microanalysis in the Electron Microscope." North Holland-Publ., Amsterdam.
Chitwood, B. G., and Chitwood, M. B. (1950). "An Introduction to Nematology." Univ. Park Press, Baltimore, Maryland. Reprinted.
Clarke, A. J., and Hennessy, J. (1976). Nematologica 22, 190-195.
Clarke, A. J., and Perry, R. N. (1977). Nematologica 23, 350-368.
Clarke, A. J., and Perry, R. N. (1980). Parasitology (in press).
Clarke, A. J., Perry, R. N., and Hennessy, J. (1978). Nematologica 24, 384-392.
Croll, N. A., and Viglierchio, D. R. (1969). Proc. Helminthol. Soc. Wash. 36, 1-9.
Croll, N. A., Slater, L., and Smith, J. M. (1972). Exp. Parasitol. 31, 356-360.
Dankwarth, L. (1971). Z. Zellforsch. 113, 581-609.
Diamond, J. M. (1971). Fed. Proc. Fed. Am. Soc. Exp. Biol. 30, 6-13.
Diamond, J. M., and Bossert, W. H. (1967). J. Gen. Physiol. 50, 2061-2083.
Dropkin, V. H., Martin, G. C., and Johnson, R. W. (1958). Nematologica 3, 115-126.
Duchâteau-Bosson, G., Florkin, M., and Jeuniaux, C. (1961). Arch. Int. Physiol. Biochim. 69, 97-116.
Duggal, C. L. (1978). J. Zool. London 186, 39-46.
Duncan, C. J., ed. (1976). Calcium Biol. Syst., Symp. Soc. Exp. Biol. 30.
Ellenby, C. (1968). Experientia 24, 84.
Ellenby, C. (1974). J. Exp. Biol. 61, 773-779.
Ellenby, C., and Gilbert, A. B. (1957). Nature (London) 180, 1105-1106.
Ellenby, C., and Gilbert, A. B. (1958). Nature (London) 182, 925-926.

Ellenby, C., and Perry, R. N. (1976). *J. Exp. Biol.* **64,** 141-147.

Ellenby, C., and Smith, L. (1964). *Nematologica* **10,** 342-343.

Ellis, W. G. (1937). *J. Exp. Biol.* **14,** 340-350.

Fairbairn, D. (1961). *Can. J. Zool.* **39,** 153-162.

Forman, J. C., Garland, L. G., and Morgan, J. L. (1976). *Symp. Soc. Exp. Biol.* **30,** 193-218.

Greenaway, P. (1971). *J. Exp. Biol.* **54,** 199-214.

Harpur, R. P., and Popkin, J. S. (1965). *Can. J. Biochem.* **43,** 1157-1169.

Harpur, R. P., and Popkin, J. S. (1973). *Can. J. Physiol. Pharmacol.* **51,** 79-90.

Harris, J. E., and Crofton, H. D. (1957). *J. Exp. Biol.* **34,** 116-130.

Hill, A. E. (1975a). *Proc. R. Soc. London Ser. B.* **190,** 99-114.

Hill, A. E. (1975b). *Proc. R. Soc. London Ser. B.* **190,** 115-134.

Hobson, A. D., Stephenson, W., and Beadle, L. C. (1952a). *J. Exp. Biol.* **29,** 1-21.

Hobson, A. D., Stephenson, W. and Eden, A. (1952b). *J. Exp. Biol.* **29,** 22-29.

Kämpfe, L. (1962). *Parasitol. Schriftenr.* **14,** 142-170.

Krogh, A. (1939). "Osmotic Regulations in Aquatic Animals." Cambridge Univ. Press, London and New York.

Kümmel, G., Dankwarth, L., Braun-Schubert, G., and Gertz, K. H. (1969). *Z. Vgl. Physiol.* **84,** 118-134.

Lang, M. A., and Gainer, H. (1969). *J. Gen. Physiol.* **53,** 323-341.

Lee, D. L. (1960). *Parasitology* **50,** 241-246.

Lee, D. L., and Atkinson, H. J. (1976). "The Physiology of Nematodes," 2nd ed. Macmillan, New York.

Lee, H. F., Chen, I. L., and Lin, R. P. (1973). *J. Parasitol.* **59,** 289-298.

Lockwood, A. P. M. (1960). *Comp. Biochem. Physiol.* **2,** 241-289.

Marks, C. F., Thomason, I. J., and Castro, C. E. (1968). *Exp. Parasitol.* **22,** 321-337.

Mikkelson, R. B. (1976). *In* "Biological Membranes" (D. Chapman and D. F. Wallach, eds.), Vol. 3, pp. 153-190. Academic Press, New York.

Myers, R. F. (1966). *Nematologica* **12,** 579-586.

Narang, H. K. (1972). *Parasitology* **64,** 253-268.

Newall, D. R. (1980). Volume and Ionic Regulation in *Enoplus communis, Enoplus brevis* (Bastian) and *Globodera rostochiensis.* Ph. D. Thesis, University of Newcastle upon Tyne, England.

Oschman, J. L., and Berridge, M. J. (1971). *Fed. Proc. Fed. Am. Soc. Exp. Biol.* **30,** 49-56.

Paltridge, R. W., and Janssens, P. A. (1971). *Comp. Biochem. Physiol. B* **40,** 503-513.

Pannikar, N. K., and Sproston, N. G. (1941). *Parasitology* **33,** 214-223.

Pantin, C. F. A. (1931). *J. Exp. Biol.* **8,** 63-72.

Pantin, C. F. A. (1964). "Notes on Microscopical Technique for Zoologists." Cambridge Univ. Press, London and New York.

Perry, R. N. (1977). *Parasitology* **75,** 215-231.

Perry, R. N., Clarke, A. J., and Hennessy, J. (1980). *Rev. Nématol.* (in press).

Potts, W. T. W. (1958). *J. Exp. Biol.* **35,** 749-764.

Prosser, C. L. (1973). *In* "Comparative Animal Physiology" (C. L. Prosser, ed.), 3rd ed., pp. 1-78. Saunders, Philadelphia, Pennsylvania.

Read, C. P., Douglas, L. T., and Simmons, J. E. (1959). *Exp. Parasitol.* **8,** 58-75.

Roggen, D. R., Raski, D. J., and Jones, N. O. (1967). *Nematologica* **13,** 1-16.

Rojas, E., and Tobias, J. M. (1965). *Biochim. Biophys. Acta* **94,** 394-404.

Schepfer, W. H. (1932). *Rev. Suisse. Zool.* **39,** 89-194.

Schmidt, J. M., Todd, K. S., and Levine, N. D. (1974). *J. Nematol.* **6,** 27-29.

Schoffeniels, E. (1960). *Arch. Int. Physiol.* **68,** 696-698.

Schoffeniels, E. (1967). "Cellular Aspects of Membrane Permeability." Pergamon, Oxford.

Searcy, D. G., Kisiel, M. J., and Zuckerman, B. M. (1976). *Exp. Aging Res.* **2**, 293-301.

Simon, W., Morf, W. E., and Ammann, D. (1977). *In* "Calcium Binding Proteins and Calcium Function" (R. H. Wasserman, R. A. Corradino, E. Carafoli, R. H. Kretsinger, P. H. Maclennan, and F. L. Siegal, eds.), pp. 50-62. North-Holland Publ., Amsterdam.

Smith, I. C. P. (1971). *Chimia* **25**, 349-360.

Stephenson, W. (1942). *Parasitology* **34**, 253-265.

Todd, K. S., Jr., Levine, N. D., and Andersen, F. L. (1970). *Proc. Helminthol. Soc. Wash.* **37**, 57-63.

Truter, M. R. (1976). *Symp. Soc. Exp. Biol.* **30**, 19-40.

Tucker, L. E. (1977a). *J. Exp. Biol.* **71**, 49-66.

Tucker, L. E. (1977b). *J. Exp. Biol.* **71**, 81-93.

Tucker, L. E. (1977c). *J. Exptl. Biol.* **71**, 95-110.

Vanfleteren, J. R. (1975). *Nematologica* **21**, 413-424.

Viglierchio, D. R. (1974). *Trans. Am. Microsc. Soc.* **93**, 325-338.

Viglierchio, D. R., Croll, N. A., and Gortz, J. H. (1969). *Nematologica* **15**, 15-21.

Walker, J. L. (1971). *Anal. Chem.* **43**, 89A-93A.

Wall, B. J. (1970). *J. Insect Physiol.* **16**, 1027-1042.

Wallace, H. R. (1956). *Ann. Appl. Biol.* **44**, 274-282.

Wallace, H. R. (1957). *Ann. Appl. Biol.* **45**, 251-255.

Wallace, H. R. (1966). *Nematologica* **12**, 57-69.

Wallace, H. R. (1971). *In* "Plant Parasitic Nematodes" (B. M. Zuckerman, W. F. Mai, and R. A. Rohde, eds.), Vol. 1, pp. 257-280. Academic Press, New York.

Weinstein, P. P. (1952). *Exp. Parasitol.* **1**, 363-376.

Wilson, R. A., and Webster, L. A. (1974). *Biol. Rev.* **49**, 127-160.

Womersley, C. (1978). "Physiological and Biochemical Aspects of Anhydrobiosis in some Free-Living and Plant-Parasitic Nematodes." Ph.D. Thesis, University of Newcastle Upon Tyne, England.

Wright, D. J. (1973). "Aspects of Nitrogenous Excretion in the Free-Living Nematode *Panagrellus redivivus* (Goodey, T., 1945)." Ph.D. Thesis, University of Newcastle upon Tyne, England.

Wright, D. J., and Newall, D. R. (1976). *In* "The Organization of Nematodes" (N. A. Croll, ed.), pp. 163-210. Academic Press, New York.

Wright, K. A. (1963). *J. Morphol.* **112**, 233-259.

Wright, K. A., and Chan, J. (1973). *Tissue & Cell* **5**, 373-380.

Wright, K. A., and Hope, W. D. (1968). *Can. J. Zool.* **46**, 1005-1011.

Zuckerman, B. M., Himmelhoch, S., Nelson, B., Epstein, J., and Kisiel, M. (1971). *Nematologica* **17**, 478-487.

7

Nematode Energy Metabolism

R. BOLLA

Department of Biology
University of Missouri—St. Louis
St. Louis, Missouri 63121

I.	Introduction	166
	A. General Problems	166
	B. Metabolic Patterns	166
	C. Current Status	166
II.	Energy Storage Molecules	167
	A. Fatty Acids	167
	B. Carbohydrates	168
III.	Utilization of Energy Reserves	168
	A. Animal-Parasitic Nematodes	168
	B. Free-Living and Plant-Parasitic Nematodes	169
IV.	End Products of Metabolism	171
V.	Glycolysis and the Tricarboxylic Acid Cycle	173
	A. Animal-Parasitic Nematodes	173
	B. Free-Living and Plant-Parasitic Nematodes	176
VI.	Other Pathways of Energy Metabolism	178
	A. Glycerol Production	178
	B. Glyoxylate Cycle	179
	C. Phosphogluconate Pathway	182
	D. Gluconeogenesis	183
	E. β-Oxidation of Fatty Acids	183
	F. Protein Metabolism for Energy Production	185
VII.	Energy Production and Regulation of Metabolism	185
VIII.	Conclusions	189
	References	189

NEMATODES AS BIOLOGICAL MODELS
VOLUME 2

I. INTRODUCTION

A. General Problems

The small size of most nematodes, the complexity of their life cycle, and, until recently, the inability to maintain free-living and parasitic species *in vitro* on a synthetic defined medium to obtain synchronous populations, have made it difficult to study energy metabolism in these animals. Most metabolic studies, therefore, have been on entire populations of nematodes containing a mixture of developmental stages, rather than on individual stages or on individual tissues. For these reasons, the most detailed studies of nematode metabolism have been conducted on adults of larger animal-parasitic nematodes. In addition, developmental studies of energy metabolism and its regulation have been almost entirely confined to studies with parasitic nematodes (Barrett, 1976; Barrett and Beis, 1975; Rubin and Trelease, 1975, 1976; Körting and Fairbairn, 1971).

B. Metabolic Patterns

Because of the extreme adaptability of nematodes, it is difficult to make generalizations about cellular energy production in these animals. Many animal-parasitic nematodes demonstrate complete aerobic metabolism during the free-living phases of their life cycle, but shift to anaerobic metabolism as parasites (Saz *et al.*, 1971; Barrett and Beis, 1975; Körting and Fairbairn, 1971). Free-living nematodes may, under normal conditions, use aerobic metabolic pathways coupled to cytochrome-mediated electron transport for energy production, but when environmental conditions become restricted these nematodes may switch to anaerobic pathways. Because of this adaptability some of these nematodes may survive anaerobically for extended periods.

Regardless of the pathway used to generate cellular energy, nematodes generally catabolize energy-producing substrates stepwise for substrate-linked phosphorylation of high-energy compounds. Anaerobically, pyruvate and acetaldehyde may be used as electron acceptors, or aerobically, electrons may be moved along an electropotential gradient involving flavoproteins and cytochromes. Some nematodes parasitic in an anaerobic environment have evolved mechanisms for flavoprotein-mediated electron transport which do not require molecular oxygen for the production of cellular energy (Bryant, 1975).

C. Current Status

The use of nematodes as model systems for genetic and aging studies has added impetus to investigations of energy metabolism and its regulation in a variety of free-living nematodes. This chapter is focused on a comparison of the metabolic pathways used for energy metabolism and control as influenced by nematode species, environmental condition, and developmental stage. An emphasis is placed on species of animal-parasitic nematodes that have both free-living and parasitic stages in their life cycle. These nematodes may be excellent model systems for the study of substrate and gene-level regulations of eukaryotic metabolism. Where possible, energy metabolism and its regulation are discussed in free-living and plant-parasitic nematodes.

II. ENERGY STORAGE MOLECULES

A. Fatty Acids

The concentration and type of energy storage molecules in nematodes are affected by environmental factors, habitat, and the developmental stage (Krusberg *et al.*, 1973; Krusberg, 1971, 1967; Lee and Atkinson, 1976; Von Brand, 1973). The lipid concentration in adult and larval stages of most free-living nematodes and of most plant-parasitic nematodes ranges from 23 to 40% of the total dry weight (Nicholas, 1975; Lee and Atkinson, 1976; Barrett *et al.*, 1971; Krusberg, 1971; Sivapalan and Jenkins, 1966). This concentration is five- to ten-fold greater than in adult animal-parasitic nematodes, (Greichus and Greichus, 1966; Fairbairn, 1969; Von Brand, 1973). The embryonic stages of *Ascaris,* which has primarily aerobic metabolism (Barrett *et al.*, 1970), and the aerobic free-living larval stages of several other animal-parasitic nematodes have lipid concentrations more comparable to free-living and plant-parasitic nematodes than to the adult parasitic stages of their own life cycle (Nicholas, 1975; Von Brand, 1973; Wilson, 1965; Barrett and Beis, 1975; Barrett, 1968). Barrett (1968) suggests that this difference in lipid concentration reflects the nematode's habitat and type of metabolism.

Even-numbered carbon fatty acids ranging from C_{12} to $C_{20:5}$ have been identified in several species; however, C_{18} to C_{20} fatty acids predominate (Lee and Atkinson, 1976; Krusberg, 1971; Sivapalan and Jenkins, 1966). In *Panagrellus redivivus* unsaturated fatty acids amount for about 96% of the neutral lipid fraction and 87% of the phospholipid fraction (Sivapalan and Jenkins, 1966; Rothstein, 1970). Linoleic and linolenic acids are the major unsaturated fatty acids and stearic acid is the major saturated fatty acid in this nematode. *Ditylenchus triformis, D. dipsaci, Pratylenchus penetrans, Aphelenchoides ritzemabosi,* and *Tylenchorhynchus claytoni* have a similar fatty acid composition (Krusberg, 1967, 1971). Vaccenic and oleic acids are the major C_{18} fatty acids in *D. triformis, D. dipsaci,* several species of *Meloidogyne,* and *Turbatrix aceti* (Krusberg, 1967, 1971; Krusberg *et al.*, 1973; Fletcher and Krusberg, 1973; Castillo and Krusberg, 1971). Branched-chain fatty acids have been identified in *T. aceti* but not in several other free-living and plant-parasitic nematodes

(Fletcher and Krusberg, 1973; Krusberg, 1967, 1971; Krusberg *et al.*, 1973). Triglycerides and free fatty acids are the major fat storage molecules of animal-parasitic nematodes (Von Brand, 1973; Saz, 1969); free-living and plant-parasitic nematodes contain lower concentrations of triglycerides and free fatty acids, but higher concentrations of phospholipids than do their animal-parasitic counterparts (Cole and Krusberg, 1967; Castillo and Krusberg, 1971; Fairbairn, 1969; Magat *et al.*, 1972; Von Brand, 1973). Unlike most free-living and plant-parasitic nematodes, animal-parasitic nematodes contain odd-numbered carbon fatty acids as well as branched-chain fatty acids (Fairbairn, 1969; Von Brand, 1973).

B. Carbohydrates

Carbohydrate reserves are present in nematodes mainly as glycogen, although significant amounts of trehalose and free glucose have been reported (Magat *et al.*, 1972; Barrett *et al.*, 1970; Roberts and Fairbairn, 1965; Cooper and Van Gundy, 1970, 1971; Von Brand, 1973; Tracey, 1958; Krusberg, 1971). Free glucose generally constitutes from 0.1 to 0.8% of tissue solids and trehalose from 0.1 to 4.8% (Lapp and Mason, 1978; Von Brand, 1973; Krusberg, 1971). In *Aphelenchoides* spp. glycogen is about 85% of the tissue carbohydrate (Krusberg, 1971; Cooper and Van Gundy, 1970, 1971). In *Ascaris,* however, the glycogen content ranges from 20% of the total carbohydrate in male reproductive tissue to nearly 90% in the muscle of female worms (Fairbairn and Passey, 1957). In most nematodes glycogen concentration is highest in the hypodermis, the noncontractile regions of muscle, intestinal cells, and epithelial cells of the reproductive system (Dropkin and Acedo, 1974; Lee and Atkinson, 1976). The pseudocoelomic fluid of most nematodes, although containing small amounts of glycogen, is generally enriched in mono- and disaccharides (Von Brand, 1973; Lee and Atkinson, 1976).

III. UTILIZATION OF ENERGY RESERVES

A. Animal-Parasitic Nematodes

The use of stored glycogen and lipid for the production of cellular energy in nematodes is influenced by the species, the life-cycle stage, and the availability of oxygen. This is particularly true in animal parasitic nematodes that have an alternation of life cycle between free-living aerobic larval stages and adults parasitic in microaerobic or anaerobic environments. Changes in energy metabolism have been demonstrated during the embryogenesis and development of *Ascaris* (Barrett, 1976; Beis and Barrett, 1975; Fairbairn, 1969; Rubin and Trelease, 1975; Ward and Fairbairn, 1970). The glycogen content in the developing eggs of this nematode decreases from 15% dry weight at day 4 to 2% at day 12. Over the next 8 days, the glycogen concentration increases threefold as infective larvae differentiate. The change in glycogen content is paralleled from days 0–10 by a decrease in triglyceride concentration; from days 10–25, as the carbohydrate concentration increases, the triglyceride concentration continues to decrease (Barrett, 1976; Beis and Barrett, 1975; Barrett *et al.*, 1970; Rubin and Trelease, 1975; Ward and Fairbairn, 1970). There is, therefore, a resynthesis of carbohydrate from triglycerides during the developmental change from aerobic larval metabolism to anaerobic metabolism as a parasitic adult.

A similar shift in energy metabolism has been demonstrated during the development of *Strongyloides ratti* and *Nippostrongylus brasiliensis*. The infective third-stage larvae of these species are nonfeeding and rely entirely on endogenous energy stores for metabolic energy (Barrett, 1968; 1969a,b; Roberts and Fairbairn, 1965; Körting and Fairbairn, 1971). The adults, which are parasitic in the intestine of rats, metabolize exogenous energy-producing substrates acquired from the host. The infective larvae have been shown to have low concentrations of glycogen but high levels of free fatty acids as compared to the adults (Barrett, 1968). The total lipid content of *N. brasiliensis* infective larvae decreases about 0.9% per day as the larvae age *in vitro* (Wilson, 1965). *Strongyloides ratti* third-stage infective larvae metabolize lipids at a rate of 1.3% per day during 12 days *in vitro* (Barrett, 1968). In the adults of *N. brasiliensis* the production of cellular energy from lipids may be of limited importance since glycogen has been shown to be the primary energy storage molecule (Roberts and Fairbairn, 1965). Barrett (1968) suggests that triglycerides in adult parasitic nematodes may function as energy reserves for gametes and as a mechanism for the adaptation of the nematode to a parasitic mode of life at higher temperatures.

B. Free-Living and Plant-Parasitic Nematodes

Glycogen is the principal energy storage molecule in free-living and plant-parasitic nematodes (Krusberg, 1971; Dropkin and Acedo, 1974; Barrett *et al.*, 1971; Nicholas, 1975), and is used for energy production when the nematodes are maintained under adverse environmental conditions. Thus, *Aphelenchus avenae* and *Caenorhabditis* spp., when maintained microaerobically or anaerobically on a nutritionally limited medium, rapidly metabolize glycogen stores (Cooper and Van Gundy, 1970, 1971). *Caenorhabditis* spp., which apparently lack the ability to synthesize glycogen from neutral lipid, die after about 80 hr of anaerobic starvation when glycogen stores are depleted. *Aphelenchus avenae,* on the other hand, is capable of entering a cryptobiotic state after about 120 hr of oxygen stress, thereby curtailing energy metabolism. Upon recovery from anaerobiosis, this nematode resynthesizes carbohydrate from neutral lipid (Cooper and Van Gundy, 1970). When *A. avenae* and *Caenorhabditis* spp. are starved aerobically, lipid reserves are the main source of cellular energy. These reserves decrease linearly for about 10 days in both *A. avenae* and *Caenorhabditis* spp. from 33% dry weight and 36% dry weight to 5% and 8% dry weight, respectively. *Aphelenchus avenae* does not use glycogen during periods of aerobic starvation. *Caenorhabditis* spp., however, use glycogen at a relatively rapid rate when starved aerobically (Cooper and Van Gundy, 1970, 1971). When *A. avenae* becomes anhydrobiotic, glycogen is rapidly used for the production of glycerol and trehalose. Glycogen concentrations decrease from 45 μg/mg dry weight in active *A. avenae* in an aerobic environment of high relative humidity to 3 μg/mg dry weight in anhydrobiotic *A. avenae*. This nematode resynthesizes glycogen at the expense of lipid stores upon rehydration (Crowe *et al.*, 1977).

Similar situations exist in other free-living and plant-parasitic nematodes. In *P. redivivus* and *T. aceti,* carbohydrate and lipid reserves decrease when the nematodes are starved aerobically (Barrett *et al.,* 1971). Barrett *et al.* (1971) also observed the incorporation of radioactive carbon from [U-^{14}C] palmitate into glycogen by both *T. aceti* and *P. redivivus,* indicating active glyconeogenesis. Van Gundy *et al.* (1967) reported that there was a rapid decrease in total body lipid concentration in *M. javanica* and *Tylenchulus semipenetrans* when these nematodes were incubated aerobically at 27°C. Lipids were not metabolized, however, either at lower incubation temperatures or under microaerobic or anaerobic conditions.

In an ultrastructural study of glycogen and lipid metabolism by developmental stages of *M. incognita,* Dropkin and Acedo (1974) observed an accumulation of lipid droplets in the intestine, hypodermis, and muscle of motile second-stage larvae. Small but significant concentrations of α- and β-glycogen were also present in the hypodermis and noncontractile regions of the muscle. Two to three days after invasion of host plant roots the number and size of lipid droplets decreased, and extensive deposits of glycogen of nematode origin were observed in the hypodermis and intestine. After infection, *M. incognita* undergoes an initial growth and molting period followed by the development of a sessile adult female, or of an adult male that leaves the plant. Preceding molting, glycogen deposits were metabolized. Lipid droplets reappeared at about 8 days postinfection and were abundant from 12 days postinfection onward. Eggs of this nematode also contained abundant glycogen stores (Dropkin and Acedo, 1974).

Much less information is available on the metabolism of endogenous protein stores by nematodes for the generation of cellular energy. The small size of most nematodes has made it difficult to study the protein content of specific tissues. Since the major protein concentration is cuticular collagen (Von Brand, 1973; Lee and Atkinson, 1976), it is difficult to estimate the portion of total body protein that might be available for the production of cellular energy. If losses of total protein are observed with starvation, cuticular molting must be considered as a possible source of this loss.

IV. END PRODUCTS OF METABOLISM

Although some species of nematodes may metabolize energy-producing carbon substrates completely to CO_2 and water, many species excrete partially metabolized products such as amino acids, peptides, ethanol, carbohydrates, intermediates of metabolic pathways, and short-chain volatile fatty acids (Saz *et al.*, 1971; Saz and Lescure, 1966; Fairbairn, 1954; Von Brand, 1973; Haskins and Weinstein, 1957; Rogers, 1952; Lee and Atkinson, 1976; Myers and Krusberg, 1965; Cooper and Van Gundy, 1971). It is evident that some of the excreted products, if catabolized to completion, would produce additional cellular energy. This suggests that when abundant exogenous nutritional substrate is available less energy is expended by the nematode in running a minimally efficient, partially wasteful, metabolic pathway than in running a highly efficient complete pathway that would require the synthesis of additional enzymes at a large energy cost to the animal. Many parasitic nematodes, however, live in microaerobic or anaerobic environments that are restrictive to complete oxidative metabolism. These nematodes must, therefore, evolve maximally efficient but wasteful substitute metabolic pathways.

The major end products of energy metabolism in obligative or facultative anaerobic animal-parasitic nematodes are short-chain volatile fatty acids such as α-methyl butyric and α-methyl valeric acids and organic acids such as lactate, pyruvate, and succinate (Ellison *et al.*, 1960; Greichus and Greichus, 1966; Saz, 1969, 1970; Bryant, 1975). Organic acid end products of carbohydrate and lipid metabolism are also excreted by several plant-parasitic nematodes, but usually not by free-living nematodes (Von Brand, 1973; Nicholas, 1975; Lee and Atkinson, 1976). Specific metabolic end products will be considered in discussions of individual energy-producing metabolic pathways.

The excretion of amino acids has been observed in several nematodes, probably as a means of removing toxic ammonia produced from protein and nucleic acid metabolism (Miller and Roberts, 1974; Myers and Krusberg, 1965; Nicholas *et al.*, 1960; Rothstein, 1963). *Caenorhabditis briggsae* and *D. triformis* do not excrete urea and therefore probably compensate, in part, by excreting amino acids (Myers and Krusberg, 1965; Nicholas *et al.*, 1960). Rothstein and co-workers have identified several radioactive essential and nonessential amino acids as excretion products of bacteria-free *C. briggsae* cultured *in vitro* on either [^{14}C]formate, [^{14}C]serine, [^{14}C]acetate, [^{14}C]glucose, [^{14}C]malate, or $NaH^{14}CO_3$. The concentration of amino acids excreted was greater than the concentration retained in the tissues (Rothstein, 1965). Of the total nitrogen excreted by surface-sterilized *D. triformis, D. dipsaci, D. myceliophagus, M. incognita, A. rutgersi,* and *P. penetrans* in 24 hr, 23 to 41% is excreted as amino acid (Myers and Krusberg, 1965; Balasubramanian and Myers, 1971). *Ditylenchus triformis* excretes 53–169% of its free amino acid pool in 24 hr and synthesizes several radioactive nonessential amino acids *de novo* from [2-^{14}C]acetate. However, no essential amino acids are produced (Myers and Krusberg, 1965). *Meloidogyne incognita* and *A. rutgersi* synthesize both essential and nonessential amino acids from acetate for excretion (Balasubramanian and Myers, 1971). Even though *C. briggsae* and *A. rutgersi* have been shown to synthesize essential amino acids *de novo,* these amino acids are still required for *in vitro* growth and reproduction on a defined culture medium (Vanfleteren and Roets, 1973; Balasubramanian and Myers, 1971). These observations further support the hypothesis that amino acid synthesis is a primary means of removing toxic ammonia.

Several studies suggest that *de novo* synthesis of amino acids in nematodes does not differ from that in other animals (Rothstein and Mayoh, 1964a,b; Rothstein, 1963, 1965; Nicholas *et al.*, 1960; Myers and Krusberg, 1965; Rothstein and Tomlinson, 1962). Synthesis of [^{14}C]serine from [^{14}C]formate by *C. briggsae* apparently occurs by a simple one-carbon transfer to glycine; however, decarboxylation of serine to produce glycine does not occur. Rothstein and co-workers suggest further that the radioactive glutamic and aspartic acid synthesized from [^{14}C]formate are synthesized by a one-carbon transfer to glycine to produce serine, which can then be converted by serine dehydrase to [3-^{14}C]pyruvate. The subsequent decarboxylation of pyruvate to acetate could lead, via the tricarboxylic acid cycle, to the production of radioactive keto acids. [^{14}C]Glutamic acid and [^{14}C]aspartic acid could then be produced by transamination. Production of [^{14}C]glycine from [4-^{14}C]aspartic acid in *C. briggsae* could occur via the glyoxylate cycle (Rothstein, 1963; Rothstein and Mayoh, 1964a,b). Transaminases involving keto acids, glutamate, pyruvate, and aspartate, as well as enzymes for amino acid oxidation, reductive amination, and for amino acid decarboxylation, have been directly and indirectly demonstrated in a variety of nematodes (Scott and Whittaker, 1970; Haskins and Weinstein, 1957; Pollak and Fairbairn, 1955; Polyakova, 1962; Rogers, 1952).

Although amino acid excretion may represent one mechanism of nitrogen detoxification in some nematodes, ammonia is generally the main excretory product of nitrogen metabolism. Ammonia excretion may account for 20–70% of the total nitrogen excreted (Rogers, 1969; Rothstein, 1965). This is not unexpected since most animals living in an aquatic or semiaquatic environment are ammonotelic (Campbell, 1973).

Urea is not produced in several free-living and plant-parasitic nematodes suggesting the absence of a functional ornithine-urea cycle for the detoxification of ammonia. Although *Pelodera strongyloides* does not excrete urea, both urea and uric acid are found in the tissues of this nematode, along with the enzymes uricase and xanthine oxidase. This suggests the presence of a complete purine degradation pathway with the excretion of ammonia into an aquatic environment (Scott and Whittaker, 1970).

Urea production may, however, be a major mechanism for ammonia detoxification in some animal-parasitic nematodes. Urease and arginase activity have been reported in several species (Rogers, 1969), and the addition of ornithine, arginine, or citrulline to an *in vitro* culture medium for some species results in increased urea production. This cycle is, however, not present in all species of animal-parasitic nematodes, and when it is present it may have adaptive significance (Rogers, 1969).

Toxic ammonia may also be excreted by the synthesis of various amides, secondary aliphatic amines, and volatile amines (Rogers, 1969; Von Brand, 1973; Nicholas, 1975).

V. GLYCOLYSIS AND THE TRICARBOXYLIC ACID CYCLE

Metabolism of carbohydrate via the Embden–Meyerhof pathway occurs in most nematode species (Von Brand, 1973; Nicholas, 1975; Saz, 1969; Lee and Atkinson, 1976). The metabolism of phosphoenolpyruvate (PEP) produced in this pathway depends, however, on the species of nematode, its nutritional state, and the partial pressure of oxygen in the environment in which the nematode is maintained.

A. Animal-Parasitic Nematodes

Many animal-parasitic nematodes exhibit facultative anaerobiosis during some portion of their life cycle. Thus, in some of these nematodes, adult stages that are parasitic in environments of low oxygen tension may have an anaerobic metabolism, whereas developmental stages that may be free-living may produce energy aerobically. In some cases the parasitic adults are capable of using oxygen if available (Von Brand, 1973).

Adult *Ascaris* muscle contains a complete enzyme sequence for the metabolism of glucose to lactate, but *Ascaris* does not excrete lactate as a major end product of carbohydrate metabolism (Barrett and Beis, 1973). The activities of pyruvate kinase (PK) and lactate dehydrogenase (LDH) in *Ascaris* muscle are low relative to the activity of phosphoenolpyruvate carboxykinase (PEPCK). *Ascaris*, therefore, apparently produces oxaloacetate by PEPCK-catalyzed CO_2 fixation into PEP. The oxaloacetate can then be converted to malate by a malate dehydrogenase (MDH)-catalyzed reaction, requiring the dephosphorylation of either inosine triphosphate or guanosine triphosphate and the oxidation of reduced nicotinamide adenine dinucleotide (NAD) (Barrett and Beis, 1973; Saz 1969, 1972). The mitochondria are permeable to the malate, which apparently enters the mitochondria (Papa *et al.*, 1970), where one-half of the malate is metabolized to fumarate by fumarate hydratase. Fumarate is then reduced to succinate by fumarate reductase (reverse reaction of succinate dehydrogenase) coupled to the oxidation of reduced NAD and to the flavoprotein-coupled production of ATP (Bryant, 1970, 1975; Saz and Lescure, 1966). The other half of the malate entering the mitochondria is converted to pyruvate by mitochondrial malic enzyme. This reaction is coupled to the reduction of NAD^+. The reduced NAD produced is available for oxidation in the fumerate reductase-catalyzed reaction, indicating that the dismutation reactions are coupled. The succinate and pyruvate produced by these reactions can permeate the mitochondria and can enter the cytoplasm, where they can be metabolized to the major volatile fatty acids excreted by *Ascaris* (Saz and Lescure, 1966, 1967).

Developing eggs and larval forms of *Ascaris* are free-living in an oxygen-rich environment and appear to use aerobic metabolic pathways for metabolism of PEP. The activities of LDH, PEPCK, NAD-linked malic enzymes, and fumarate reductase are lower in the developmental stages than in the adult (Barrett, 1976). A complete sequence of tricarboxylic acid cycle enzymes (Barrett, 1976) and a variety of functional cytochromes (Kmetec *et al.*, 1963; Costello *et al.*, 1963; Oya *et al.*, 1963) have been reported in developing *Ascaris* eggs. It appears therefore, that early developmental stages metabolize PEP through a functional tricarboxylic acid cycle coupled to cytochrome-mediated electron transport (Beis and Barrett, 1975). This is further supported by the observation that ATP production in these larval forms is coupled to the oxidation of malate, 2-ketoglutarate, and reduced NAD (Barrett, 1976).

A pathway comparable to that in the adult *Ascaris* may be present in the adults of several animal-parasitic nematodes (Von Brand, 1973; Umezurike and Anya, 1978; Saz, 1969; Van den Bossche *et al.*, 1969, 1971; Ward and Huskisson, 1978; Fairbairn, 1954; Ward *et al.*, 1969; Langer and Smith, 1971), as well as in some larval forms (Ward *et al.*, 1969). Thus evidence has been reported to suggest the PEP succinate pathway in parasitic adults of *Haemonchus contortus* (Ward and Huskisson, 1978) and in *Trichinella spiralis* larvae (Ward *et al.*, 1969). *Haemonchus contortus* larvae, however, have a full complement of tricarboxylic acid cycle enzymes (Ward and Schofield, 1967; Van den Bossche *et al.*, 1969). In addition, a particulate fraction, prepared from third-stage *H. contortus* larvae, oxidized reduced cytochrome *c*, NADH, and succinate by a cyanide-sensitive pathway (Ward and Schofield, 1967). It appears that free-living *H. contortus* larvae metabolize carbohydrate through the tricarboxylic acid cycle coupled to a cytochrome-mediated electron transport, but that adults of this nematode use a PEP succinate pathway comparable to that in *Ascaris* (Ward and Schofield, 1967; Van den Bossche *et al.*, 1969).

Strongyloides ratti is the best example of metabolic regulation and adaptation in nematodes. This nematode has a genetically determined alternation of generations (Bolla and Roberts, 1968), consisting of a parthenogenetic parasitic female alternating with a free-living larval generation and a free-living generation of adult males and females. The parthenogenetic female is parasitic in the rat intestine, whereas the free-living adults and larvae live in soil associated with the host feces. All developmental and adult stages of this nematode contain enzymes necessary for the catabolism of glucose to PEP (Körting and Fairbairn, 1971). First-stage larvae, which must stand an initial period of anaerobiosis, and parasitic females both lack NAD-dependent, ADP-activated isocitrate dehydrogenase. In the parasitic female, aconitase activity and PK activity are low and PEPCK activity is high (PK/PEPCK activity ratio = 0.143). The absence of NAD-dependent, ADP-regulated isocitrate dehydrogenase, the rate-limiting enzyme of the tricarboxylic acid cycle, is evidence that this cycle is nonfunctional in these facultative anaerobic stages. This is marked in contrast to the nonfeeding third-stage larvae and the free-living adults, which have a complete enzyme sequence of the tricarboxylic acid cycle along with a complete cytochrome system (Körting and Fairbairn, 1971). These free-living stages of the life cycle have a PK/PEPCK ratio of 2.23 in the third-stage larvae and 1.75 in the free-living adults, and produce nine times more CO_2 from $[U\text{-}^{14}C]$palmitate than do the parasitic females. LDH activity is present in all life-cycle stages of *S. ratti,* but the activity is reduced in the adult forms (Körting and Fairbairn, 1971). It appears that regulation of energy metabolism during the development of *S. ratti* adapts this nematode for maximal energy production by the development of enzyme pathways capable of complete aerobic metabolism of carbohydrate and lipid to CO_2, whereas parasitic adults have a metabolism similar to that of *Ascaris*. The first-stage larvae have enzyme patterns characteristic of a transitional stage between anaerobic and aerobic metabolism.

Dictyocaulus viviparus, Obeliscoides cuniculi, Syphacia muris, and *N. brasiliensis* (Vaatstra, 1969; Lee and Fernando, 1971; Van den Bossche *et al.,* 1971; Saz *et al.,* 1971) have an active PEPCK but lack malic enzyme. Aerobically adult *N. brasiliensis* accumulates lactate and small amounts of pyruvate, and excretes CO_2 as an end product of carbohydrate metabolism (Saz *et al.,* 1971). Succinate is, however, the major anaerobic metabolic end product. PK and LDH activities are of a similar order of magnitude in this nematode. PEPCK is present but its activity is low relative to PK (PK/PEPCK = 1.86). Adult *N. brasiliensis* incorporates greater than 1% of the radioactivity from exogenously supplied $NaH^{14}CO_3$ into glycogen (Saz *et al.,* 1971), suggesting that PEPCK functions mainly for gluconeogenesis. Anaerobic succinate production probably occurs via PEPCK-catalyzed CO_2 fixation into PEP, without the subsequent mitochondrial dismutation of malate via malic enzyme. Therefore, the malate entering the mitochondria is apparently metabolized to fumarate and the fumarate is reduced to succinate in a fumarate reductase-catalyzed reaction. Energy production is probably by the coupling of this latter reaction to a mitochondrial flavoprotein dehydrogenase-catalyzed phosphorylation of ADP.

Another variation of the PEP succinate pathway is exhibited by the filarial parasite *Setaria cervi* and by *Strongylus brevicaudata* (Anwar *et al.*, 1977; Umezurike and Anya, 1978). Although these nematodes excrete primarily lactate and succinate, the PK activity is low (PK/PEPCK in *S. brevicaudata* = 0.19, and in *S. cervi* = 0.40). It has been proposed that carbohydrate metabolism is via the PEP succinate pathway with the subsequent reduction of pyruvate (produced in the malic enzyme-catalyzed reaction) to lactate by LDH coupled to the oxidation of NADH. ATP synthesis would in these nematodes involve flavoprotein-mediated phosphorylation of ADP coupled to fumerate reduction, as occurs in *Ascaris*. The reduction of pyruvate, produced directly from anaerobic glycolysis to lactate, is only a minor pathway in these nematodes (Anwar *et al.*, 1977; Umezurike and Anya, 1978).

Rhabdias bufonis, parasitic in amphibian lungs, apparently has a regulatory mechanism for a situation-dependent shift between the PEPCK-catalyzed carboxylation of PEP to oxaloacetate and the PK-catalyzed carboxylation of PEP to pyruvate. This shift may be regulated by enzyme competition between PEPCK and PK for the available substrate (Anya and Umezurike, 1978). This again stresses that, through evolution, facultative anaerobiosis has developed in many animal-parasitic nematodes as a mechanism of adaptation for the maximum efficiency of energy production from the available substrate.

B. Free-Living and Plant-Parasitic Nematodes

Many free-living and plant-parasitic nematodes metabolize glucose to PEP by the classical Embden–Meyerhof pathway. However, as in the animal-parasitic nematodes, the metabolism of PEP depends upon species and environment (Nicholas, 1975; Lee and Atkinson, 1976).

Ditylenchus spp. have been shown to metabolize radioactive glucose to $^{14}CO_2$. Activities of several glycolytic enzymes, including hexokinase (which has a maximum affinity for glucose, followed by—in decreasing order—fructose, galactose, and mannose), phosphoglucomutase, and phosphopyruvate hydratase, have been reported in *D. dipsaci* and *D. triformis* (Krusberg, 1960). Ells (1969) reported the enzymes that catalyze reactions from fructose 1,6-diphosphate to pyruvate in *T. aceti*, but was unable to demonstrate specific glucokinase, hexokinase, phosphorylase, phosphoglucomutase, or phosphofructokinase. *T. aceti*, as well as several other free-living nematodes, however, uses carbohydrate and lipid reserves for energy production under adverse nutritional conditions. This suggests that many free-living nematodes, including *T. aceti*, have a complete glycolytic enzyme sequence and are capable of complete anaerobic glycolysis (Cooper and Van Gundy, 1970, 1971; Barrett *et al.*, 1970).

The necessity of the oxidative metabolism of PEP through the tricarboxylic acid cycle coupled to terminal electron transport for ATP production is evident in several free-living and plant-parasitic nematodes. These nematodes will survive anaerobically but require oxygen for motility, growth, development and reproduction, (Nicholas, 1975; Rohde, 1971; Atkinson, 1973; Lee and Atkinson, 1976). Thus it has been reported that both *C. briggsae* and *T. aceti* will survive for several days in a nonmotile state when maintained anaerobically or when 1×10^{-4} *M* cyanide is added to the culture medium (Rothstein and Tomlinson, 1962; Barrett *et al.*, 1971), but that aerobic conditions are required for *in vitro* reproduction, development, and growth to occur.

Direct and indirect evidence has been reported for a complete tricarboxylic acid cycle in several free-living and plant-parasitic nematodes (Castillo and Krusberg, 1971; Rothstein, 1965; Rothstein and Tomlinson, 1962; Rothstein *et al.*, 1970; Barrett *et al.*, 1971; Hieb and Dougherty, 1966; Section IV, this chapter). It is unclear, however, if this cycle is completely or only partially functional. *Caenorhabditis briggsae* has been shown to excrete a number of essential and nonessential amino acids whose synthesis is associated with the transamination of keto acid intermediates of the tricarboxylic acid cycle (see Section IV). Several enzymes of this cycle have also been identified in both *T. aceti* and *P. redivivus* (Hieb and Dougherty, 1966; Barrett *et al.*, 1971). Although ketoglutarate dehydrogenase activity was not directly identified in these nematodes, the evidence that both *T. aceti* and *P. redivivus* excrete radioactive CO_2 when incubated with [*U*-^{14}C]palmitate (Barrett *et al.*, 1971) indicates that this enzyme must be present, since the production of CO_2 would require β-oxidation of palmitate coupled to a functional tricarboxylic acid cycle. Further evidence for mitochondrial oxidative metabolism in *T. aceti* has been reported by Rothstein *et al.* (1970). These investigators observed that the morphology of mitochondria isolated from *T. aceti* is comparable to that of mammalian mitochondria. In addition, mitochondrial fractions isolated from this nematode supported oxidation of all tricarboxylic acid cycle intermediates coupled to NADH reduction, and was stimulated by ADP and dinitrophenol. Cyanide and azide inhibited these oxidations. The P:O ratio, with succinate as a substrate, was 1.3 to 1.5. Cytochromes a, b, $c + c_1$ and an additional heme protein, in considerable excess, have also been identified in mitochondria from *T. aceti*, and although not directly identified, all the experimental evidence points to cytochrome a_3 as the terminal oxidase (Rothstein *et al.*, 1970).

Krusberg (1960) has identified citrate-condensing enzyme, isocitrate dehydrogenase, and malate dehydrogenase in both *D. triformis* and *D. dipsaci*, and fumerase in *D. triformis*. Succinate dehydrogenase and 2-ketogluterate dehydrogenase could not be demonstrated in these nematodes. However, the evidence presented is not sufficient to conclude definitely that these enzymes are absent. The presence of cytochromes *c* and $a + a_3$ and activities of both cytochrome oxidase and NADH-cytochrome *c* oxidase in *D. triformis* suggest a functional cytochrome-mediated oxidation of reduced substrates for the production of ATP. LDH activity was insignificant in *D. triformis* and could not be identified in *D. dipsaci*, further suggesting that the main pathway of energy metabolism in these nematodes involves the mitochondrial oxidation of reduced substrates via the tricarboxylic acid cycle. Although malic enzyme has been reported from both of these nematodes, the available evidence suggests that this enzyme is involved in gluconeogenesis rather than in carbohydrate catabolism (Krusberg, 1960).

Both *T. aceti* and *P. redivivus* have been reported to contain both PEPCK and malic enzyme activities (Barrett *et al.*, 1971). Although the major role of these enzymes may be for gluconeogenesis, their presence, along with the ability of these nematodes to survive anaerobically or in the presence of cyanide, suggests that they may be able to adapt to adverse conditions by directing their metabolic pathways toward anaerobiosis for survival. A similar metabolic regulation for the adaptation to adverse conditions is exhibited by *A. avenae*. In environments devoid of oxygen, this nematode uses typical anaerobic pathways to convert 75–80% of its glycogen stores to ethanol for excretion coupled to the oxidation of NADH (Cooper and Van Gundy, 1970).

VI. OTHER PATHWAYS OF ENERGY METABOLISM

A. Glycerol Production

Glycerol is synthesized as an end product of metabolism in some free-living nematodes. This production depends upon the conditions under which the nematodes are maintained. Rothstein (1969) and Liu and Rothstein (1976) observed that glycerol, glucose, and trehalose were synthesized from acetate and were excreted by *C. briggsae* and *T. aceti* maintained in a complete nutritional medium. *Panagrellus redivivus* also excreted some products of glycerol-synthetic pathways when similarly maintained (Rothstein, 1969). However, only minimal amounts of glycerol were excreted by these nematodes maintained in unsupplemented buffer solutions. Decreased glycerol excretion under starvation conditions did not result in increased tissue retention of glycerol, but was directly related to decreased glycerol synthesis.

Liu and Rothstein (1976) proposed a pathway of glycerol synthesis involving the metabolism of oxaloacetate produced in the tricarboxylic acid cycle via a gluconeogenic pathway. This involves the PEPCK-catalyzed decarboxylation of oxaloacetate and the subsequent metabolism of PEP to glyceraldehyde 3-phosphate via enzymes of the glycolytic sequence operating in the reverse direction. Following the triose phosphate isomerase-catalyzed conversion of glyceraldehyde 3-phosphate to dihydroxyacetone phosphate (DHAP), DHAP can be reduced to glycerol 1-phosphate, and then dephosphorylated to glycerol. The activity of glycerol phosphate dehydrogenase in the reducing direction with DHAP as a substrate was sevenfold greater in nematodes from a complete nutrient medium than in those maintained in buffer (Liu and Rothstein, 1976). Substantial increases in hexokinase, fructose-1,6-diphosphatase, and glucose-6-phosphate dehydrogenase were also observed in those nematodes maintained in a complete medium. This suggests that any glucose synthesized by gluconeogenesis would be rapidly converted to triose phosphates for the synthesis of glycerol. A portion of the glycerol synthesized under either experimental condition is used for triglyceride synthesis.

The glycerol synthesis pathway is apparently so active in *C. briggsae* raised on a nutrient medium that glycerol production becomes a major pathway of carbohydrate metabolism in this nematode. Liu and Rothstein (1976) have proposed that this pathway is regulated by induction and repression of genes responsible for the synthesis of the enzymes catalyzing glycerol production. The advantage to the nematodes of glycerol excretion is not clear. It might be suggested, however, that in a nutrient-rich environment the synthesis of large amounts of glycogen would be energetically disadvantageous and that excretion of glycerol would be an efficient mechanism for removing end products of metabolism.

During the early stages of anhydrobiosis, glycogen and lipid reserves decrease in *A. avenae,* with a nearly quantitative increase in tissue glycerol and trehalose (Crowe *et al.,* 1977; Madin and Crowe, 1975). Glycerol and trehalose synthesis in *A. avenae* may occur by a pathway similar to that proposed for *C. briggsae.* However, direct metabolism of glycogen to glycerol could occur from DHAP synthesized by the aldolase-catalyzed cleavage of fructose 1,6-diphosphate. This direct pathway, which would bypass the tricarboxylic acid cycle would be energetically disadvantageous to the nematode. During the initial steps of entry into anhydrobiosis, *A. avenae* catabolizes glycogen and lipid at about the same rate, suggesting that catabolism of glycogen to glycerol is mediated through the tricarboxylic acid cycle for maximum energy production. It has been suggested that glycerol and trehalose synthesis for anhydrobiosis may function to protect the nematode from dehydration and to supply substrates for gluconeogenesis following rehydration (Crowe *et al.,* 1977; Madin and Crowe, 1975).

B. Glyoxylate Cycle

In plants and microorganisms, the glyoxylate cycle is used to bypass acetyl-CoA from the tricarboxylic acid cycle for carbohydrate biosynthesis (White *et al.,* 1978). Organisms lacking this cycle are unable to use acetyl-CoA from oxidation of fatty acids for gluconeogenesis, since two carbon atoms are lost as CO_2 in the tricarboxylic acid cycle. In this cycle, citrate synthase catalyzes citrate formation by condensation of acetyl-CoA with oxaloacetate. Isocitrate is then cleaved by isocitrate lyase into glyoxylate and succinate. Malate is synthesized from glyoxylate by the malate synthase-catalyzed condensation of glyoxylate with a second molecule of acetyl-CoA. The malate is then metabolized via MDH to oxaloacetate for reentry into the cycle. The succinate produced can be used for a variety of biosynthetic processes, including gluconeogenesis and amino acid synthesis. In microorganisms and plants the enzymes of the glyoxylate cycle are localized in catalase- and peroxidase-rich organelles (glyoxysomes or peroxisomes).

Although *T. aceti,* which uses acetate as a nutrient substrate, might be expected to have a glyoxylate cycle, it was not until Rothstein and Mayoh (1964a,b, 1965, 1966) identified isocitrate lyase and malate synthase activity in *P. redivivus. T. aceti, Rhabditis anomala,* and *C. briggsae* that the glyoxylate cycle was reported in higher eukaryotes. These investigators initially observed the synthesis of radioactive glyoxylate and succinate from L-[1,5-^{14}C]isocitrate by these nematodes. It was subsequently observed that tissue homogenates primarily produced radioactive malate from [1-^{14}C]acetyl-CoA plus glyoxylate, but when these extracts were incubated with [1-^{14}C]acetyl-CoA in the absence of glyoxylate, citrate was the major radioactive product and malate synthesis was decreased (Rothstein and Mayoh, 1964a,b). Reiss and Rothstein (1974) have isolated isocitrate lyase from *T. aceti.* The synthesis of radioactive glycine from [4-^{14}C]aspartate by *C. briggsae* (Rothstein, 1963; Rothstein and Tomlinson, 1962; Rothstein and Mayoh, 1964b) and the incorporation of radioactivity from [U-^{14}C]palmitate into glycogen in *T. aceti* and *P. redivivus* (Barrett et al., 1971) further indicate the presence of a complete glyoxylate cycle in these free-living nematodes.

Isocitrate lyase and malate synthase activity have also been identified in *C. elegans* (Colonna and McFadden, 1975; Patel and McFadden, 1977), in developing eggs of *Ascaris* (Barrett et al., 1971; Barrett and Beis, 1975; Patel and McFadden, 1978; Rubin and Trelease, 1976), and in various life-cycle and developmental stages of *S. ratti* (Körting and Fairbairn, 1971). Malate synthase activity could not be identified in the free-living adults of *S. ratti* and was insignificant in third-stage larve (Körting and Fairbairn, 1971). These latter stages were also incapable of gluconeogenesis from [^{14}C]palmitate.

Beginning at about the fifteenth day of development of *Ascaris* eggs, the activity of both isocitrate lyase and malate synthase increases concomitantly with the initiation of carbohydrate resynthesis from triglycerides. Enzyme activity then declines as dormancy is reached (Barrett et al., 1970; Rubin and Trelease, 1976). The activities of other gluconeogenic enzymes increase in parallel to the increased activity of the glyoxylate cycle enzymes during *Ascaris* development (Barrett *et al.*, 1970).

Efforts have recently been made to identify glyoxysomes in free-living nematodes shown to have enzymes of the glyoxylate cycle and in embryonated *Ascaris* eggs. Aueron and Rothstein (1974) reported that a subcellular fraction of *T. aceti*, isolated by methods developed for mitochondrial isolation from rat liver, contained the mitochondrial enzyme fumarase and enzymes characteristic of glyoxysomes. These investigators were unable to separate glyoxysomal and mitochondrial enzyme activities using several other separation techniques. Using methods that would be expected to result in the leakage of enzymes from glyoxysomes but not from mitochondria, these authors suggest the possible presence of peroxisome-like organelles in *T. aceti*. McKinley and Trelease (1978a) reported a similar cobanding of marker enzymes of the glyoxylate and tricarboxylic acid cycle at a density of 1.204 g/cm^3 following isopycnic centrifugation of mitochondrial fractions of *T. aceti*. These activities could not be separated by treating the mitochondrial fractions with nitro blue tetrazolium, which increases the mitochondrial density by reacting with NADP-isocitrate dehydrogenase while not affecting glyoxysomal density (Nachlas *et al.*, 1957). In a subsequent study, McKinley and Trelease (1978b) demonstrated that in *T. aceti* the enzymes of the glyoxylate cycle, malate synthase and isocitrate lyase, are located in the mitochondrial matrix and that catalase is bound to the mitochondrial membrane. This location of catalase and isocitrate lyase in the mitochondria might explain the loss of catalase and isocitrate lyase from the mitochondrial fraction with repeated centrifugation (Liu and Rothstein, 1976). Whether the glyoxylate cycle and tricarboxylic acid cycle are present in two different types of mitochondria, or if a single type contains both enzymes, is not known. This may, however, be important in terms of the cellular regulation of acetate metabolism for maximum production of cellular energy or for gluconeogenesis.

Glyoxylate cycle enzymes are also apparently located in the mitochondria of developing *Ascaris* larvae (Rubin and Trelease, 1976). Although a subcellular fraction with an isopycnic density characteristic of glyoxysomes can be isolated from these larvae, it does not contain glyoxylate cycle enzymes (Rubin and Trelease, 1976). No microbodies were observed by the electron microscopic examination of the mitochondrial fraction containing isocitrate lyase. This observation is not unusual since glyoxylate cycle enzymes have been reported in both mitochondrial and glyoxysomal locations in yeast and protozoa (Avers, 1971; Mueller, 1969).

In subcellular fractions from *C. elegans*, the glyoxylate cycle enzymes, citrate synthase and catalase, migrate on isopycnic sucrose gradients at a density characteristic of glyoxysomes (1.25 g/cm^3), whereas the tricarboxylic acid cycle enzymes are recovered at a density of 1.18 g/cm^3. This suggests that, unlike

Ascaris larvae and *T. aceti, C. elegans* contains glyoxysome-like microbodies (Patel and McFadden, 1977, 1978). The reason for this difference is not readily apparent since comparable methods were used by these investigators for the preparation of tissue homogenates and isolation of subcellular fractions. It might be suggested, however, that *T. aceti,* which uses exogenous acetate for energy production (Nicholas, 1975), and developing *Ascaris* larvae, which use lipid storage substrates for the resynthesis of carbohydrate, have both cycles in the same cellular organelle for maximum efficiency in regulating acetate metabolism through either cycle. On the other hand, *C. elegans* may use a wider range of readily available metabolic substrates, and a physical separation of pathways may provide the most efficient mechanism for the regulation of acetate metabolism. Further studies are needed to clarify this problem.

C. Phosphogluconate Pathway

Enzymes of the hexose monophosphate shunt have been identified in several free-living, plant-parasitic and animal-parasitic nematodes (Krusberg, 1960; Panagides and Rothstein, 1973a,b; Anwar *et al.,* 1977; Körting and Fairbairn, 1971; Langer *et al.,* 1971). This pathway generally functions as an alternative pathway for the oxidation of glucose to CO_2, pentoses, and reduced NADP, and can account for the complete oxidation of glucose.

An interesting function of this cycle has been reported in *T. aceti* (Panagides and Rothstein, 1973a,b). This nematode synthesizes and excretes free [^{14}C]ribitol when maintained *in vitro* on a medium supplemented with [1-^{14}C]glucose or [1-^{14}C]acetate. The labeling patterns of the ribitol produced indicate that it is synthesized in the hexose monophosphate shunt (Panagides and Rothstein, 1973b). NADP-dependent ribulose reductase activity catalyzing the reduction of D-ribulose to ribitol has been identified in this nematode; however, ribulose 5-phosphate is not a substrate for this enzyme. Both D- and L-xylulose are also reduced in this nematode by an NADH- or NADPH-dependent reductase, but xylitol is not excreted. Low levels of ketopentose reductase activity were also found in homogenates of *C. briggsae* and *P. redivivus,* but only *P. redivivus* excreted any ribitol. The biological significance of ribitol production coupled to NADPH oxidation is unknown. The only obvious function would be the oxidation of NADPH produced in the hexose monophosphate shunt. This oxidation, however, would be of no advantage since the NADPH produced is necessary for fatty acid synthesis in these nematodes (Rothstein and Gotz, 1968; Rothstein, 1970). It might be suggested that the excretion of ribitol serves as a regulatory mechanism on the hexose monophosphate shunt or on fatty acid biosynthesis by removing NADPH. Further study is required to determine the biological function of this pathway.

D. Gluconeogenesis

The presence of enzymes of gluconeogenic pathways, the glyoxylate cycle, the tricarboxylic acid cycle, and fatty acid oxidation in a variety of nematodes (Von Brand, 1973; Nicholas, 1975; Saz and Lescure, 1967) indicates a capability for gluconeogenesis. Developing *Ascaris* larvae are dependent on gluconeogenesis for the development of viable infective larvae (Section III, A); the parasitic nematode *Cooperia punctata* also requires gluconeogenic pathways for survival (Ridley *et al.*, 1977). This nematode, which lives in the rumen of cattle, apparently uses the volatile fatty acid end products of symbiotic rumen bacteria metabolism, such as acetate, butyrate, propionate, and succinate, as substrates for cellular energy production. Ridley *et al.* (1977) demonstrated the incorporation and metabolism of [^{14}C]propionate to protein, lipid, glucose, and respired CO_2 by this nematode. These investigators propose a pathway of propionate metabolism via CO_2 fixation into propionyl-CoA to yield methylmalonyl-CoA, which can be isomerized to succinyl-CoA. This succinyl-CoA can then enter the tricarboxylic acid cycle for the production of oxaloacetate. The subsequent decarboxylation of oxaloacetate to PEP and the reduction of PEP to triose phosphate would allow for the conversion of two molecules of propionate to one molecule of glucose. This proposed pathway requires at least a partially functional tricarboxylic acid cycle; however, no investigations have been done to identify either enzymes or intermediates of this cycle in *C. punctata*.

E. β-Oxidation of Fatty Acids

β-Oxidation of long-chain fatty acids coupled to the tricarboxylic acid cycle and electron transport yields large amounts of cellular energy. In mammalian systems, complete oxidation of one molecule of palmitate ($C_{16:0}$) produces 131 molecules of ATP (White *et al.*, 1978). β-Oxidation occurs in the mitochondrial matrix and requires the activation of the fatty acid by the formation of acyl-CoA fatty acids coupled to ATP pyrophosphate cleavage in the cell cytoplasm. This activation is catalyzed by a group of chain-length-specific acyl-CoA synthetases and is followed by the carnitine acyltransferase-catalyzed oxidation of the fatty acid acyl-CoA to synthesize a fatty acyl carnitine ester. This ester readily permeates the mitochondria and is transported to the mitochondrial matrix, where the acyl group is transferred from carnitine to intermitochondrial CoA by a second carnitine acyltransferase. Once in the mitochondria the fatty acid is oxidized through the β-oxidation cycle in which each turn of the cycle requires four enzymatic reactions to yield reduced FAD, reduced NAD$^+$, acetyl-CoA and a 2-carbon shortened acyl fatty acid. The first reaction, coupled to the reduction of FAD and catalyzed by acyl-CoA dehydrogenase, yields 2,3-trans unsaturated derivatives of the fatty acid. The unsaturated bond is then hydrated by enol-CoA hydratase, forming a 3-hydroxy fatty acid. The hydroxy fatty acid is then dehydrogenated to a 3-keto derivative, coupled to NAD reduction, by β-hydroxyacyl-CoA dehydrogenase. The 3-keto fatty acid reacts with a second molecule of acetyl-CoA to yield acetyl-CoA and a 2-carbon shortened fatty acid in a reaction catalyzed by thiolase. During one turn of this cycle, one molecule of FADH and one molecule of NADH are produced per 2-carbon unit oxidized. These reduced cofactors can then enter the electron transport chain for energy production. The acetyl-CoA can be used for the synthesis of additional cellular energy or as a precursor for gluconeogenesis, can be metabolized to products for excretion, or can itself be excreted. The β-oxidation cycle degrades odd-numbered fatty acids and branched-chained fatty acids to acetyl-CoA and propionyl-CoA. The propionyl-CoA is metabolized via methylmalonyl-CoA to succinyl-CoA for entry into the tricarboxylic acid cycle or for other metabolic pathways including gluconeogenesis.

The β-oxidation of fatty acids has been demonstrated in a variety of nematodes that use fatty acid stores for energy production during periods of aerobic starvation (Von Brand, 1973; Nicholas, 1975) or during development (Barrett et al., 1970; Fairbairn, 1969; Wilson, 1965; Körting and Fairbairn, 1971). Barrett et al. (1970) reported a long-chain acyl-CoA synthetase, acetyl-CoA acyl transferase, and all other mitochondrial matrix enzymes of β-oxidation in P. redivivus and T. aceti. These nematodes were also shown to metabolize aerobically [^{14}C]palmitate to CO_2 and glycogen.

Enzymes of β-oxidation have also been demonstrated in embryonating eggs of Ascaris (Barrett et al., 1970; Ward and Fairbairn, 1970), in adult male and female Ascaris (Barrett and Beis, 1973), and in various stages of the life cycle of S. ratti (Körting and Fairbairn, 1971). Specific chain length acyl-CoA synthetases, including a short-chain synthetase capable of acylating 2-methylbutyric and 2-methylvaleric acids, have been identified in 15-day-old Ascaris larvae (Ward and Fairbairn, 1970). The activity of the β-oxidation cycle enzymes in developing Ascaris egg changes coincident with developmental changes in lipid use. Short-chain acyl-CoA synthetase is most active during the developmental period when volatile fatty acids are used most rapidly. Long-chain acyl-CoA synthetase is most active when carbohydrate resynthesis is maximal and the enzymes of the glyoxylate cycle and glyconeogenesis have maximal activity (Barrett et al., 1970). Enzymes of β-oxidation are active in S. ratti infective larvae, which use lipid reserves for energy metabolism (Barrett, 1969a,b), and in free-living adults, which have been shown to metabolize [^{14}C]palmitate to CO_2 (Körting and Fairbairn, 1971). These enzymes either are lacking or are minimally active in parasitic females and first-stage larvae.

F. Protein Metabolism for Energy Production

Few studies have been done on the use of endogenous protein as a source of energy in nematodes. During starvation of the larval stages of the animal-parasitic nematodes *N. brasiliensis, Cooperia punctata,* and *Ancylostoma caninum,* large losses of total protein have been reported (Wilson, 1965; Eckert, 1967; Clark, 1969). In *N. brasiliensis* nonfeeding infective larvae, total nitrogen is lost at a rate of 2.17% of total dry weight per day (Wilson, 1965). It is evident that in this nematode and in *C. punctata* the loss of total protein results from the use of protein for energy production. However, the protein loss in *A. caninum* larvae is probably due to the loss of the cuticle during exsheathment (Wilson, 1965; Eckert, 1967; Clark, 1969).

Both *C. briggsae* and *D. triformis* have been shown to use protein during starvation (Rothstein, 1965; Myers and Krusberg, 1965). A 2–4% decrease in total protein during starvation of *D. triformis* suggests that protein degradation may be a means of energy production in this species.

VII. ENERGY PRODUCTION AND REGULATION OF METABOLISM

The efficiency and amount of cellular energy produced depend on the metabolic cycle used by individual species of nematode. In turn, the type of cycle used in these highly adaptable animals depends on the availability of oxygen. At one extreme are those nematodes having a PK/PEPCK ratio greater than one and having active LDH for the anaerobic synthesis of lactate or ethanol with a net production of 2 moles of ATP per mole of glucose metabolized. At the other end of the spectrum are the free-living nematodes that have an active tricarboxylic acid cycle coupled to classical electron transport pathways. In these species there is a possible maximum of 36–38 moles of ATP produced per mole of glucose catabolized. Developmental stages of some animal-parasitic nematodes that have a complete tricarboxylic acid cycle and some form of electron transport for the oxidation of NADH or FADH coupled to phosphorylation of ADP would be included in this latter group. Those free-living nematodes and the free-living developmental stages of animal-parasitic nematodes with complete glyoxylate cycles would, however, produce less cellular energy per mole of glucose metabolized. These forms, which shunt some of the carbon atoms from glucose into the glyoxylate pathway, decrease the amount of cellular energy available from glucose by bypassing the main energy-producing portion of the tricarboxylic acid cycle. Assuming the metabolism of succinate, produced from the isocitrate lyase cleavage of isocitrate, through the terminal steps of the tricarboxylic acid cycle to oxaloacetate, about 10 moles of ATP would be synthesized per mole of acetyl-CoA entering the cycle. This is a maximum estimation and may not reflect the true energy yield, since experimental evidence suggests that succinate is used for both gluconeogenesis and amino acid synthesis. The actual energy yield may rather be about 4–6 moles of ATP per mole of acetyl-CoA entering the cycle, since this cycle is primarily a biosynthetic pathway.

The animal-parasitic nematodes that operate the PEP succinate pathway described for *Ascaris* produce only limited amounts of cellular energy per mole of glucose metabolized. Although the exact energy yield from this pathway is not known, it may be estimated that 2–4 moles of ATP may be produced per mole of carbohydrate metabolized.

Maximal energy production would occur in those nematode species that metabolize fatty acids by β-oxidation coupled to the tricarboxylic acid cycle and electron transport, since approximately 17 moles of ATP are produced for each 2-carbon unit of fatty acid oxidized. This pathway is, therefore, of extreme energetic value to nematodes under aerobic starvation conditions, during non-feeding developmental stages, or when the exogenous substrate is physically limited, as in developing *Ascaris* eggs. Oxygen is, however, required to maximal energy production of β-oxidation (White *et al.*, 1978). It is possible either that adult parasitic nematodes have sufficient oxygen available in the environment, or that the oxygen bound by the pseudocoelomic fluid hemoglobin or by the muscle myoglobin (Lee and Smith, 1965) is available for use in lipid metabolism.

Regardless of the pathway of metabolism used by various species of nematodes, the pathways that have evolved are apparently sufficient for the success of the nematodes. Thus, animal-parasitic nematodes, living in an anaerobic or microaerobic environment rich in exogenous substrate, apparently do not require a metabolic pathway that is more efficient in energy production than that of glycolysis coupled to the PEP succinate pathway. Free-living nematodes on the other hand, which may experience periods of starvation and anaerobiosis, require a more efficient mechanism of energy production, and a mechanism such as the glyoxylate pathway, for conserving cellular energy. The flux of both carbohydrate and lipid through metabolic pathways in the different nematode species must be highly regulated to allow adaptation to a variety of environmental conditions.

Metabolic rate is regulated by end product feedback and by cellular concentrations of [ATP] and [NAD]. In aerobic systems [NAD]/[NADH] is generally adjusted to 750–1000 and [ATP]/[AMP] is regulated at about 500 (White *et al.*, 1978). Relatively small changes in [NAD] and [ATP] result in relatively large changes in [NADH] and [AMP] and subsequently large inhibition or stimulation of energy metabolism. In the glycolytic pathway the major regulatory enzymes are phosphofructokinase, hexokinase, and PK. These enzymes are regulated by end product feedback and by [ATP]/[AMP] and [NAD]/[NADH]. This regulation affects only glycolysis and does not affect gluconeogenesis pathways that involve other enzymes of glycolytic sequence.

The glyoxylate cycle in plants and bacteria is regulated by the concentration of PEP, acting as a negative modulator of isocitrate lyase, and by the activity of aconitate hydratase. Thus large accumulations of PEP repress the movement of carbon atoms through this pathway and regulate gluconeogenesis.

The flux of acetyl-CoA through the tricarboxylic acid cycle is regulated by cellular concentrations of ATP and by the rate of electron flow through electron transport pathways. The cycle can proceed no faster than the rate at which the cell uses the ATP generated by the oxidation of reduced NAD. The entry of citrate into the tricarboxylic acid cycle via citrate synthase is regulated by both [NADH] and [ATP] (White *et al.*, 1978). This is the committed step and hence is the rate-limiting step of both this and the glyoxylate cycle. ATP regulation is mediated through GTP inhibition of 2-ketogluterate dehydrogenase and the coincident accumulation of succinyl-CoA. Succinyl-CoA acts as a negative modulator of citrate synthase by binding to the enzyme and effecting an increase in the K_m of the enzyme for acetyl-CoA. As a consequence of these regulatory mechanisms the tricarboxylic acid cycle is under fine regulatory control. Since β-oxidation is generally coupled to the tricarboxylic acid cycle and to the glyoxylate cycle (when present), the regulation of β-oxidation would involve mechanisms regulating the pathways to which it is coupled. Thus cellular concentrations of NAD and ATP have major regulatory roles in fatty acid oxidation.

The enzymes of any metabolic pathway cay be divided into two classes: those enzymes catalyzing reactions that are near thermodynamic equilibrium, and those enzymes catalyzing reactions far removed from equilibrium. The latter group generally provides a focus of metabolic control since they catalyze the rate-limiting steps in metabolic pathways (Crabtree and Newsholme, 1972). An estimation of the catalytic activities of these nonequilibrium enzymes therefore provides an estimate of the catalytic rate of the pathways. Changes in the activity of these rate-limiting enzymes will alter the rate of substrate flux through the pathway, whereas changes in the activity of equilibrium-reaction enzymes will have little effect on metabolic rate unless their activity becomes insignificant.

Metabolic regulation has been best studied in animal-parasitic nematodes (Von Brand, 1973; Bryant, 1970, 1975; Saz, 1969) and, in particular, in the developmental stages and adults of *Ascaris* (Barrett and Beis, 1975; Barrett, 1976; Ward and Fairbairn, 1970). In developing *Ascaris* larvae the maximum catalytic capacity of the nonequilibrium-reaction enzymes of glycolysis, β-oxidation, the TCA cycle, and gluconeogenesis corresponds to the onset of carbohydrate and lipid use in the early stages of development and the onset of resynthesis of carbohydrate from triglycerides later in development (Barrett, 1976; Beis and Barrett, 1975). Concomitant with changes in the catalytic capacity of these rate-limiting enzymes are changes in [ATP]/[ADP] and [NAD]/[NADH]. [ATP]/[ADP] is low during active development, increases as dormancy is reached, and decreases again when the egg is activated for further development. Similar but opposite changes in [NAD]/[NADH] occur coincidentally. When the [ATP]/[ADP] ratios are high, as in the dormant egg, an increase and accumulation of the intermediates of the TCA cycle occur. This buildup of intermediates may have a regulatory function in glycolysis. Metabolic control by concentrations of ATP and NAD may be lost in dormant eggs and regained when activation occurs (Beis and Barrett, 1975).

PEP metabolism in nematodes having the PEP succinate type of pathway is apparently regulated by the catalytic activities of PK as compared to PEPCK in the direction of CO_2 fixation. In *N. brasiliensis* and *R. bufonis* this apparently occurs by substrate competition between PK and PEPCK (Saz *et al.*, 1971; Anya and Umezurike, 1978). By a comparison of the activity of PEPCK to PK, a valid estimate of the nature of the end products of metabolism and of the possible metabolic pathway can be made. This is readily apparent when applied to the activities of these enzymes in the developmental and life-cycle stages of *S. ratti*. The free-living third-stage larvae and adults of this nematode, which have an oxidative metabolism, have PK/PEPCK ratios of 2.2 and 1.8, respectively. The parasitic adults, which apparently use the PEP succinate pathway, have a PK/ PEPCK ratio of 0.14 (Körting and Fairbairn, 1971). First-stage larvae, which may be a transition stage between the parasitic and free-living types of metabolism and which may produce lactate as a major end product (Körting and Fairbairn, 1971), have a PK/PEPCK ratio of 1.6. The mechanism of metabolic regulation in *S. ratti* is not clear; however, the enzyme patterns reported in the developmental and life-cycle stages suggest that (1) regulation may be by product feedback inhibition; (2) there may be repression and derepression of the genes coding for enzymes of the pathways; or (3) both may occur.

Those nematodes having the PEP succinate pathway must have an additional regulation at the level of the mitochondrial dismutation reaction. That is, there must be a regulation of malate flux in order to maintain a balance between NADH and fumerate production since this pathway is the only source of mitochondrial ATP production for the nematode (Landsperger and Harris, 1976). A study by Landsperger and Harris (1976) suggests that the NAD malic enzyme is a product feedback-inhibited enzyme and would afford the needed regulation. The reader is referred to reviews by Bryant (1970, 1975) for a further discussion of metabolic regulation in parasitic nematodes.

Metabolic regulation has not been well investigated in free-living nematodes, although substrate and enzyme-level regulation appears to be similar to that in other systems (Nicholas, 1975). The regulation of metabolism that occurs when *C. briggsae, T. aceti,* and *P. redivivus* are cultured in a nutritionally rich medium and when *A. avenae* enters anhydrobiosis is interesting to consider. As discussed earlier (Section VI,A), these nematodes shift their metabolism toward a less efficient use of carbohydrate and produce glycerol. It has been suggested (Liu and Rothstein, 1976) that the shift mainly to glycerol metabolism in *C. briggsae* may involve repression and derepression of gene function. This proposal is novel in terms of the regulation of metabolic pathways in higher eukaryotic organisms and certainly merits additional investigation. The regulation of metabolic pathways in nematodes during periods of starvation or during entry into an anhydrobiotic state is another area that deserves further study.

VIII. CONCLUSIONS

Pathways of energy metabolism in nematodes are extremely variable and reflect both the adaptability and diversity of the animals. Thus it is difficult to make valid comparisons between species and even between developmental stages within a species. For this reason the study of energy production and metabolic regulation is in an undeveloped state when compared to what is known in other animals. This metabolic pathway adaptability may, however, be a feature of the nematode that can be exploited to the utmost for investigating eukaryotic metabolic regulation. Thus nematodes such as *S. ratti* that have both free-living and parasitic generations with characteristic patterns of metabolic enzymes, or nematodes such as *C. briggsae* that alter their metabolic enzyme pattern to fit the nutritional state of the environment, may be excellent model systems to study metabolic regulation. In this sense, the experimental evidence that suggests that some regulation may involve repression and depression of the genome may renew and encourage interest in studying metabolic regulation in nematodes. Such studies will be facilitated by the recent advent of *in vitro* culture for several nematodes, and by the development of microbiochemical techniques.

REFERENCES

Anwar, N., Ansari, A. A., Ghatak, S., and Murti, C. R. K. (1977). *Z.Parasitenkd.* **51**, 275–283.
Anya, A. O., and Umezurike, G. M. (1978). *Parasitology* **76**, 21–27.
Atkinson, H. J. (1973). *J. Exp. Biol.* **59**, 255–274.
Aueron, F., and Rothstein, M. (1974). *Comp. Biochem. Physiol. B* **49**, 261–271.
Avers, C. (1971). *Sub-Cell Biochem.* **1**, 25–37.
Balasubramanian, M., and Myers, R. F. (1971). *Exp. Parasitol.* **29**, 330–336.
Barrett, J. (1968). *Nature (London)* **218**, 1267–1268.
Barrett, J. (1969a). *Parasitology* **59**, 859–875.
Barrett, J. (1969b). *Parasitology* **59**, 3–17.
Barrett, J. (1976). *In* "Biochemistry of Parasites and the Host Parasite Relationship" (H. Van den Bossche, ed.), pp. 117–123. North-Holland Publ., Amsterdam.
Barrett, J., and Beis, I. (1973). *Comp. Biochem. Physiol. B* **44** 751–761.

Barrett, J., and Beis, I. (1975). *Dev. Biol.* **42**, 181–187.

Barrett, J., Ward, C. W., and Fairbairn, D. (1970). *Comp. Biochem. Physiol.* **35**, 577–586.

Barrett, J., Ward, C. W., and Fairbairn, D. (1971). *Comp. Biochem. Physiol. B* **38**, 279–284.

Beis, I., and Barrett, J. (1975). *Dev. Biol.* **42**, 188–195.

Bolla, R. I., and Roberts, L. S. (1968). *J. Parasitol.* **54**, 849–855.

Bryant, C. (1970). *Adv. Parasitol.* **8**, 139–172.

Bryant, C. (1975). *Adv. Parasitol.* **13**, 35–69.

Campbell, J. W. (1973). *In* "Comparative Animal Physiology" (C. L. Prosser, ed), 3rd ed., pp. 279–306. Saunders, Philadelphia, Pennsylvania.

Castillo, J. M., and Krusberg, L. R. (1971). *J. Nematol.* **3**, 284–288.

Clark, F. E. (1969). *Exp. Parasitol.* **24**, 1–8.

Cole, R. J., and Krusberg, L. R. (1967). *Exp. Parasitol.* **21**, 232–239.

Colonna, W. J., and McFadden, B. (1975). *Arch. Biochem. Biophys.* **170**, 608–619.

Cooper, A. F., and Van Gundy, S. D. (1970). *J. Nematol.* **2**, 305–315.

Cooper, A. F., and Van Gundy, S. D. (1971). *J. Nematol.* **3**, 205–214.

Costello, L. C., Oya, H., and Smith, W. (1963). *Arch. Biochem. Biophys.* **103**, 345–351.

Crabtree, B., and Newsholme, E. A. (1972). *Biochem. J.* **126**, 49–58.

Crowe, J. H., Madin, K. A. C., and Loomis, S. H. (1977). *J. Exp. Zool.* **201**, 57–64.

Dropkin, V. H., and Acedo, J. (1974). *J. Parasitol.* **60**, 1013–1021.

Eckert, J. (1967). *Z. Parasitenkd.* **29**, 209–241.

Ellison, T., Thomson, W. A. B., and Strong, F. M. (1960). *Arch. Biochem. Physiol.* **91**, 247–254.

Ells, H. A. (1969). *Comp. Biochem. Physiol.* **29**, 689–701.

Fairbairn, D. (1954). *Exp. Parasitol.* **3**, 52–63.

Fairbairn, D. (1969). *In* "Chemical Zoology" (M. Florkin and B. T. Scheer, eds.), Vol. 3, pp. 361–378. Academic Press, New York.

Fairbairn, D., and Passey, R. F. (1957). *Exp. Parasitol.* **6**, 566–574.

Fletcher, C. L., and Krusberg, L. R. (1973). *Comp. Biochem. Physiol. B* **45**, 159–165.

Greichus, A., and Greichus, Y. A. (1966). *Exp. Parasitol.* **19**, 85–90.

Haskins, W. T., and Weinstein, P. P. (1957). *J. Parasitol.* **43**, 19–24.

Hieb, W. F., and Dougherty, E. C. (1966). *Nematologia,* **12**, 93.

Kmetec, E., Beaver, P. C., and Bueding, E. (1963). *Comp. Biochem. Physiol.* **9**, 115–120.

Körting, W., and Fairbairn, D. (1971). *J. Parasitol.* **57**, 1153–1158.

Krusberg, L. R. (1960). *Phytopathology* **50**, 9–22.

Krusberg, L. R. (1967). *Comp. Biochem. Physiol.* **21**, 83–90.

Krusberg, L. R. (1971). *In* "Plant Parasitic Nematodes" (B. M. Zuckerman, W. F. Mai, and R. A. Rohde, eds.), Vol. 2, pp. 213–234. Academic Press, New York.

Krusberg, L. R., Hussey, R. S., and Fletcher, C. L. (1973). *Comp. Biochem. Physiol. B* **45**, 335–341.

Landsperger, W. J., and Harris, B. G. (1976). *J. Biol. Chem.* **251**, 3599–3602.

Langer, B. W., and Smith, W. J. (1971). *Comp. Biochem. Physiol. B* **40**, 833–840.

Langer, B. W., Smith, W. J., and Theodorides, V. J. (1971). *J. Parasitol.* **57**, 485–486.

Lapp, D. F., and Mason, S. L. (1978). *J. Parasitol.* **64**, 645–650.

Lee, D. L., and Atkinson, H. J. (1976). "Physiology of Nematodes," 2nd ed. MacMillan, New York.

Lee, D. L., and Smith, M. H. (1965). *Exp. Parasitol.* **16**, 392–424.

Lee, E. H., and Fernando, M. A. (1971). *Int. J. Biochem.* **2**, 403–408.

Liu, A., and Rothstein, M. (1976). *Comp. Biochem. Physiol. B* **54**, 233–238.

McKinley, M. P., and Trelease, R. M. (1978a). *Biochem. Biophys. Res. Commun.* **81**, 434–438.

McKinley, M. P., and Trelease, R. N. (1978b). *Protoplasma* **94**, 249–261.

Madin, K. A. C., and Crowe, J. (1975). *J. Exp. Zool.* **193**, 335–342.

Magat, W. J., Hubbard, W. J., and Jeska, E. L. (1972). *Exp. Parasitol.* **32**, 102–108.

Miller, C. W., and Roberts, R. N. (1974). *Phytopathology* **54**, 1177.

Mueller, M. (1969). *Ann. N.Y. Acad. Sci.* **168**, 292–301.

Myers, R. F., and Krusberg, L. R. (1965). *Phytopathology.* **55**, 429–437.

Nachlas, M., Tsou, K., de Suza, D. E., Cheng, C., and Seligman, A. (1957). *J. Histochem. Cytochem.* **5**, 420–436.

Nicholas, W. L. (1975). "The Biology of Free-Living Nematodes." Oxford Univ. Press (Clarendon), London and New York.

Nicholas, W. L., Dougherty, E. C., Hansen, E. L., Hansen, O. H., and Moses, V. (1960). *J. Exp. Biol.* **37**, 435–443.

Oya, H., Costello, L. C., and Smith, W. N. (1963). *J. Cell Comp. Physiol.* **62**, 287–293.

Panagides, J., and Rothstein, M. (1973a). *Int. J. Biochem.* **4**, 397–406.

Panagides, J., and Rothstein, M. (1973b). *Int. J. Biochem.* **4**, 407–414.

Papa, S., Cheah, K. S., Rasmussen, H. N., Lee, I. Y., and Chance, B. (1970). *Eur. J. Biochem.* **12**, 540–543.

Patel, T., and McFadden, B. (1977). *Arch. Biochem. Biophys.* **183**, 24–30.

Patel, T., and McFadden, B. A. (1978). *Exp. Parasitol.* **44**, 72–81.

Pollak, J. K., and Fairbairn, D. (1955). *Can. J. Biochem. Physiol.* **33**, 307–316.

Polyakova, O. I. (1962). *Biokhimiya* **27**, 430–436.

Reiss, U., and Rothstein, M. (1974). *Biochemistry* **13**, 1796–1800.

Ridley, R. K., Slonka, G. F., and Leland, S. E. (1977). *J. Parasitol.* **63**, 348–356.

Roberts, L. S., and Fairbairn, D. (1965). *J. Parasitol.* **51**, 129–138.

Rogers, W. P. (1952). *Aust. J. Sci. Res. B* **5**, 210–222.

Rogers, W. P. (1969). *In* "Chemical Zoology" (M. Florkin and B. T. Scheer, eds.), Vol. 3, pp. 379–428. Academic Press, New York.

Rohde, R. A. (1971). *In* "Plant Parasitic Nematodes" (B. M. Zuckerman, W. F. Mai, and R. A. Rohde, eds.), Vol. 2, pp. 235–246. Academic Press, New York.

Rothstein, M. (1963). *Comp. Biochem. Physiol.* **9**, 51–59.

Rothstein, M. (1965). *Comp. Biochem. Physiol.* **14**, 541–552.

Rothstein, M. (1969). *Comp. Biochem. Physiol.* **30**, 641–648.

Rothstein, M. (1970). *Int. J. Biochem.* **1**, 422–428.

Rothstein, M., and Gotz, P. (1968). *Arch. Biochem. Biophys.* **126**, 131–140.

Rothstein, M., and Mayoh, H. (1964a). *Arch. Biochem. Biophys.* **108**, 134–142.

Rothstein, M., and Mayoh, H. (1964b). *Biochem. Biophys. Res. Commun.* **14**, 43–47.

Rothstein, M., and Mayoh, H. (1965). *Comp. Biochem. Physiol.* **16**, 361–365.

Rothstein, M., and Mayoh, H. (1966). *Comp. Biochem. Physiol.* **17**, 1181–1188.

Rothstein, M., and Tomlinson, G. (1962). *Biochem. Biophys. Acta.* **63**, 471–480.

Rothstein, M., Nicholls, F., and Nicholls, P. (1970). *Int. J. Biochem.* **1**, 695–705.

Rubin, H., and Trelease, R. N. (1975). *J. Parasitol.* **61**, 577–588.

Rubin, H., and Trelease, R. N. (1976). *J. Cell. Biol.* **70**, 374–383.

Saz, H. J. (1969). *In* "Chemical Zoology" (M. Florkin and B. T. Scheer, eds.), Vol. 3, pp. 329–360. Academic Press, New York.

Saz, H. J. (1970). *J. Parasitol.* **56**, 634–642.

Saz, H. J. (1972). *In* "Comparative Biochemistry of Parasites" (H. Van den Bossche, ed.), pp. 33–47. Academic Press, New York.

Saz, H. J., and Lescure, O. L. (1966). *Comp. Biochem. Physiol.* **18**, 845–857.

Saz, H. J., and Lescure, O. L. (1967). *Comp. Biochem. Physiol.* **22**, 15–28.

Saz, H. J., and Vidrine, A. (1959). *J. Biol. Chem.* **234**, 2001–2005.

Saz, D. K., Bonner, T. P., Karlin, M., and Saz, H. J. (1972). *J. Parasitol.* **57**, 1159–1162.

Scott, H. L., and Whittaker, F. H. (1970). *J. Nematol.* **2**, 193–203.

Sivapalan, P., and Jenkins, W. R. (1966). *Proc. Helminthol. Soc. Wash.* **33**, 149–157.

Tracey, M. V. (1958). *Nematologica* **3**, 179–183.

Umezurike, G. M., and Anya, A. O. (1978). *Comp. Biochem. Physiol. B* **59**, 147–151.

Vaatstra, W. J: (1969). *Hoppe-Seyler's Z. Physiol. Chem.* **350**, 701–709.

Von Brand, T. (1973). "Biochemistry of Parasites." Academic Press, New York.

Van den Bossche, H., Vanparijs, O. F. J., and Thienpont, D. (1969). *Life. Sci.* **8**, 1047–1054.

Van den Bossche, H., Schaper, J., and Borgers, M. (1971). *Comp. Biochem. Physiol. B* **38**, 43–52.

Vanfleteren, J. R., and Roets, D. E. (1973). *Nematologica* **18**, 325–338.

Van Gundy, S. D., Bird, A. F., and Wallace, H. R. (1967). *Phytopathology* **57**, 559–571.

Ward, C. W., and Fairbairn, D. (1970). *Dev. Biol.* **22**, 366–387.

Ward, C. W., and Schofield, P. J. (1967). *Comp. Biochem. Physiol.* **23**, 335–359.

Ward, C. W., Castro, G. A., and Fairbairn, D. (1969). *J. Parasitol.* **55**, 67–71.

Ward, P. F. V., and Huskisson, N. S. S. (1978). *Parasitology* **77**, 255–271.

White, A., Handler, P., Smith, E. L., Hill, R. L., and Lehman, I. R. (1978) "Principles of Biochemistry," 6th ed. McGraw-Hill, New York.

Wilson, P. A. G. (1965). *Exp. Parasitol.* **16**, 190–194.

8

Longevity and Survival in Nematodes: Models and Mechanisms

A. A. F. EVANS

Imperial College at Silwood Park
Ascot Berkshire SLS 7DE, England

C. WOMERSLEY*

Department of Zoology
University of Newcastle
Newcastle-upon-Tyne NE1 4LP, England

I.	Introduction	193
II.	Model Nematodes for Studying Longevity and Survival	195
	A. Tylenchid Nematodes	195
	B. Rhabditid Nematodes	200
	C. Other Free-Living Nematodes	201
	D. Comparison of Models	202
III.	Physiological Aspects of Desiccation Survival	202
	A. Theoretical Considerations	202
	B. Previous Work with Anhydrobiotes	203
	C. Physiological Changes in Nematodes	204
	D. Perspectives and Challenges	207
IV.	Conclusions	208
	References	208

I. INTRODUCTION

All organisms have a natural life span which arises from the particular life history strategy that they have evolved in interaction with their other attributes

---

* Present address: Department of Nematology, University of California, Riverside, California 92521.

193

NEMATODES AS BIOLOGICAL MODELS
VOLUME 2

such as morphology, physiology, and behavior. Such attributes are continuously molded by the organism's environment, which is likely to range from the fairly uniform and predictable to the rather harsh, patchy, and unpredictable. Many animals display various adaptations giving tolerance to variations in temperature, food, oxygen, salts, and water availability, and the length of the life span (normal longevity) may be altered if these are suboptimal. Thus rhabditid nematodes live longer when cultured in axenic media rather than in bacteria (Croll *et al.*, 1977; Otter, 1933) or when cultured at suboptimal temperatures, but this is properly the area of studies on aging (see this volume, Chapters 1 and 2).

Nematodes have attracted attention because they have adaptations for survival in the dormant state which apparently alter normal aging by slowing or ceasing development, thereby conferring a much increased life span.

These ideas are reflected in the concepts developed by students of nematode survival mechanisms. Van Gundy (1965) recognized three main categories under which survival could be discussed; (1) factors in the life cycle (e.g., mode of reproduction, egg structure, larval modification); (2) metabolic factors (e.g., lipid or carbohydrate storage, respiratory patterns); and (3) hypobiotic factors (covering dormancy and cryptobiosis). In his discussion of hypobiosis, Van Gundy followed Keilin (1959), who emphasized the metabolic state of an animal by using *hypobiosis* to characterize both dormancy (hypometabolism) and cryptobiosis (ametabolism). Gordon (1973) attributed hypobiosis in nematodes of veterinary importance to either immunological or physiological origins.

Cooper and Van Gundy (1971b) again placed these concepts in the context of the metabolism of an organism where senescence, quiescence, and cryptobiosis described decreasing levels of metabolic activity. Evans and Perry (1976) suggested that, in accordance with recent evidence showing developmental processes to be the principal factor controlling an animal's response, dormancy in nematodes was of two forms: (1) quiescence and (2) diapause. Anhydrobiosis was seen as an extreme form of *quiescence* (a suspension of normal activity which is readily reversible upon return to favorable conditions), rather than an expression of *diapause* (which is mediated by endogenous factors controlling the animal's further development). This view of anhydrobiosis was also expressed by Crowe and Madin (1974), although they used dormancy as the term to describe factors under developmental control. Mansingh (1971) used a similar approach in proposing that insect dormancy be classed as either (1) hibernation, (2) estivation, or (3) athermopause. Each of these classes could be subdivided into one of three categories (quiescence, oligopause, or diapause) representing a sequence of evolutionary development increasing in intensity of dormancy.

This chapter concentrates on the ability of some nematodes to survive various degrees of desiccation, and discusses the use of certain free-living nematodes as models for study and the physiological mechanisms by which survival may be achieved. Nematodes living in permanent water (marine or freshwater) probably have little need to survive desiccation, so this ability is best developed in those groups of nematodes that live near, at, or above the soil surface. Nielsen (1967) lists 28 species abundant in habitats known to be subjected to severe desiccation, e.g., moss cushions; the ability to survive desiccation is a property widely shared among the different taxonomic groups, some of which are now discussed.

II. MODEL NEMATODES FOR STUDYING LONGEVITY AND SURVIVAL

A. Tylenchid Nematodes

1. Anguina tritici

For the past 200 years *Anguina tritici* has been known for the remarkable ability of its second larval stage to remain viable, though desiccated, for long periods. Infective second-stage larvae enter young wheat plants from the soil, penetrating between leaf sheaths to the growing point. After flower initiation they invade the differentiating tissues of the floret, inducing galls within which they develop to adults and then mate. Eggs hatch after 2–5 days, and the L1 stage larvae molt to the L2 stage, which accumulate while the gall is still green, but which prepare for cryptobiosis as the host plant ripens. The mature gall is usually smaller than a wheat grain, brown to black in color, and packed with cryptobiotic second-stage larvae some of which are capable of remaining invasive for 32 years (Limber, 1973). Under field conditions the galls either fall to the ground or are harvested with the grain (Southey, 1972). From the galls large numbers of clean, age-synchronized nematodes can be obtained for experimentation under greenhouse conditions.

2. Ditylenchus spp.

The stem and bulb nematode *D. dipsaci* has been known in a number of more or less host-specific biological races for about 100 years. Under favorable conditions rapid breeding produces large populations that quickly kill host tissue. Aggregations of desiccation-resistant larvae ("eelworm wool") are produced when senescence or death of host tissue makes further reproduction impossible (e.g., in seed pods of field beans or decaying bulbs of daffodils). These aggregations contain predominantly the L4 stage (Perry, 1977), which shows survival abilities similar to *Anguina tritici* larvae. In addition, there is evidence that these larvae may vary in infectivity, entering a diapause in response to autumn chilling that lasts for up to 36 weeks (Evans and Perry, 1976). Eelworm wool is variable from one population to another, but large amounts may be collected at one time from fairly uniform material. Some populations of *D. dipsaci* have been cultured on fungi in the laboratory (Viglierchio, 1971), and if such cultured nematodes showed desiccation resistance, progress into as well as revival from the cryptobiotic state could be studied. This species also resists freezing damage when desiccated (Bosher and McKeen, 1954).

Two other species, *Ditylenchus destructor* and *D. myceliophagus*, are taxonomically close and morphologically similar to *D. dipsaci* and may be cultivated on fungi, but only *D. myceliophagus* shows some desiccation resistance (see Section III,C,1). Most isolates show a preference for the mycelium of the cultivated mushroom *Agaricus bisporus* and have a generation time of about 18 days at 23°C (Cayrol, 1964; Evans and Fisher, 1969). Large populations can be obtained using mass culture methods (Evans, 1970) that yield a mixture of adults and larvae.

3. *Aphelenchus avenae*

Although first described by Bastian in 1865 and studied intermittently since then as a possible plant pathogen or biological control agent (Christie and Arndt, 1936; Rhoades and Linford, 1959), it is only in the last 10 years that the potential use of *Aphelenchus avenae* as a model stylet-bearing nematode has begun to be appreciated and exploited. Accordingly, a comprehensive—though not exhaustive—review of aspects of its behavior, physiology, and experimental uses is presented here. Additional information on taxonomy, morphology, life history, and population dynamics may be found elsewhere (Thorne, 1961; Mankau and Mankau, 1963; Hooper, 1974). It must suffice to note that both parthenogenetic and amphimictic isolates have been studied from all over the world and that considerable ecological differences can be found between them, especially regarding temperature optima and reproduction rates (Evans and Fisher 1970b; Dao, 1970) (Table 1).

The culturing of *A. avenae* is possible on a wide range of soil fungi, but plant-pathogenic forms are preferred (Townshend, 1964; Mankau and Mankau, 1963) and the organism may be able to utilize fungi more effectively than other nematodes (Kondrollochis, 1977). Feeding by some populations can reduce the intensity of plant disease symptoms (Klink and Barker, 1968), whereas in other instances, symptoms were aggravated (Chin and Estey, 1966). Reproduction while feeding on tomato plants was observed in a Canadian population of *A. avenae* (Chin and Estey, 1966); this ended years of controversy about its parasitic status by demonstrating that at least some populations may be facultative plant pathogens.

Prior to the use of axenic culture, little was known of the nutritional requirements of *A. avenae*. Reproduction on tobacco callus (Barker and Darling, 1965) was apparently much slower than on the most suitable fungi. Individuals in a Brownhill population could grow larger, but not reproduce faster, when feeding on fungi cultured on nitrogen-rich agar (Evans and Fisher, 1970a). Axenic culture has shown that reproduction is best when the chemically defined *Caenorhabditis briggsae* minimal medium is supplemented with chick embryo extract (Hansen *et al.*, 1970). Human serum (Tarakanov, 1975) gave similar results. Population growth rate may be increased using a higher CO_2 concentration in the gas phase (Buecher *et al.*, 1974), but the nutrition of *A. avenae* has yet to be investigated systematically in the way it has been for other nematodes. Monoxenic, mass-culture methods using soil fungi growing on a sterile, cooked-wheat substrate (Evans, 1970) allowed production of large quantities of mixed populations for physiological and biochemical investigations.

Feeding on fungi involved location of the hyphae, penetration of the cell by stylet thrusting, and immediate, rapid ingestion of cytoplasm without extracorporeal digestion (Fisher and Evans, 1967). In axenic culture a range of activities associated with perforation of membranes was performed. When the stylet tip remained protruded after stylet thrusting, ingestion sometimes followed for about a minute. Chemical rather than tactile stimuli may be the more important in promoting feeding activities; ingestion occurred in a solution of glucose plus amino acids, whereas only stylet thrusting occurred in glucose (Fisher, 1975).

Among most parthenogenetic populations males are relatively rare, varying from 1 in 10^5 to 1 in 10^4 (Hechler, 1962) at temperatures up to 30°C. However, Dao (1970) found that the sex ratio in two populations could be influenced by temperature; a few males occurred in both populations at 25°C, a predominance of males at 32°–33°C, and only a slight excess of females after growth at 29°C for 1 month in a Netherlands population. A similar temperature effect has also been investigated in the Californian isolate NC-32 (Hansen *et al.*, 1972); only males develop at 30°C in this population which, at temperatures up to 28°C, produced only females. The change to maleness could be induced at 28°C by including ethanol (1–2%) or CO_2 (5%) in the axenic medium, or by monoxenic culture, whereas mitomycin C (an inhibitor of DNA synthesis) prevented the masculinizing effect of 30°C (Hansen *et al.*, 1973; Buecher *et al.*, 1974). Similar treatment of the Brownhill isolate at 30°C failed to induce maleness, although treatment at 32°C with 10% CO_2 did. Furthermore, ethanol at 2% was toxic to this isolate, but population increase was stimulated by growth under 10% CO_2 (Buecher and Hansen, 1975).

Following the observation that reproduction of *A. avenae* was reduced below 5% O_2 and inhibited by intermittent anaerobiosis (Cooper *et al.*, 1970), respiration rates were studied by Cooper and Van Gundy (1970). During starvation,

TABLE I

Characteristics of *Aphelenchus avenae* Populations

Population name	Main features	References
Brownhill (Australian)	Parthenogenetic, $n=8$; ♂ rare	Fisher (1972); Evans and Fisher (1970b)
Perth (Australian)	Amphimictic, $n=8$, ♂ = 50%	Fisher (1972); Evans and Fisher (1970b)
NC-32 (Californian)	Parthenogenetic; $n=9$; ♂ = 0% below 28°C, 100% at 30°C	Cooper and Van Gundy (1970); Hansen *et al.*, 1970; Fisher and Triantaphyllou (1976)

initial rates of respiration (5.6–5.8 μl O_2/mg/hr at 27°C) and glycogen content (8%) were maintained at the expense of stored neutral lipids. An oxygen debt was incurred during 16 hr of anaerobiosis during which glycogen was depleted. Longer periods of anaerobiosis progressively delayed the time at which the initial respiration rate was resumed until, after 120 hr or more, respiration immediately following anaerobiosis could not be detected and lipid utilization had ceased. After 24 hr of aeration, however, normal respiration was resumed and glycogen levels recovered. Survival after 20 days under anaerobic conditions was above 80%. Since reproduction of *A. avenae* was inhibited by intermittent anaerobiosis and was much reduced below 5% O_2 (Cooper *et al.*, 1970), it was interesting to observe that this was also the oxygen concentration below which aerobic respiration (with lipid catabolism) could not be sustained. During short periods (12–16 hr) of anaerobiosis, glycogen catabolism yielded lactic acid, but during longer periods, ethanol was produced (Cooper and Van Gundy, 1971a). There was also evidence that under normal aerobic conditions ethanol or propanol could be taken up and metabolized, having a lipid-sparing effect during starvation. Respiration was also studied by Marks (1971) in conjunction with nematicide effects on *A. avenae* (see below).

Respiration rate at 26°C was measured by Cartesian diver to estimate the calorific equivalent of energy respired per day (assuming 1 kcal = 4.825 liters O_2) during growth and development of individuals of *A. avenae* (de Soyza, 1973). This was used in the construction of an energy budget which, although imprecise, suggested that net production efficiency was greatest in the fourth larval stage.

When Arrhenius plots of respiration rates at different temperatures were made for *A. avenae* (NC-32), a decreased slope was observed above 20°C which was interpreted as a change in phase of polar lipids from the liquid–crystal to a more liquid state above 20°C (Lyons *et al.*, 1975). However, there were no ill effects on reproduction in this population, as was the case in *Caenorhabditis elegans*. Studies on British populations of *A. avenae* yield Arrhenius plots with breaks at 25°C, whereas an amphimictic isolate from Malawi had a nonintersecting discontinuity below 20°C (A. H. W. Mendis and A. A. F. Evans, unpublished results).

Little has yet been done to study enzymes in *A. avenae*. Polygalacturonase and cellulase activities were found in homogenates, but these enzymes were not released to an incubating medium (Barker, 1966). A number of dehydrogenases, esterases, phosphatases, and a peroxidase were demonstrated qualitatively using staining in gels following electrophoresis (Evans, 1971a,b; Dickson *et al.*, 1970, 1971). Enzymes involved in folate metabolism are present in the NC-32 strain (Platzer, 1974). Enzymes of the glycolytic and tricarboxylic acid (TCA) cycle pathways, together with the classical mammalian electron transport chain, have been found in mitochondria-rich fractions from *A. avenae* (Evans and Raison; A. A. F. Evans and M. Kondrollochis, unpublished results).

Serological studies with *A. avenae* showed it was only distantly related to *Panagrellus* and *Diplogaster* (El Sherif and Mai, 1968), but the presence in *A. avenae* of antigens cross-reacting with *Neoaplectana* and *Dictyocaulus* (Andreeva *et al.*, 1975) has led to speculation that vaccines from free-living nematodes may protect against animal parasitic nematodes (Ershov *et al.*, 1974; Busalev *et al.*, 1976).

Nematicide studies have frequently employed *A. avenae* as a convenient test organism (Castro and Thomason, 1971; Evans, 1973). The alkyl halide nematicides ethylene dibromide (EDB) and 1,2-dibromo-3-chloropropane readily moved into and out of nematodes, as did water molecules (Marks *et al.*, 1968); other large or charged organic molecules (glucose, glycine, sodium acetate) move more slowly (Castro and Thomason, 1973). These nematicides were concentrated within living nematodes at the expense of the external solution, but selective permeability was lost on death.

Such studies emphasized that water movement, with consequent change in size, is the primary event in osmoregulation (see also Myers, 1966; Wright and Newell, this volume, Chapter 6). Further studies showed that ethylene dibromide probably reacts with heme proteins of the respiratory pathway and with serine residues of an esterase or protease in causing the death of *A. avenae* (Castro and Belser, 1978). This conclusion is supported by earlier studies on intact adults and larvae showing that anaerobiosis gave some protection to intoxication by EDB in nonmolting stages (Evans and Thomason, 1971), but that nematodes in a molt were the most susceptible stage. However, respiration during intoxication was almost unchanged (Marks, 1971), and this, together with the marked cyanide insensitivity of respiration in some isolates of *A. avenae* suggests that an oxidase other than cytochrome $a + a_3$ may be present (A. A. F. Evans, unpublished results).

Organophosphate and carbamate nematicides have effects on *A. avenae*. Feeding and reproduction were disrupted by low concentrations of thionazin (Kondrollochis, 1972), and mortality to phorate was enhanced by anaerobiosis (cf. alkyl halides; Evans and Kondrollochis, 1972). Behavioral effects (abnormal stylet thrusting and bulb pumping, body coiling) were seen in aldicarb and phorate, but exposure for short periods is rarely lethal (Keetch, 1974). The toxicity of phorate was related to the ability of *A. avenae* to absorb more and metabolize less, and also to the production of metabolites more toxic than phorate (Le Patourel and Wright, 1976a) which were optically active (Le Patourel and Wright, 1976b). The toxicity of aldicarb (an oxime carbamate) was principally related to the accumulation of toxic metabolites (Batterby *et al.*, 1977). Because these compounds are active in nematodes, their site of action has been sought. Acetyl cholinesterase was demonstrated in the nerve ring (Wright and Awan, 1976), but other neurotransmitter substances are also present in the nervous system. Dopamine and dopamine decarboxylase have been demonstrated, the former in the nerve ring (Wright and Awan, 1978), and 5-hydroxytryptophan, epinephrine, and serotonin caused vulval contractions in *A. avenae* and other species (Croll, 1975).

Most data on the survival of *A. avenae* during desiccation have been anecdotal e.g., Bovien, 1937, but recent work (Crowe and Madin, 1974, 1975; Madin and Crowe, 1975; Crowe *et al.*, 1977) presages a new departure in this field which is discussed below (Section III).

B. Rhabditid Nematodes

Within the order Rhabditida are many species which feed predominantly on microorganisms that thrive on rapidly degradable substrates such as rotting wood, dung, fallen leaves, carrion, and decaying fruits. Because these substrates often last for only a few days or weeks, animals utilizing them have adopted a strategy of rapid exploitation of a transient resource in what is often a relatively unpredictable habitat and in which a short life cycle and high fecundity are useful features. High local densities are often reached in populations of such nematodes. Typically, all active stages are susceptible to unfavorable conditions, especially desiccation or temperature extremes, but in a few families a resistant larval stage, the *dauer larva,* may be formed which is more tolerant of unfavorable conditions, including desiccation.

Few studies have been made on dauer larvae, despite knowledge of their existence for many years (Maupas, 1899; Rogers and Sommerville, 1963), but they offer models for morphological, physiological, and biochemical studies on resistant stages, and for studies on control of development. A few of the better-known examples are discussed below and are later evaluated in comparison with other models.

1. Caenorhabditis elegans and C. briggsae

The dauer stage of *C. elegans,* described in detail by Cassada and Russell (1975), is formed as a facultative response to unfavorable culture conditions prior to the second molt. These larvae apparently continue in a state similar to the premolt lethargus, covered with a sheath of L2 cuticle, unable to feed and decreasingly active, but having formed a cuticle with a "striated" layer (Popham and Webster, 1978) which is not seen in normally developing stages. They can survive without food for more than 70 days (Klass and Hirsh, 1976), can tolerate high temperatures and anaerobiosis longer than adults (Anderson, 1978), are more resistant to chemicals and desiccation, and can wave most of the body in the air from projections on the substrate (Cassada and Russell, 1975). In axenic cultures of *C. briggsae,* dauer formation occurred despite adequate nutrition, suggesting that it was possibly a response to crowding or accumulating metabolic products (Yarwood and Hansen, 1969).

Normal development is resumed shortly after the return of favorable conditions. A convenient diagnostic test for dauer stage estimation is resistance to 1%

sodium dodecyl sulfate (Cassada and Russell, 1975). However, desiccation resistance has apparently not been investigated experimentally despite numerous qualitative observations, a surprising omission reminiscent of that noted by Rogers and Sommerville (1963) in work on infective larvae, the analogous stage in the life cycle of many parasitic nematodes. Glycerol was excreted by *C. briggsae* after incubation in a culture medium containing [^{14}C]acetate, whereas in water, glucose is the major product of acetate metabolism (Rothstein, 1972). Endogenous glycerol and small amounts of glucose and trehalose were found labeled after incubation of *C. briggsae* in a growth medium containing [2^{14}C]acetate (Liu and Rothstein, 1976).

2. Other Rhabditids

Other possible models may be found among less well-studied rhabditids. Bovien (1937) observed that the dauer stages of *Diplogaster stercorarius* found under the elytra of a dung beetle (*Aphodius* sp.) tended to float in water, could move on a dry slide without difficulty, and were resistant to heat. However, they were only moderately resistant to desiccation compared to *Anguina* and *Aphelenchus*, being capable of surviving 6 hr on a dry slide, but not longer. The dauer stages, when moving, seemed to leave minute oil drops behind them. *Mesodiplogaster lheritieri* dauer larvae also had oily cuticles (Grootaert, 1976). *Rhabditis dubia* and *Neoaplectana bibionis* have a facultative ability to form dauer larvae. These lasted 12 months in water and survived in moist soil but not in soil that had been allowed to dry out (Bovien, 1937). *Rhabditis maupasi* populations studied in forest soils had between 33 and 77% dauer larvae and 0 to 7% adults, and were of low density and patchy distribution (Sohlenius, 1973a).

Dauer larvae of *Pelodera strongyloides* withstood repeated exposure to 37°C, but adults could not (Yarwood and Hansen, 1968). They showed strong galvanotaxis at currents of 3 μA, compared to 32 μA for nondauers, and no response for adults (Whittaker, 1969). *Panagrellus* spp. have frequently been used as models in comparative studies with other rhabditid species, but only Lees (1953) studied survival in *P. silusiae* during desiccation. Rapid drying allowed only a few minutes of survival, but slower water loss while embedded in an agar gel allowed revival after 2 months. *Acrobeloides nanus* in declining cultures produced a higher proportion of short larvae that could resist adverse conditions (Sohlenius, 1973b).

C. Other Free-Living Nematodes

Plectus rhizophilus was nominated by Nielsen (1967) as the most drought-resistant free-living nematode known to him, surviving violent temperature fluctuations in its moss-cushion habitat. However, *Plectus* species, although distributed worldwide, are slow breeders [60 days, egg to egg, for *P. parietinus* (Maggenti, 1961)].

D. Comparison of Models

Survival abilities in desiccated nematodes range from years (L2 of *A. tritici*) to weeks (dauer larvae of rhabditids), but this contrasts with the ease of culture and generation time. For most of the year all the individuals comprising a population of *A. tritici* are in the cryptobiotic L2 stage, which is the bridge between one generation and the next. *Anguina* has therefore been under strong pressure during its evolution as a pathogen of grass and wheat in the "fertile crescent" of the Middle East to develop the greatest possible potential for survival. As a model it is well suited to studies on survival of larvae during, and recovering from, desiccation (Bhatt and Rohde, 1970; Womersley, 1978). It is less well suited to studies on larvae preparing to enter desiccation within developing galls, since these occur naturally only once per season and L2 stages are mixed with adults and younger stages. *Ditylenchus dipsaci* shows similar survival abilities to *A. tritici,* and the existence of races offers the chance of studying intraspecific variability.

Aphelenchus avenae is already confirmed as a versatile laboratory tool although it lacks some properties of an ideal model. Of these, perhaps the ability to create and manipulate mutants for genetic analysis is most critical. However, the existence worldwide of many naturally occurring populations differing in their biological, physiological, and ecological attributes, yet sharing many features in common, suggests their value for use in revealing the physiology and biochemistry of adaptations for survival in very different habitats. This is particularly relevant as a prospect for examining the physiology of desiccation survival (Section III).

Rhabditid nematodes have a limited ability to survive desiccation; even the dauer larvae, having a striated layer in the cuticle, like tylenchids, tolerate low relative humidities for only a few months. However, no serious study of the dauer larva has been initiated, and insights into control of dauer stage induction could come from the study of mutants showing aberrant behavior (Cassada and Russell, 1975).

Other models could be sought or exploited. For example, *Plectus rhizophilus* might have rather unique properties developed to a high degree, which will make it worth culturing despite difficulties.

III. PHYSIOLOGICAL ASPECTS OF DESICCATION SURVIVAL

A. Theoretical Considerations

The cells of an organism consist of a range of organic and inorganic chemicals in an aqueous environment that is highly organized and under close control to maintain homeostasis. Low molecular weight substrates for intermediary metabolism are generally kept at low concentration, many molecules are stored as polymers of carbohydrate, lipid, or protein, and most enzymes are bound to membranes—all features that avoid high osmotic pressures within cells (Hochachka and Somero, 1973; Atkinson, 1977). Similarly, inorganic ions may be transported into or out of cells to regulate their water content. Oxygen and carbon dioxide are dissolved in cell water, and toxic wastes are converted to soluble forms for excretion. Because of its central role, the nature of water within cells has been of interest and is now generally conceived of as existing in two different forms—free or bulk water and bound water. Bulk water represents the cytoplasmic solvent that can vary in quantity in many cells without causing damage. Bound water, by contrast, is vital to the maintenance of structural integrity of many cell components, e.g., the macromolecules of chromosomes and membranes, which are liable to denaturation in its absence (see Crowe and Clegg, 1973).

Lack of water therefore threatens normal metabolism, detoxification and excretion, and diffusion of oxygen and carbon dioxide, and leads to increased salt concentrations in cells. Ultimately, bound water may be removed from vital cell components and death ensues. It is not surprising, therefore, that some animals have various metabolic as well as structural adaptations that help to avoid or minimize the effects of severe desiccation.

B. Previous Work with Anhydrobiotes

Physiological and metabolic mechanisms have been studied most intensively in organisms other than nematodes; these are extensively discussed by Crowe and Clegg (1973) and Crowe and Madin (1974) and are only briefly considered here. Evans and Perry (1976) reviewed the information on nematodes.

Slow drying facilitates survival in tardigrades, rotifers, and nematodes, and the rate of water loss is controlled in high humidities. Some animals such as the *D. dipsaci* L4 stages can even do this at low humidities. Aggregations of animals survive better than individuals, but the ability to revive is decreased by the passage of time or by subjection to repeated dehydration–rehydration cycles. Cuticle permeability decreases with water loss, and during this period of slow drying morphological and physiological changes occur—e.g., tardigrades shrink and form a "tun" and nematodes coil into a spiral, shrinking the cuticle between the annulations. These changes cannot occur if the animal is anesthetized.

Desiccated brine shrimps (*Artemia* spp.) contain much glycerol which is suggested to (1) substitute for bound water in sensitive macromolecules; (2) inhibit free radical formation; and (3) protect reactive groups from proximity effects (such as cross-linking between adjacent parts of a dried protein molecule). Glycerol may also provide protection from damage during shrinkage because of its incompressibility.

Trehalose, a nonreducing disaccharide of glucose, is synthesized in response to desiccation in many anhydrobiotes, generally at the expense of glycogen. Reducing sugars react with amino groups of proteins, yielding "melanoidins" by the so-called browning reaction, so trehalose may represent a "safe," nonreactive, carbohydrate store.

Theoretical considerations and experimental work suggest that other chemicals may effectively substitute for bound water in desiccating cells. Inositol in its various forms has a suitable shape to replace bound water; it is nonvolatile and nontoxic, and was the best of a range of chemicals for protecting bacterial cells from the effects of desiccation (Webb, 1965). Glycerol was also protective at some relative humidities (RH) but polyhydroxy, straight-chain compounds (unlike hydroxycyclohexanes, with a ring structure) were toxic.

Lipids were implicated in desiccation survival by Van Gundy (1965) and more recently by Bird and Buttrose (1974), who suggested that lipid droplets within *A. tritici* larvae coalesced to form a water-impermeable sheath. The prime importance of lipids may, however, be their food-reserve function, providing energy for the metabolic preparations as desiccation commences, rather than their role in any specific mechanism involved in desiccation survival. Thus, more "opaque" females (those containing lipids and carbohydrates) of *Helicotylenchus dihystera* and *Scutellonema cavenessi* survived desiccation than did transparent females lacking lipids (Demeure *et al.*, 1978).

Prehydration of anhydrobiotes in high RH allows potentiation of dry tissues for rapid resumption of metabolism when free water is restored (e.g., Bhatt and Rohde, 1970), but ionic regulation in anhydrobiotes has not previously been studied.

C. Physiological Changes in Nematodes

1. Changes during Preparation for Desiccation

In old cultures of fungi on agar, Mankau and Mankau (1963) noticed adults and larvae of *A. avenae* coiling or clustering and becoming dormant. Townshend (1964) found many survivors after 12 months at 15°–20°C in dried fungal cultures on agar. The factors affecting induction of anhydrobiosis in *A. avenae* were studied by Crowe and Madin (1975). When aggregates of nematodes (principally L4 and adult stages) were dried at 97% RH for at least 72 hr, recovery after exposure to 0% RH exceeded 90%, provided the aggregates weighed at least 90 mg wet weight. Such pellets were kept for up to 18 months with little decrease in viability, and the nematodes had all coiled into tight spirals. Chemical analyses for lipid, glycogen, glycerol, trehalose, and glucose contents of slowly drying *A. avenae*, made on a dry weight basis at intervals up to 72 hr (Table II), showed that lipid content rapidly declined in the first 24 hr and had dropped to 60% of the

TABLE II

Analyses of Lipid and Carbohydrate before (B), during (D), and after (A) desiccation[a,b]

	Lipid	Glycogen	Glucose	Trehalose	Glycerol	Inositol		References
						Bound	Total	
Aphelenchus avenae								
B	120	60	10	20	0	—	—	Madin and Crowe (1975)
→ D [D]c	→ 80 [128]c	→ 10 [3]	→ 10 [3–8]	→ 120 [125]	→ 55 [40–50]	—	—	Crowe et al. (1977)
→ [A]c	[51]c	[40]	[3–8]	[35]	[0]	[0]	—	
Anguina tritici								
B	406	11.1	0.6–3.1	67–116	0.1–5.3	3.2–5.0	7.2–8.0	Womersley (1978)
D 24 hr	—	—	2.4–3.0	4.6–5.6	2.0–7.4	0.4–1.1	—	
↑ A 48 hr	—	—	0.7–2.7	2.9–3.8	—	0	—	
Ditylenchus dipsaci								
B	—	—	8.8–16.8	10.5–14.9	5.4–9.6	0	1.2–2.0	Womersley (1978)
D	383	—	1.3–4.4	24–65	0.5–5.2	0–0.7	2.1–3.4	
A	—	—	—	—	—	—	—	
Ditylenchus myceliophagus								
B	307	—	12.0–16.9	1.9–6.1	0–10.7	0–0.7	1.9–3.0	Womersley (1978)
Panagrellus redivivus								
B	231	252	12.0–64.8	3.1–9.1	10.1–81.0	0.3–1.6	2.3–2.9	Womersley (1978)
Turbatrix aceti								
B	219	227	17–80	5.7–9.0	3.0–4.9	0.7–1.5	2.3–3.5	Womersley (1978)

[a] Results are given in micrograms per milligram dry weight.

[b] From Crowe et al. (1977)—results after 24-hr revival—and Womersley (1978)—analyses for lipid and glycogen made on independent samples.

[c] All numbers in brackets are from Crowe et al. (1977).

205

original value by 72 hr (Madin and Crowe, 1975). Glycogen similarly dropped to only 10% of the original value, but glucose concentration was unchanged. Concomitantly, the trehalose content increased from 20 to 120 μg/mg dry weight, whereas glycerol, undetectable in the nematode before drying began, rose to 55 μg/mg dry weight. Survival in dry air was strongly correlated with both glycerol and trehalose concentrations.

Hydrated active *A. avenae* had adenosine triphosphate (ATP) concentrations of 2.23 μg/mg dry weight. After desiccation ATP rose to 4.59 μg/mg dry weight, indicating a higher energy charge in the anhydrobiotic state (Willett *et al.*, 1978).

A comparative analysis of carbohydrate and lipid contents of nematodes resistant and susceptible to desiccation revealed that *A. tritici* lost its small amount of glycerol when desiccated (Table II), but had far more bound and total inositol than other species (Womersley, 1978). Trehalose levels were also much higher than other species except *D. dipsaci,* and the glucose content was low. *Ditylenchus dipsaci* decreased the glucose and glycerol content upon desiccation, but total inositol increased. Inositol and glycerol levels in fresh *D. myceliophagus* were similar to *D. dipsaci,* but trehalose was lower. The other two free-living nematodes, *Panagrellus redivivus* and *Turbatrix aceti,* had a large glucose content, and the inositol content was again similar to desiccated *D. dipsaci.* Glycerol levels in *P. redivivus* were variable but sometimes exceeded the level in desiccated *A. avenae* (Table II). Ribitol, fructose, and sorbitol were also measured but were similarly sparse in all cases except for 5 μg/mg sorbitol in *T. aceti.* Incubation in 5% inositol solutions improved the survival of *A. tritici, D. dipsaci,* and *D. myceliophagus* during subsequent desiccation and revival (Womersley, 1979a).

2. Changes with Revival from Desiccation

The recovery of *A. avenae* was slowest in those batches stored longest at 0% RH, but it was improved by rehydration at high humidity (97% RH for 24 hr) before transfer to water (Crowe and Madin, 1975). Ability to survive reexposure to dry air was rapidly lost (50% survival after 22 min in water following 10 days at 0% RH). Reproduction following revival after 18 months of desiccation confirmed the viability of many individuals. Hydration began during storage at 97% RH, but uptake was very rapid from water (Crowe *et al.*, 1977). Within an hour the nematodes were almost completely rehydrated and metabolic changes were in progress. Glycerol was released to the water within 15 min, and concentrations within the worms dropped 90% in the first hour. Trehalose and lipid concentrations decreased, glucose remained constant, and glycogen gradually increased. Respiration reached a maximum rate of 25 μl/mg dry weight/hr about 1 hr after rehydration, slowing to 7.2–9.0 after 24 hr (Bhatt and Rohde, 1970).

Analyses of *A. tritici* larvae 24 hr after revival showed trehalose at only 5% of the level in desiccated larvae, and it decreased further by 48 hr (Womersley, 1978). Bound inositol also decreased, and had almost disappeared by 48 hr. Glucose remained unchanged and glycerol was slightly increased at 24 hr.

Concentrations of ATP in larvae of *A. tritici* from galls 3–5 years old were similar (about 1×10^{-10} g/larva). Within 40 min of revival the ATP concentration had doubled, but it declined again under starvation (Spurr, 1976).

Analyses of sodium, potassium, magnesium, and calcium during desiccation showed that salt loss through the cuticle was less than during subsequent revival in distilled water, and that only the loss of potassium, magnesium, and calcium was slightly regulated (Womersley, 1979b). (See also this volume, Chapter 6.)

D. Perspectives and Challenges

The chemical changes accompanying desiccation and revival in nematode anhydrobiotes help to account for the benefits of slow drying at high RH and perhaps offer an explanation of why nematodes such as *Haemonchus contortus* can dry as slowly as *D. dipsaci*, but nevertheless survive less well (Ellenby, 1969). Despite slow drying their internal chemistry may not be rearranged suitably for maximum protection from desiccation.

Results with nematodes show similarities with other model systems in that glycogen is decreased, trehalose is synthesized, and other hexoses are relatively unchanged. It is difficult to comment on the candidate chemicals for bound-water replacement before much more experimental work is done. Indeed, the distinction between free and bound water will remain only a conceptual one until the two types can be assayed, as has been attempted by Andronikashvili *et al.* (1969), and it can be demonstrated that bound water may be replaced. It seems probable that glycerol is synthesized *de novo* in *A. avenae*, possibly via a gluconeogenesis pathway such as that postulated for *C. elegans* (Liu and Rothstein, 1976), at the expense of neutral lipids. The origin of inositol in *A. tritici* and *D. dipsaci* is less certain. They may obtain much inositol through feeding on plant tissue, and it may not be entirely coincidental that seed-borne nematodes are anhydrobiotes to various degrees.

It is recognized that chemical analyses were performed on populations; analyses of individuals are as yet impossible for technical reasons. Thus, the causal connection between survival and concentrations of glycerol and trehalose has yet to be established.

Other theories regarding water-replacement chemicals require direct testing on the nematode models, especially those relating to the postulated survival-enhancing abilities of glycerol, which has an additional function as a cryoprotectant in some insects. Furthermore, regulation of inorganic ions and their interactions with water-replacement chemicals need to be investigated.

IV. CONCLUSIONS

Analytical work on nematode models has just begun, but it is already clear that some free-living species are useful for investigating anhydrobiosis. Desiccated *A. avenae* containing glycerol show a chemical disposition similar to *Artemia,* and they have, moreover, all the advantages of a convenient laboratory animal.

Larvae of *A. tritici,* on the other hand, apparently employ the most effective water-replacement chemical inositol (Webb, 1965). What is not yet clear is the comparative advantage of glycerol vis-à-vis inositol or other cyclitols and alditols as a replacement for bound water. It could be argued that if an "ideal" replacement for bound water does exist, then *A. tritici* larvae possess it; indeed, so vital is the L2 stage to the survival of this species that any extra metabolic cost to synthesize inositol (if necessary) would have to be met. However, only one of the many available populations of *A. avenae* has yet been analyzed, and its survival was tested for only 18 months. A physiological challenge similar to that met by *A. tritici* might be faced by *A. avenae* in desert soils, and their study might reveal populations where inositol is preferred over glycerol. In addition, populations of this species showing a range of survival abilities could be compared to discover those features of metabolism most closely related to survival.

Dauer larvae of rhabditid nematodes are clearly different and withstand desiccation for only relatively short periods. Methods now exist to examine why survival is not more prolonged.

Finally, an understanding of the means by which metabolism is controlled while entering anhydrobiosis has implications for economic nematology. Nematodes entering cryptobiosis also show increased resistance to pesticides, drugs, anaerobiosis, or starvation, suggesting that these properties may have a common metabolic basis (Evans and Perry, 1976). The power to manipulate the physiology of parasitic nematodes might allow them to be kept in a susceptible state in which drug action is most effective. Free-living nematodes offer models for investigating such possibilities.

REFERENCES

Atkinson, D. E. (1977). "Cellular Energy Metabolism and Its Regulation." Academic Press, New York.

Anderson, G. L. (1978). *Can J. Zool.* **56,** 1786–1791.

Andreeva, G. N., Ermolin, G. A., and Tarakanov, V. I. (1975). *Tr. Vses. Inst. Gel'mintol.* **22,** 3–7.

Andronikashvili, E. L., Mrevilishvili, G. M., and Privalov, P. L. (1969). *In* "Water in Biological Systems" (L. P. Kayushin, ed.). Plenum, New York.

Barker, K. R. (1966). *Proc. Helminthol. Soc. Wash.* **33,** 134–138.

Barker, K. R., and Darling, H. M. (1965). *Nematologica* **11,** 162–166.

Bastian, H. C. (1865). *Trans. Linn. Soc. London* **25,** 73–184.

Batterby, S., Le Patourel, G. N. J., and Wright, D. J. (1977). *Ann. Appl. Biol.* **86**, 69–76.
Bhatt, B. D., and Rohde, R. A. (1970). *J. Nematol.* **2**, 277–285.
Bird, A. F., and Buttrose, M. S. (1974). *J. Ultrastruct. Res.* **48**, 177–189.
Bosher, J. E., and McKeen, W. E. (1954). *Proc. Helminthol. Soc. Wash.* **21**, 113–117.
Bovien, P. (1937). *Vidensk. Medd. Dan. Naturhist. Foren. Khobenhavn* **101**, 1–114.
Buecher, E. J., and Hansen, E. L. (1975). *Nematologica* **20**, (1974), 371–372.
Buecher, E. J., Yarwood, E., and Hansen, E. L. (1974). *Proc. Soc. Exp. Biol. Med.* **146**, 299–301.
Busalev, V. M., Ershov, V. S., and Tarakanov, V. I. (1976). *Byull. Vses. Inst. Gel'mintol.* **17**, 26–29.
Cassada, R. C., and Russell, R. L. (1975). *Dev. Biol.* **46**, 326–342.
Castro, C. E., and Belser, N. O. (1978). *Nematologica* **24**, 37–44.
Castro, C. E., and Thomason, I. J. (1971). *In* "Plant Parasitic Nematodes" (B. M. Zuckerman, W. F. Mai and R. A. Rohde, eds.), Vol. 2, pp. 289–296. Academic Press, New York.
Castro, C. E., and Thomason, I. J. (1973). *Nematologica* **19**, 100–108.
Cayrol, J-C. (1964). *Nematologica* **10**, 361–368.
Chin, D. A., and Estey, R. H. (1966). *Phytoprotection* **47**, 66–72.
Christie, J. R., and Arndt, R. C. (1936). *Phytopathology* **26**, 698–701.
Cooper, A. F., and Van Gundy, S. D. (1970). *J. Nematol.* **2**, 305–315.
Cooper, A. F., and Van Gundy, S. D. (1971a). *J. Nematol.* **3**, 205–214.
Cooper, A. F., and Van Gundy, S. D. (1971b). *In* "Plant Parasitic Nematodes" (B. M. Zuckerman, W. F. Mai and R. A. Rohde, eds.) Vol. II, pp. 297–318. Academic Press, New York.
Cooper, A. F., Van Gundy, S. D., and Stolzy, L. H. (1970). *J. Nematol.* **2**, 182–188.
Croll, N. A. (1975). *Can. J. Zool.* **53**, 894–903.
Croll, N. A., Smith, J. M., and Zuckerman, B. M. (1977). *Exp. Aging Res.* **3**, 175–189.
Crowe, J. H., and Clegg, J. S. (1973). "Anhydrobiosis." Dowden, Huchinson and Ross, Inc., Stroudsburg, Pennsylvania.
Crowe, J. H., and Madin, K. A. C. (1974). *Trans. Am. Microsc. Soc.* **93**, 513–524.
Crowe, J. H., and Madin, K. A. C. (1975). *J. Exp. Zool.* **193**, 323–334.
Crowe, J. H., Madin, K. A. C., and Loomis, S. H. (1977). *J. Exp. Zool.* **201**, 57–63.
Dao D., F. (1970). *Meded. Landbouwhogesch. Wageningen.* **70**, 1–181.
Demeure, Y., Reversat, G., Van Gundy, S. D., and Freckman, D. W. (1978). *Nematropica* **8**, 7–8.
de Soyza, K. (1973). *Proc. Helminthol. Soc. Wash.* **40**, 1–10.
Dickson, D. W., Sasser, J. N., and Huisingh, D. (1970). *J. Nematol.* **2**, 286–293.
Dickson, D. W., Huisingh, D., and Sasser, J. N. (1971). *J. Nematol.* **3**, 1–16.
Ellenby, C. (1969). *Symp. Soc. Exp. Biol.* **23**, 83–97.
El Sherif, M., and Mai, W. F. (1968). *Nematologica* **14**, 593–595.
Ershov, V. S., Berezhko, V. K., Kashinsky, A. D., Naumycheva, M. I., and Nisenbaum, I. A. (1974). *In* "Parasitic Zoonoses" (E. J. L. Soulsby, ed.), pp. 343–348. Academic Press, New York.
Evans, A. A. F. (1970). *J. Nematol.* **2**, 99–100.
Evans, A. A. F. (1971a). *Int. J. Biochem.* **2**, 72–79.
Evans, A. A. F. (1971b). *Int. J. Biochem.* **2**, 262–264.
Evans, A. A. F. (1973). *Ann. Appl. Biol.* **75**, 469–473.
Evans, A. A. F., and Fisher, J. M. (1969). *Nematologica* **15**, 395–402.
Evans, A. A. F., and Fisher, J. M. (1970a). *Nematologica* **16**, 295–304.
Evans, A. A. F., and Fisher, J. M. (1970b). *Aust. J. Biol. Sci.* **23**, 507–509.
Evans, A. A. F., and Kondrollochis, M. (1972). *Int. Symp. Europ. Soc. Nematol., 11th,* Abstr.
Evans, A. A. F., and Perry, R. N. (1976). *In* "The Organization of Nematodes" (N. A. Croll, ed.), pp. 383–424. Academic Press, New York.
Evans, A. A. F., and Thomason, I. J. (1971). *Nematologica* **17**, 243–254.

Fisher, J. M. (1972). *Nematologica* **18**, 179–189.

Fisher, J. M. (1975). *Nematologica* **21**, 358–364.

Fisher, J. M., and Evans, A. A. F. (1967). *Nematologica* **13**, 425–428.

Fisher, J. M., and Triantaphyllou, A. C. (1976). *J. Nematol.* **8**, 248–255.

Gordon, H. (1973). *Adv. Vet. Sci. Comp. Med.* **17**, 395–437.

Grootaert, P. (1976). *Biol. Jahrb. Dodonaea.* **44**, 191–202.

Hansen, E., Buecher, E. J., and Evans, A. A. F. (1970). *Nematologica* **16**, 328–329.

Hansen, E. L., Buecher, E. J., and Yarwood, E. A. (1972). *Nematologica* **18**, 253–260.

Hansen, E. L., Buecher, E. J., and Yarwood, E. A. (1973). *Nematologica* **19**, 113–116.

Hechler, H. C. (1962). *Proc. Helminthol. Soc. Wash.* **29**, 162–167.

Hochachka, P. W., and Somero, G. N. (1973). "Strategies of Biochemical Adaptation." Saunders, Philadelphia, Pennsylvania.

Hooper, D. J. (1974). C.I.H. Descriptions of Plant-Parasitic Nematodes. Set 4, No. 50., 4pp. Commonwealth Inst. Helminthol., St. Albans, U.K.

Keetch, D. P. (1974). *Nematologica* **20**, 107–118.

Keilin, D. (1959). *Proc. R. Soc. London Ser B* **150**, 149–191.

Klass, M., and Hirsh, D. (1976). *Nature (London)* **260**, 523–525.

Klink, J. W., and Barker, K. R. (1968). *Phytopathology* **58**, 228–232.

Kondrollochis, M. (1972). *Int. Symp. Eur. Soc. Nematol., 11th,* Abstr.

Kondrollochis, M. (1977). *Nematologica* **23**, 260–263.

Lees, E. (1953). *J. Helminthol.* **27**, 95–103.

Le Patourel, G. N. J., and Wright, D. J. (1976a). *Pestic. Biochem. Physiol.* **6**, 296–305.

Le Patourel, G. N. J. and Wright, D. J. (1976b). *Comp. Biochem. Physiol. C* **53**, 73–74.

Limber, D. P. (1973). *Proc. Helminthol. Soc. Wash.* **40**, 272–274.

Liu, A., and Rothstein, M. (1976). *Comp. Biochem. Physiol. B* **54**, 233–238.

Lyons, J. M., Keith, A. D., and Thomason, I. J. (1975). *J. Nematol.* **7**, 98–104.

Madin, K. A. C., and Crowe, J. H. (1975). *J. Exp. Zool.* **193**, 335–342.

Maggenti, A. R. (1961). *Proc. Helminthol. Soc. Wash.* **28**, 118–130.

Mankau, R., and Mankau, S. K. (1963). *In* "Soil Organisms" (J. Doeksen and J. Van der Drift, eds.), pp. 271–280. North-Holland Publ., Amsterdam.

Mansingh, A. (1971). *Can. Entomol.* **103**, 983–1009.

Marks, C. F. (1971). *J. Nematol.* **3**, 113–118.

Marks, C. F., Thomason, I. J., and Castro, C. E. (1968). *Exp. Parasitol.* **22**, 321–337.

Maupas, E. (1899). *Arch. Zool. Exp. Gen.* **7**, 563–628.

Myers, R. F. (1966). *Nematologica* **12**, 579–586.

Nielsen, C. O. (1967). *In* "Soil Biology" (A. Burges and F. Raw, eds.), pp. 197–211. Academic Press, New York.

Otter, G. W. (1933). *Parasitology* **25**, 296–307.

Perry, R. N. (1977). *Parasitology* **74**, 139–148.

Platzer, E. G. (1974). *Comp. Biochem. Physiol. B* **49**, 3–13.

Popham, J. D., and Webster, J. M. (1978). *Can. J. Zool.* **56**, 1556–1563.

Rhoades, H. L., and Linford, M. B. (1959). *Plant Dis. Rep.* **43**, 323–328.

Rogers, W. P., and Sommerville, R. I. (1963). *Adv. Parasitol.* **1**, 109–177.

Rothstein, M. (1972). *In* "Biology of Nematodes: Current Studies" (Behme *et al.*, eds.), pp. 37–47. MSS Information Corp., New York.

Sohlenius, B. (1973a). *Pedobiologia* **13**, 368–375.

Sohlenius, B. (1973b). *Oikos* **24**, 64–72.

Southey, J. (1972). C.I.H. Descriptions of Plant-Parasitic Nematodes Set 1, No. 13., 4pp. Commonwealth Inst. Helminthol., St. Albans, U.K.

Spurr, H. W. (1976). *J. Nematol.* **8**, 152–158.

Tarakanov, V. I. (1975). *Mater. Nauchn Konf. Vses. Obs. Gel'mintol.* **27,** 152–161.

Thorne, G. (1961). "Principles of Nematology." McGraw-Hill, New York.

Townshend, J. L. (1964). *Can. J. Microbiol.* **10,** 727–737.

Van Gundy, S. D. (1965). *Annu. Rev. Phytopathol.* **3,** 43–68.

Viglierchio, D. R. (1971). *Nematologica* **17,** 386–392.

Webb, S. J. (1965). "Bound Water in Biological Integrity." Thomas, Springfield, Illinois.

Whittaker, F. W. (1969). *Proc. Helminthol. Soc. Wash.* **36,** 40–42.

Willett, J. D., Freckman, D. W., and Van Gundy, S. D. (1978). *J. Nematol.* **10,** 301–302.

Womersley, C. (1978). "Physiological and Biochemical Aspects of Anhydrobiosis in Some Plant-Parasitic and Free Living Nematodes." Ph.D. Thesis, The University of Newcastle upon Tyne, England.

Womersley, C. (1980a). *Ann. Appl. Biol.* (In Press).

Womersley, C. (1980b). *Parasitology.* (In Press).

Wright, D. J., and Awan, F. A. (1976). *Nematologica* **22,** 326–331.

Wright, D. J., and Awan, F. A. (1978). *J. Zool.* **185,** 477–489.

Yarwood, E. A., and Hansen, E. L. (1968). *J. Parasitol.* **54,** 133–136.

Yarwood, E. A., and Hansen, E. L. (1969). *J. Nematol.* **1,** 184–189.

9

The Nematode Cuticle and Its Surface

ALAN F. BIRD

C.S.I.R.O., Institute of Biological Resources
Division of Horticultural Research
Adelaide, 5001 South Australia

I. Introduction 213
II. Whole Cuticle Structure 214
    A. Variation in Cuticle Structure throughout the Nematoda 214
    B. Proposal for a Basic Nomenclature 216
III. The Epicuticle 219
IV. Dynamic State of the Epicuticle 225
    A. Changes in Ultrastructure 225
    B. Interaction with Microorganisms 227
V. Development of the Epicuticle 232
VI. Summary and Conclusions 233
    References 234

I. INTRODUCTION

A number of review articles on the nematode cuticle have been published in the last decade. The most recent of these are those of Bird (1976, 1979a) and Lee and Atkinson (1976). These authors, while emphasizing the complexity and variability of nematode cuticles, support the use of a simplified nomenclature of cuticle structure which divides the cuticle into three regions or zones—namely, cortical, median, and basal. It is obvious that many exceptions to this fundamental pattern occur, and I shall mention some of these below. However, I think that they are adaptations to survival in changing environments, particularly where parasitism is involved.

In particular, I propose to consider the structure and functions of the surface or epicuticle of the cortical zone, for it is here that reactions similar to those

NEMATODES AS BIOLOGICAL MODELS
VOLUME 2

occurring at cell surfaces and in cell membranes are thought to occur in a wide range of "helminth" organisms (Lumsden, 1975). At the moment, particularly for the Nematoda, these ideas require more experimental evidence to establish them as facts. However, the use of sensitive techniques currently employed by membrane physicists and chemists to isolate, label, analyze, measure, and observe interactions taking place in cell membranes have in many instances yet to be used on the nematode epicuticle. There is no doubt that the free-living bacterial-feeding nematodes such as those belonging to the genus *Caenorhabditis,* and in particular *C. elegans* (Riddle, 1978), are the experimental models of choice for this purpose.

II. WHOLE CUTICLE STRUCTURE

A. Variation in Cuticle Structure throughout the Nematoda

Nematodes exhibit great variability in cuticle morphology. This is often, but not always, associated with parasitism. Cuticles may range from being nonexistent, as in adult females of *Bradynema* (Riding, 1970) where the hypodermal membrane covers the surface of the nematode in the form of differently shaped microvilli, to being very thin in comparison to body diameter, as in the parasitic larvae of *Reesimermis nielseni* (Poinar and Hess, 1976) and *Romanomermis culicivorax* (Poinar and Hess, 1977), to being complex multilayered structures, as are found in many parasitic and some free-living forms (Bird, 1976).

In the former instances, the cuticles have been modified or completely reduced, thus permitting the uptake of nutrients from the hemocoel of their insect hosts. These nutrients have been shown by Rutherford *et al.* (1977), using radioisotope tracer techniques on parasitic larval stages of *Mermis nigrescens*, to include glucose and amino acids. The sites for glucose transport are thought to lie beneath the cuticle, whereas those for amino acids are thought to be located in the cuticle.

Morphological evidence for the movement of much larger molecules through the cuticles of the parasitic larvae of the insect-parasitic nematodes *Reesimermis nielseni* and *Romanomermis culicivorax* has been provided by Poinar and Hess (1976, 1977). In both these larvae the cuticle consists of three membranous structures overlying hypodermal cells whose outer surface is highly convoluted to form microvilli which are thought to have active absorptive surfaces. These membranous cuticular structures have pores in them with a diameter of 7–11 nm through which ferritin particles (4–5 nm in diameter) can pass.

At the other extreme, as may be expected, there is great variability in cuticular morphology ranging from the multilayered cuticles of nematodes parasitic in mammals to the multilayered cuticles of nematodes parasitic in insects. In the adult cuticles of *Strongylus equinus* and *Ascaris suum* (Bird, 1958; Bird and Deutsch, 1957), the cuticles are 50–80 μm thick. The rather complex and bizarre forms, such as the adult cuticles of *Nippostrongylus brasiliensis,* have a fluid-filled layer containing hemoglobin and esterase traversed by fibrils of collagen (Lee, 1965). The adult female cuticle of *Mermis nigrescens* (Lee, 1970) contains canals that run from the surface into the cuticle, and has two layers of giant fibers that spiral around the nematode, below which is a thick layer made up of a network of fibers. Another complex cuticle is that of *Acanthonchus duplicatus,* which contains canals that connect the hypodermis to the surface of the cuticle (Wright and Hope, 1968). Yet another is that of the marine nematode *Deontos-toma californicum* (Siddiqui and Viglierchio, 1977), which like *Mermis* contains canals that run from its surface into the cuticle, but which differs markedly from the cuticle of *Mermis* in many other respects.

Nematode cuticles that appear to vary considerably from the basic three-zoned pattern mentioned above are quite common and I am sure that many more will be described in future years. This is to be expected, because these structures reflect a host of adaptations that have evolved in these animals over millions of years in response to their various changing environments.

No doubt there are many chemical differences in these various structures, but only the cuticles from the larger ascarids have been dissected and analyzed (Bird, 1957; McBride and Harrington, 1967; Fujimoto and Kanaya, 1973). It seems that in these large ascarids there is some lipid on the surface, and at least two structural proteins have been identified. The first of these has a high proline and alanine content but is relatively low in glycine. It does not give the characteristic wide-angle X-ray diffraction pattern of collagen, and is not hydrolyzed by col-lagenase. It is located in the cortical zone and has been named "cuticulin" by Fujimoto and Kanaya (1973). The second, which generally is thought to be located in the fiber layers of the basal zone, is hydrolyzed by collagenase and does have the characteristic wide-angle X-ray diffraction pattern of collagen, although it differs from vertebrate collagen in a number of respects. The collagen subunits of this protein are thought to be covalently linked by disulfide cross-bridges, giving rise to a folded polypeptide chain.

The chemical analysis of cuticles of nematodes of microscopic size, such as free-living forms and the larval stages of parasitic ones, is fraught with the technical difficulties of isolating these structures from their owners in a state both free from underlying hypodermal debris and intact.

There must always be some concern that any physical or chemical method that is employed might also remove part of the cuticle itself. However, what has already been done for the egg shell can presumably also be done for the cuticle. The egg shells of *Meloidogyne incognita* obtained by ultrasonic disintegration have been compared both chemically and structurally with those from which larvae had hatched normally (Bird and McClure, 1976). They were found to be similar, although the outermost lipoprotein layer appeared to have been stripped away during the process of ultrasonic disintegration.

In our experiments the active larvae that had hatched were separated from the shells by allowing them to migrate through mesh of 20 μm pore size and in both cases shells were separated, after washing on sieves, by differential centrifugation using a potassium tartrate density gradient. Presumably, cuticles of free-living nematodes such as *C. elegans* could be obtained and compared at different stages of development using similar techniques.

B. Proposal for a Basic Nomenclature

There seems to be general agreement that the nematode cuticle either consists of or has evolved from a three-layered or zoned structure. The problem that arises and tends to lead to confusion is that different authors insist on giving different names to these zones. Provided that these synonyms can be equated there is no problem. In order to assist the reader, a diagram (Fig. 1) illustrating these structures and the various names that they have been called is included, with a plea to future authors not to invent any more!

The surface of the cuticle consists of a triple-layered structure common to all nematodes for which the term epicuticle is proposed (Lee and Atkinson, 1976; De Grisse, 1977). This is synonymous with external cortical zone and is of equivalent dimensions to what is known as the outer epicuticle in the generalized arthropod cuticle (Neville, 1975), which has a thickness of 17 nm. Neville (1975) points out that he uses the term outer epicuticle in preference to that of cuticulin to avoid confusion, because although this term has been used synonymously for the 17 nm outermost trilaminar membrane in the Insecta, it has also been used for impregnated and tanned lipid substances. Furthermore, as mentioned above, the term cuticulin has been used for a protein isolated from *Ascaris suum* cuticles. I do not think the terms exocuticle, mesocuticle, and endocuticle proposed by De Grisse (1977) for the remaining zones of the nematode cuticle are valid since they are commonly used for regions in the insect cuticle that differ markedly in morphology, dimensions, and chemical composition from nematode cuticles.

The next zone moving inward (Fig. 1) is the cortical zone, often referred to as the internal cortical layer and, in *Meloidogyne* (Johnson and Graham, 1976), as the outer median layer. The next is the median zone, which is known in *Meloidogyne* as the inner median layer. The innermost of the three zones is, of course, the basal zone, which lies just above the hypodermis.

Nematodes of the genus *Caenorhabditis* provide an excellent model biological system for studies on the structure and formation of the nematode cuticle. The structure of the adult cuticle (Zuckerman *et al.*, 1973) fits easily into the pattern of the generalized cuticle outlined above. The surface membrane (Fig. 2) is

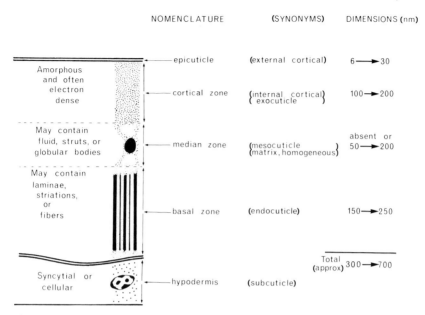

NOMENCLATURE	(SYNONYMS)	DIMENSIONS (nm)

Amorphous and often electron dense

epicuticle — (external cortical) — 6 ——► 30

cortical zone — (internal cortical / exocuticle) — 100 ——► 200

May contain fluid, struts, or globular bodies

median zone — (mesocuticle / matrix, homogeneous) — absent or 50 ——► 200

May contain laminae, striations, or fibers

basal zone — (endocuticle) — 150 ——► 250

Total (approx) 300 ——► 700

Syncytial or cellular

hypodermis — (subcuticle)

Fig. 1.   Diagram of a typical nematode cuticle of the type common to many free-living and larval forms, showing the relative positions of the epicuticle and the cortical, median, and basal zones or layers and some of the synonyms used for these areas.

equivalent to the epicuticle, the external and internal cortical layers to the cortical zone, the fluid-filled layer containing the supporting struts the median zone, and the fiber and basal layers jointly are equivalent to a common basal zone.

The altered cuticle of the dauer larva, a resistant larval stage (Fig. 3), resembles the cuticle of the infective stage of many nematodes in possessing a striated basal zone. However, the median zone depicted in the model of a typical infective larva (Bird, 1971) is missing. Thus the dauer of *C. elegans* has a cuticle that consists, from the outermost inward, of the following zones (Popham and Webster, 1978): the epicuticle, which is about 8 nm thick (outer cortical zone), a cortical zone which is about 84 nm thick (median zone), and a basal zone of variable thickness (basal zone).

The basal zone of the dauer larva of *C. elegans* contains a striated or striped layer about 84 nm thick, resembling that found in the cuticles of many infective larvae and thought to be made up of lattice of regularly arranged vertical rods or striations (Bird, 1971). Popham and Webster (1978) have provided evidence that supports their contention that these striations are the cut surfaces of a network of two sets of interconnecting sheets or laminae. They point out that in transverse and saggital sections every line has the same electron density, which is what one would expect if sheets or laminae were sectioned. Popham and Webster (1978),

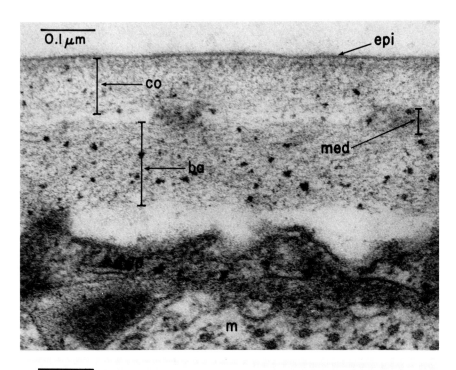

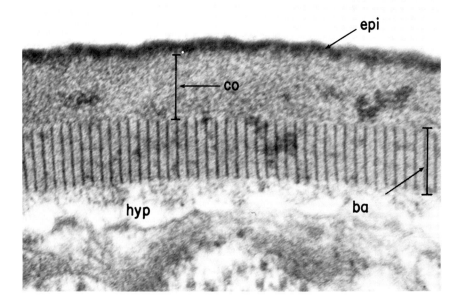

propose that to achieve this effect if rods were present it would be necessary for all of them to be cut in exactly the same manner. However, sections that show silver to gold interference colors are usually about 100 nm thick, so that were rods present, each section would include about five of them superimposed; thus, it is very doubtful if their electron density would vary from section to section if they were of similar thickness. By tilting oblique sections and recording and observing the same area at different axes and degrees of tilt away from the horizontal, it is possible to illustrate these longitudinal and circumferential laminae. Popham and Webster (1978) conclude that the striated layer of the cuticle of dauer larvae of *C. elegans* consists of blocks of polymerized protein approximately 84 nm long by 14.5 nm wide by 4.5 nm thick, separated by sheets of osmiophilic lipoprotein which intersect at right angles to each other. Presumably similar structures are found in the numerous other instances known throughout the Nematoda of striated basal layers with both inter- and intraspecific differences occurring in the dimensions of the protein blocks.

The most consistent structural pattern in the cuticle of nematodes occurs at its surface and this structure, which can most conveniently be called the epicuticle, will now be considered.

III. THE EPICUTICLE

The outermost triple-layered structure or epicuticle (external cortical layer) is an essential component of almost all nematode cuticles. Its thickness varies over a wide range both within and between different species (Bird, 1976). Different thicknesses reported by different workers for the same stage of the same species can be attributed to a variation in preparative procedures, such as fixation and staining on the one hand or failing to calibrate electron microscope magnification on the other.

Measurements of frozen sections of adjacent *Ascaris suum* cuticle have shown decreases in width of up to one-eighth because of fixation in formaldehyde. However, most preparative techniques for electron microscopy are so standardized at present that it is doubtful if much variation due to fixation occurs. Overstaining and the cutting of oblique sections certainly add to the difficulties in resolving the triple-layered structure of the epicuticle.

Fig. 2. (*top*)Transverse section of adult *Caenorhabditis elegans* cuticle (×150,000), showing epicuticle (epi), cortical zone (co), median zone (med), basal zone (ba), hypodermis (hyp), and somatic muscle (m).

Fig. 3. (*bottom*)Transverse section of dauer larva of *Caenorhabditis elegans* (×150,000), showing epicuticle (epi), cortical zone (co), striated basal zone (ba), and hypodermis (hyp).

For the most accurate measurements the electron microscope should be calibrated using a reliable diffraction grating replica. Differences of as much as 10% can occur between theoretical and actual magnifications, and would certainly account for some of the variations that have been described. My own measurements show that in a young adult female of *C. elegans* the epicuticle is about 6 nm thick (Fig. 2), whereas in the dauer larva of *C. elegans* it is about 25 nm thick (Fig. 3). These measurements for the *C. elegans* female epicuticle are similar to those for the young adult female of *C. briggsae* (Zuckerman *et al.*, 1973), and those for the dauer larva of *C. elegans* are similar to those shown in the photographs of this nematode published by Popham and Webster (1978). They, however, calculate the thickness to be about 8 nm instead of about 30 nm, as shown in their photographs.

These changes that can occur in the dimensions of the epicuticle within the same species, and indeed in some forms within the same stages of a species, will be discussed below.

The nematode epicuticle may thus have the same dimensions as the cell membrane in some forms, although its great morphological variability within the phylum could lead one to believe that there is no real analogy between these structures. Martinez-Palomo (1978), as a result of his studies on cuticle formation in the microfilaria of *Onchocerca volvulus*, states quite emphatically that "the term 'membrane' should not be used when referring to the outer laminae of the microfilarial cuticle." The term outer laminae is synonymous here with epicuticle. A strong point in favor of this statement is that membrane particles, which are exposed when the hydrophobic regions of the classic lipid membrane bilayer are cleaved by freeze–fracture replication, were not seen when the epicuticle of the microfilaria of *O. volvulus* was cleaved by the freeze–etch technique (Martinez-Palomo, 1978), although membrane particles were seen when the hypodermal cell membrane was cleaved.

A membrane may be defined as a flexible intracellular structure, composed primarily of lipids and proteins, which serves as a barrier to separate aqueous compartments with different solute composition and also serves as a structural base to which certain enzymes and transport systems are bound (Lehninger, 1975). This definition seems to preclude the nematode epicuticle from being termed a membrane, because it is difficult to regard it as an intracellular structure in comparison with normal cells where cytoplasm is adjacent to the cell membrane. However, in the case of some highly specialized cells such as nerve cells, the conducting axon is surrounded by spirals of cell membranes and the cytoplasm is withdrawn from between those spirals so that the cell membranes wind tightly against each other, forming a multimembrane sheath, the so-called myelin sheath, around the axon. Could the nematode epicuticle be a highly evolved and modified cell membrane? It is interesting to note that the myelin membrane differs from various other cell membranes that have been examined by the freeze–etch technique in that it is smooth and devoid of particles (Branton and Deamer, 1972). Thus the absence of these particles in freeze–fracture replicates of the nematode epicuticle does not necessarily exclude it from being a membrane.

The nematode cuticle functions in a manner akin to cell membranes in that it exhibits selective permeability to a variety of chemical compounds, and this selectivity is lost when these animals are killed (Marks *et al.*, 1968). Furthermore, as mentioned earlier, Rutherford *et al.* (1977) have obtained evidence for the membrane transport of amino acids in the cuticle of the parasitic larva of *Mermis nigrescens*. The epicuticle of the larva of the lungworm *Protostrongylus stilesi* is similar to that of the mammalian cell membrane in that its surface carbohydrates consist of sialic acids which bear carboxyl groups that are thought to be responsible for the net negative charge at the cuticular surface (Hudson and Kitts, 1971). However, Himmelhoch *et al.* (1977) and Himmelhoch and Zuckerman (1978) have not been able to detect either sialic acids or hyaluronic acid in the surface carbohydrate of *Caenorhabditis,* since treatment with neuraminidase or hyaluronidase did not reduce the binding of cationized ferritin. Similarly, the use of proteases has failed to indicate the presence of digestible glycoproteins on the cuticle surface. The failure of these enzymes to hydrolyze their respective substrates may be due to the presence of a surface coat or glycocalyx that protects the surface of the epicuticle.

Other studies have revealed that the nematode epicuticle resembles the cell membrane in a number of ways. First it is known, as a result of staining with ruthenium red (Cook and Stoddart, 1973), that the cell membrane has a concentration of acid mucopolysaccharides at its outer surface; similar results have been obtained with *C. briggsae* (Himmelhoch and Zuckerman, 1978). Second, the cell membrane has a net negative surface charge, and a similar physiological state exists at the surface of the epicuticle of both *C. briggsae* and *C. elegans* (Himmelhoch *et al.*, 1977; Himmelhoch and Zuckerman, 1978; Zuckerman *et al.*, 1979). This net negative surface charge can be demonstrated with the aid of the electron microscope using cationized ferritin (Figs. 4 and 5). These photographs show the surface coat (glycocalyx) that lies between the surface of the epicuticle and the ferritin particles (Fig. 4) and the very dense packing of the negatively charged surface molecules (Fig. 5). This packing is considered to contrast with less dense packing of surface molecules in similarly treated cell membranes (Zuckerman *et al.*, 1979), although it is pointed out that a pattern similar to that found on the surface of *C. briggsae* epicuticle has been found on the surface of the protozoan *Leishmania donovani* (Dwyer, 1977), so that there is obviously some variation in ferritin-labeled patterns on the surface of different cell membranes.

Further support for the hypothesis that the epicuticle, or part of it, has some relationship with cell membranes can be seen in electron micrographs of sections

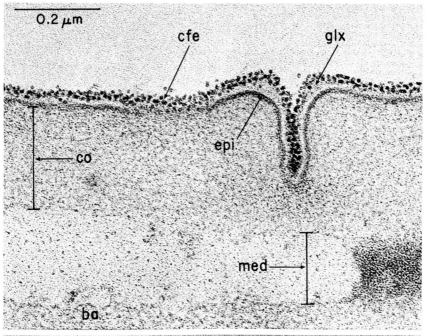

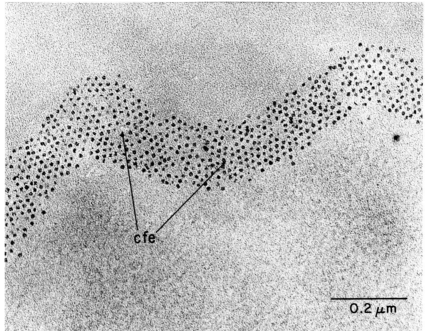

cut through cuticular invaginations such as at the anus, the oral opening, excretory pore, cuticular invagination for sensory structures such as amphids, or where the ducts of esophageal glands enter the cuticle-lined part of the digestive tract. Wright (1976) has examined these cuticular invaginations in the head regions of various nematodes in some detail. He states, "Hypodermis, continuous with that of the body wall, is a major component of the head tissues." This hypodermis shows structural and functional differentiation to cope with various specialized demands.

In many forms there is a distinct discontinuous and abrupt contact between the body wall cuticle and the esophageal cuticle. These two types of cuticles have distinctly different structures, and their epicuticles do not appear to have been examined at high resolution.

Cuticular invaginations at the vulva, excretory pore, and anus are easier to study than in the more specialized and structurally complicated cephalic region. The cuticle forms the walls of the excretory bladder in *Syphacia obvelata* (Dick and Wright, 1974) and is continuous with the epicuticle of the body wall cuticle. About half-way down the excretory sinus the epicuticle becomes discontinuous and ends. Whether or not the epicuticle merges into a cell membrane lining the rest of the excretory sinus is not clear. Similarly, the cuticle of the vulva and vagina vera of *S. obvelata* is formed by the invagination of the body wall cuticle where the basal and median zones stop at the margin of the pore, while the epicuticle continues inward and lines the vagina vera (Dick and Wright, 1974). The anal and excretory vesicles of the microfilaria of *Cardianema* sp, are partially lined by epicuticle (Johnston and Stehbens, 1973), but it is not clear whether there is any fusion with cell membranes as these invaginations extend inward. These authors do state, however, that during cuticle formation "the plasma membrane of the hypodermal cells was not seen to contribute directly to the cortex."

The anal invagination in the second-stage larvae (L2) of *Meloidogyne javanica* (Bird 1979b) does not show any distinct discontinuity between the body wall epicuticle and that lining the rectum (Figs. 6A–6D), although the other layers of the cuticles are much reduced or absent. These four photographs, all at the same magnification, show a sequence running from the rectal invagination close to the anus (Fig. 6A) to the cell membrane lining the rectal gland (Fig. 6D). The epicuticle of the L2 of *M. javanica* is about 35 nm thick (Fig. 6A) and gradually

---

Fig. 4. (*top*) Longitudinal section of the cuticle of adult *Caenorhabditis briggsae* after labeling with cationized ferritin (×120,000), showing ferritin particles (cfe), glycocalyx (glx), epicuticle (epi), cortical zone (co), median zone (med), and part of basal zone (ba). (From Himmelhoch and Zuckerman, 1978, reproduced by permission.)

Fig. 5. (*bottom*) Tangential section through the surface cuticle of adult *Caenorhabditis briggsae* after labeling with cationized ferritin (×100,000), showing densely packed ferritin particles (cfe). (From Zuckerman *et al.*, 1979, reproduced by permission.)

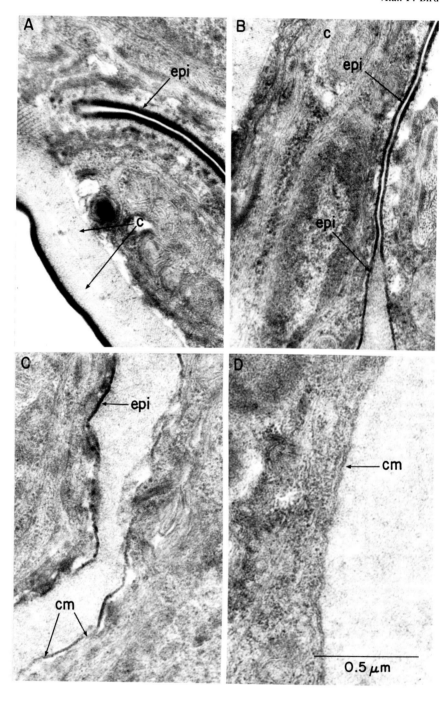

reduces in thickness as it moves inward (Figs. 6B, C), until it finally merges imperceptibly into the cell membrane lining the lumen of the rectal gland (Figs. 6C, D). In this nematode the rectal gland appears to replace the posterior gut, and there is no intestine with microvilli as in the genus *Caenorhabditis*.

The idea that perhaps the epicuticle is a specialized and highly modified type of cell membrane is at variance with the accepted picture of cellular topography. However, it does exist in a dynamic state (see below), which suggests a very close relationship with the underlying hypodermis. Extensive folding of the epicuticle of the female of *Onchocerca volvulus,* a tissue-dwelling parasite of man, which has a well-defined cuticle and an intestine with microvilli, has prompted workers studying these structures (Deas *et al.*, 1974) to state that "it is probable that a type of exchange with the external environment exists and the membrane folds described here serve as a surface amplification for such a system." The epicuticle apparently does not behave like the cell membrane in some instances. Gutman and Mitchell (1977) have shown, by a fluorescent-labeled protein that reacts with phosphorylcholine, that the internal cell membranes contain this compound. The epicuticle, on the other hand, does not appear to contain phosphorylcholine. However, this apparent difference in chemical composition may only be because the glycocalyx is acting as a barrier to the fluorescent label.

IV. DYNAMIC STATE OF THE EPICUTICLE

A. Changes in Ultrastructure

Remarkable changes in the ability of some species of nematodes to withstand environmental extremes of temperature and dehydration are linked, in the few instances where they have been examined, with morphological changes in the epicuticle (Bird and Buttrose, 1974). I suspect that these changes have not been detected more often because the required resolution of the electron microscope has not been utilized.

The second-stage larvae (L2) of the genus *Anguina* are able to survive considerable environmental extremes when they enter an anhydrobiotic state. In this dry state the L2's of *Anguina tritici* assume a characteristic coiled shape and become

Fig. 6. Saggital sections through the rectum of the L2 of *Meloidogyne javanica,* showing apparent continuity of the epicuticle with the cell membrane lining the rectal gland (×55,000). (A) Note the cuticle (c) and the epicuticle (epi) both on the surface of the nematode and lining the rectum. (B) Part of the cuticle (c) and epicuticle (epi) lining the rectum and part of the rectal gland. (C) Epicuticle (epi) lining the rectal gland and appearing to fuse with the cell membrane of the rectal gland (cm). (D) Further back from the rectum, showing the cell membrane (cm) lining the rectal gland.

closely applied to each other with their cuticles in actual contact. The surface of the cuticles of these anhydrobiotic L2's is both structurally and physiologically different from that of the hydrated and active L2's (Bird and Buttrose, 1974). These anhydrobiotic L2's can remain viable even after short exposures to extremes of temperature from as low as $-190°$ to as high as $105°C$. When hydrated, of course, these L2's cannot withstand these extremes, and one of several morphological changes that take place in them is a change in the dimensions of the surface membranes. High-resolution electron micrographs of sections cut through the cuticle surface show that the osmiophilic outermost layer of the active hydrated L2 is about half the thickness of that of the anhydrobiotic L2, whereas the innermost of these osmiophilic layers is thicker and more densely stained in the hydrated L2 (Bird and Buttrose, 1974). Survival in some nematodes is achieved by the development of dauer larvae, after the German word *dauer,* meaning continuance or *dauerhaft,* meaning durable. In the case of *C. elegans,* the dauer larva is an L2, whereas in *Bursaphelenchus lignicolus,* the Japanese pinewood nematode, the dauer larvae are L3's and L4's. These forms are produced under stress, for example, when cultures start to dry out or when the food supply becomes limited.

Again, there are both structural and physiological differences between the dauer larvae and the equivalent normal stage, and again high-resolution electron micrographs of sections cut through L4 dauer larvae and normal L4's of *B. lignicolus* show a similar twofold increase in the thickness of the outermost osmiophilic layer of the cuticle of the more resistant dauer larva compared with that of the normal larva. A similar change in surface cuticular morphology between dauer larvae and normal larvae in *C. elegans* has been reported by Cassada and Russell (1975), whose electron micrographs show that the epicuticle (outer cortical layer) of the dauer larva is thicker than that of the normal L3. The principal difference detected in cuticle structure between these two forms was that the dauer larvae had a striated basal layer whereas normal larvae did not. Cassada and Russell (1975) found that the dauer larvae of *C. elegans* were more resistant than normal larvae to a wide range of chemicals including fixatives, detergents, and anesthetics. Their resistance and survival in 1% sodium dodecyl sulfate (SDS) compared with normal larvae has been used as a method for isolating dauer larvae of *C. elegans* from the other stages. It is interesting to note that survival in 1% SDS, a chemical used for solubilizing cell membranes, is linked with a threefold thickening of the epicuticle. Kondo and Ishibashi (1978) report that the epicuticle (external cortical layer) is thicker in the dispersive (dauer) forms of *B. lignicolus* than in the propagative (normal) forms. They also state that the cuticle of the dauer form is less permeable to glutaraldehyde than that of the normal propagative form. Unlike *C. elegans,* all the stages of *B. lignicolus* have striated basal layers, although these vary in thickness in the different stages (Kondo and Ishibashi, 1978).

The changes of the fine structure of cuticles and their surfaces, which are reflected in changes in the physiology and behavior of these organisms, clearly indicates the dynamic state of the cuticle surface. It is clear that the cuticle is not just an inert exoskeleton, but that changes are constantly taking place in response to environmental fluctuations.

B. Interaction with Microorganisms

A number of interactions between bacteria or viruses and the surface of the nematode cuticle have been reported. In some instances these bacterial interactions are pathological and lead to the formation of cuticular lesions, as has been reported for the genus *Ascaris* (Weinberg and Keilin, 1912; Manter, 1929; Lubinsky, 1931; Stewart and Godwin, 1963; McKinnon and Lubinsky, 1966; Anderson *et al.*, 1971), for *Stephanurus dentatus* (Anderson *et al.*, 1973) and for *Strongylus edentatus* (Anderson *et al.*, 1978). All these nematodes are parasitic in mammals, and the bacteria responsible for the cuticular lesions are of interest not only because of their pathological effect on the nematode, but also because the nematode may act as a vector and carry certain bacteria to various organs. In the case of *S. dentatus*, the swine kidney worm (Anderson *et al.*, 1973), the nematodes form cysts in the ureter of their host. These cysts contain bacterial genera such as *Escherichia, Enterobacter,* and *Streptococcus*, which are also located on the cuticle of the nematode. Some bacteria were detected associated with a circular "knobby" cuticular lesion on this nematode. These bacteria were not identified, although their morphology under the scanning electron microscope was similar to that of coliform organisms obtained from suspensions of healthy cuticles.

Lesions on the surface of ascarids have been observed since 1912, when Weinberg and Keilin attributed their formation on *Parascaris equorum* to "large cocci." Later Manter (1929) described a similar cuticular disease in *A. suum*, when the surface of the nematode was covered with hundreds of yellowish spots. This disease was thought to be caused by a bacterium resembling *Clostridium welchii*. Two years later similar lesions were also observed on these ascarids and on the following genera: *Heterakis, Oxyuris, Toxascaris,* and *Strongylus* (Lubinsky, 1931).

Subsequently, the bacteria *Escherichia coli* and *Pseudomonas* sp. and a yeast *Candida* sp. were isolated from cuticular lesions on swine ascarids (Stewart and Godwin, 1963). These workers concluded that only the pseudomonad was capable of inducing lesions on the cuticles of healthy worms *in vitro*. More recently a light and electron microscope study of cuticular lesions on *Ascaris suum* (Anderson *et al.*, 1971) has shown that there are at least two morphologically distinct types of lesions. The first type are "small, round, and discrete with white, brownish, or yellowish coloration." This type of lesion is often found scattered

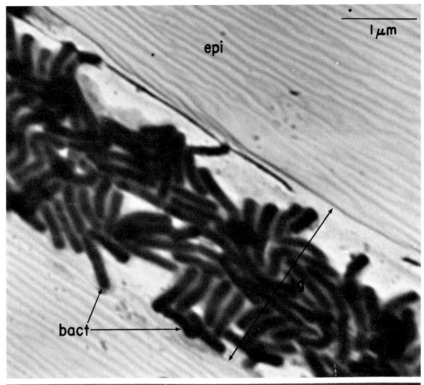

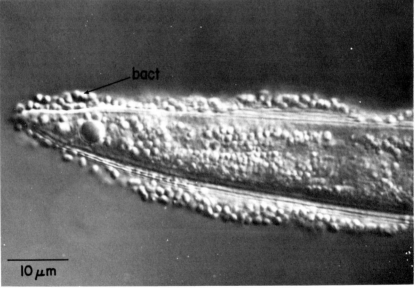

over the whole cuticular surface, in contrast to the second type, which is not widely distributed and is irregular and elongated along the circumference of the worm. Anderson *et al.* (1971) showed that both types of lesions and also healthy cuticle contained *Escherichia* sp., *Aerobacter* sp. and *Streptococcus* sp. A *Candida* sp. yeast was isolated only from the second and larger type of lesion described above. Even the smallest of these lesions have a marked effect on cuticular morphology and are colonized by numerous bacteria, but the larger lesions are associated with the destruction of the cortical zones, the formation of cavities in the median (homogeneous) zones, and hypertrophy of the innermost basal (fibrous) zone. Anderson *et al.* (1971) consider that when cuticle integrity is disrupted to this extent the effect of these lesions might ultimately be fatal.

The most recent observations on nematode cuticular lesions have been made on *Strongylus edentatus* from the horse (Anderson *et al.,* 1978). These workers have detected four distinct morphological types of lesions on the surface of this nematode: filamentous, flat, cratered, and proliferate. The first two types of lesions contained *Enterobacter aerogenes, Escherichia coli, Micrococcus* sp., *Streptococcus* sp., and *S. faecalis. S. faecalis* is found in all four lesions and is not found on the surface of healthy cuticles of this nematode, although all the other bacteria except *Proteus vulgaris* are found on healthy cuticle. However, *S. faecalis* is found on both healthy and lesioned cuticle surfaces in the male copulatory debris around the genital pore of the female nematode. Thus, Anderson *et al.* (1978) speculate that "at least some lesions of helminth cuticles may be a 'venereal disease'."

Descriptions of cuticular infections of nematodes by microorganisms are not restricted to nematodes parasitic in mammals, since a cuticle infection of *Thelastoma pterygoton* parasitic in beetles has been described by Poinar (1973). It is clear that the surface cuticle of many different species of nematodes may be inhabited by a wide range of bacteria. Many of these are not in the least pathogenic to the nematode and indeed may be of benefit. The precise nature of the chemical and physical forces that maintain this attraction between microorganism and the surface of the nematode cuticle is not known. These forces may be localized, as in the case of a pyocyanine-producing *Pseudomonas* sp. on the surface cuticle of *Ascaris suum* (Bird and Deutsch, 1957), where the bacteria are found in the transverse grooves (Fig. 7), or they may be distributed evenly throughout the entire surface, as in the case of *Corynebacterium* sp. on the cuticle of *Anguina* sp. (Fig. 8). This adhesion of the *Corynebacterium* sp. to the surface of the nematode cuticle appears to involve an attractive force of consider-

Fig. 7.   (*top*) Electron micrograph of the surface of *Ascaris suum* epicuticle (epi), showing bacteria (bact) in a transverse groove (tg) ($\times 20,000$).

Fig. 8.   (*bottom*) Nomarski interference contrast photomicrograph of the tail region of a L2 of *Anguina* sp. ($\times 1600$), showing several layers of bacteria (bact) adhering to the cuticle surface.

able magnitude because several layers of the bacterium, which project some distance into the medium in which the nematode moves, are not dislodged by this movement (Bird and Stynes, 1977). This force appears to be somewhat specific, because other genera of bacteria such as *Escherichia coli* do not adhere to the nematode cuticle, and there is some indication that a particular species of *Anguina* associates with a particular species of *Corynebacterium*.

The specific attraction between viruses and the surface cuticle of nematodes has been beautifully illustrated by Taylor and Robertson (1969, 1970a, 1970b), who showed that certain types of viruses become associated with the cuticular lining of distinct regions of their vector's anterior alimentary tract (Figs. 9 and 10). We do not know what forces are involved in maintaining or disrupting this close contact between the surface of a microorganism or a virus and that of a nematode. In the case of viruses, if only these two surfaces are involved, it has been suggested (Harrison *et al.*, 1974) that it may be either a question of adsorption and elution or the process may be more complex and involve the steric properties of the two surfaces. It is possible that other components that are secreted onto the surface of the cuticle may be involved, and it has been pointed out (Harrison *et al.*, 1974) that virus particles are separated from the surface of the esophagus by an electron-translucent layer thought to be mucus. Another hypothesis (C. E. Taylor, personal communication) is that the release of virus particles in the nematode is mediated by ionic changes created by the passage of saliva when the nematode feeds. These viruses do not appear to persist in the nematode vector after a molt and are not found in its eggs (Taylor and Robertson, 1975). It is thought that they are ''probably extracellularly attached to some part of the feeding apparatus, which is shed during a molt.''

Another interesting interaction between the nematode cuticle and a virus is that in which a cytoplasmic polyhedral virus is reported to be the causative agent in a swarming disease of the nematode *Tylenchorhynchus martini* (Ibrahim *et al.*, 1978). This swarming is thought to result from stickiness of the cuticle (Hollis, 1962) brought about by swelling and dissolution of the cuticle at its surface, and it is associated with the production of large numbers of cuticular projections (Ibrahim, 1967; Ibrahim and Hollis, 1973).

In *T. martini* viruslike inclusion bodies have been observed both on the cuticular surface and within the swarming nematode's organs, but not on normal nonswarming nematodes (Ibrahim *et al.*, 1978). These authors suggest that these inclusion bodies are similar to cytoplasmic polyhedral viruses that cause disease

Fig. 9. (*top*) Transverse section through the esophagus (es) of *Paratrichodorus pachydermus* (×43,000), showing particles of tobacco rattle virus (trv) in the lumen attached to the cuticle. (Courtesy of Dr. C. E. Taylor.)

Fig. 10. (*bottom*) Transverse section through the odontophore (od) of *Xiphinema diversicaudatum* (×95,000), showing particles of strawberry latent ring spot virus (slrsv) attached to the lumen (1). (Courtesy of Dr. C. E. Taylor.)

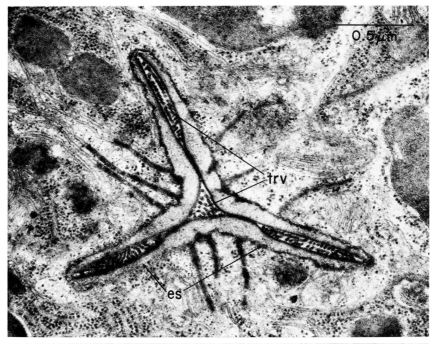

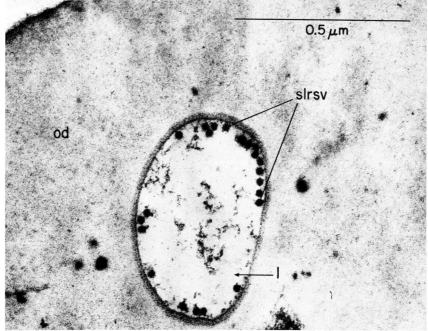

in insects. However, so far they have not been able to transmit these "viruses" from a swarming nematode to a nonswarmer.

Swarming seems to be a widespread phenomenon in the Nematoda. Ibrahim *et al.* (1978) list at least 17 species of free-living and plant-parasitic nematodes in which it has been reported, and the list is bound to increase. It remains to be seen whether or not all these changes in the surface cuticle are manifestations of viruslike diseases or whether they are brought about by changes in the nematode's environment that lead to cuticular changes that have survival value and are a reflection of the dynamic state of the cuticle surface.

V. DEVELOPMENT OF THE EPICUTICLE

Cuticle formation first takes place in the L1 in the egg. It is extremely difficult to obtain a satisfactory sequence of photographs illustrating the formation of this first larval cuticle, particularly when the hypodermal cell membrane is indistinct, as it is in the early stages of Martinez-Palomo's (1978) photographs of successive developmental sequences of *Onchocerca volvulus* microfilarial cuticle formation. I have experienced this trouble with a developing sequence of cuticle formation of the L1 in the egg of *Meloidogyne javanica*.

I am not completely convinced that the cuticle of the L1 is formed externally to the cell membrane of the hypodermis, although there seems little doubt from the work of Bonner and Weinstein (1972a), Dick and Wright (1973), and Singh and Sulston (1978) on different nematodes that in the third and fourth molts the cuticle is formed externally to the cell membrane of the hypodermis. It has been stated by Bonner and Weinstein (1972a) that the problems associated with interpreting changes in ultrastructure during cuticle formation are "compounded by the presence of an overlying old cuticle where it is often not possible to distinguish areas of new cuticle from areas of old cuticle." It was felt by these workers that studies on the formation of cuticle in the L1 during embryogenesis might overcome these problems. Unfortunately, this has not proved to be such an easy task because of the difficulties in obtaining well-fixed, embedded, and stained material from within nematode eggs. Recently, better preservation of *C. elegans* eggs has been obtained by increasing the temperature of the initial fixative to 40°C and shortening the subsequent dehydration in ethanol so that each step takes only 1–2 min (Krieg *et al.*, 1978). *Caenorhabditis elegans* has proved to be a useful model in helping to understand more about cuticle formation and molting. The great experimental advantage here is that this nematode has become so malleable in the hands of geneticists and molecular biologists that often a particular mutation is tailored to fit a particular research requirement, as in the routine use of shell-less eggs. Moreover, a single specimen may be studied throughout its life cycle. By means of Nomarski interference contrast microscopy and laser beam microsurgery Singh and Sulston (1978) studied molting in *C. elegans* and have shown the following: (1) The excretory system is not essential for molting since destruction of the excretory glands or the entire excretory system does not prevent molting. (2) The cytoplasm of the lateral median portion of the hypodermis, called by Singh and Sulston (1978) the seam cell, becomes granular because of the formation of densely packed Golgi bodies just prior to ecdysis. A similar accumulation of Golgi bodies in this region prior to the third molt in *Nippostrongylus brasiliensis* has been detected by Bonner and Weinstein (1972b). Golgi bodies are also associated with the secretion of copious amounts of a protein carbohydrate complex from the rectal gland cells of adult females of *M. javanica* (Dropkin and Bird, 1978). (3) The pharyngeal (esophageal) glands are very active just before molting and hatching, and Singh and Sulston (1978) propose that "their secretions soften and loosen the pharyngeal lining, the cuticle around the head, and perhaps the egg shell." The role of the subventral esophageal glands of the L2 of *M. javanica* in both hatching from the egg and subsequent penetration of its host-plant's root has been examined at greater length with the aid of light and electron microscopes and was found to be similar (Bird, 1967, 1968). It may be that these glands have important roles in molting, hatching, cuticle formation, and development throughout the Nematoda.

VI. SUMMARY AND CONCLUSIONS

Although there is enormous variability in the structure of cuticles throughout the Nematoda, there is nevertheless a basic pattern consisting of the epicuticle and the cortical, median, and basal layers. The surface layer or epicuticle (external cortical layer) is a trilaminate structure that varies considerably in thickness and is usually between 6 and 30 nm.

Caenorhabditis is a most suitable model for studies on these structures for the variety of reasons mentioned earlier. In this genus the epicuticle ranges from 6 to 8 nm in normal stages and from 25 to 30 nm in the more resistant dauer stage. It is postulated that the 6 to 8 nm epicuticle is a form of cell membrane; it resembles in many respects the surface structures of various other "helminth" phyla considered by Lumsden (1975) to have a great deal in common with cell membranes, both morphologically and physiologically.

The epicuticle of nematodes, like cell membranes, exhibits selective permeability, is thought to be involved in membrane transport of amino acids, has a net negative surface charge, has a concentration of acid mucopolysaccharides at its surface, can undergo structural changes between molts, and in some instances appears to merge with cell membranes where cuticular invaginations occur. I suggest that the epicuticle of nematodes originates from a cell membrane. In the course of evolution this membrane has become highly modified as the outermost part of a protective exoskeleton far removed from the cell membrane of the hypodermis which replaces it. The epicuticle's original relationship to this structure may be apparent only at the first molt, with subsequent molts reflecting its highly specialized extracellular functions. Needless to say, experimental work on intact and isolated cuticles and on cuticle formation in the embryo, such as is currently being undertaken in a number of laboratories with *C. elegans,* is needed in order to establish or disprove this hypothesis.

ACKNOWLEDGMENTS

I should like to thank B. M. Zuckerman, Pergamon Press, and C. E. Taylor for kindly permitting me to use the photographs in Figs. 4 and 5, 6, and 9 and 10, respectively.

I am most grateful to S. D. Harris for her care in preparing the figures and to my colleagues in C.S.I.R.O. for their constructive comments on the manuscript.

REFERENCES

Anderson, W. R., Madden, P. A., and Tromba, F. G. (1971). *J. Parasitol.* **57,** 1010–1014.
Anderson, W. R., Tromba, F. G., Thompson, D. E., and Madden, P. A. (1973). *J. Parasitol.* **59,** 765–769.
Anderson, W. R., Madden, P. A., and Colglazier, M. L. (1978). *Proc. Helminthol. Soc. Wash.* **45,** 219–225.
Bird, A. F. (1957). *Exp. Parasitol.* **6,** 383–403.
Bird, A. F. (1958). *Parasitol.* **48,** 32–37.
Bird, A. F. (1967). *J. Parasitol.* **53,** 768–776.
Bird, A. F. (1968). *J. Parasitol.* **54,** 475–489.
Bird, A. F. (1971). "The Structure of Nematodes." Academic Press, New York.
Bird, A. F. (1976). *In* "The Organization of Nematodes" (N. A. Croll, ed.), pp. 107–137. Academic Press, New York.
Bird, A. F. (1979a). *In* "International Seminar on Meloidogyne" (C. E. Taylor, ed.), pp. 59–84. Academic Press, New York.
Bird, A. F. (1979b). *Int. J. Parsitol.* **9,** 357–370.
Bird, A. F., and Buttrose, M. S. (1974). *J. Ultrastruct. Res.* **48,** 177–189.
Bird, A. F., and Deutsch, K. (1957). *Parasitol.* **47,** 319–328.
Bird, A. F., and McClure, M. A. (1976). *Parasitol.* **72,** 19–28.
Bird, A. F., and Stynes, B. A. (1977). *Phytopathology.* **67,** 828–830.
Bonner, T. P., and Weinstein, P. P. (1972a). *J. Ultrastruct. Res.* **40,** 261–271.
Bonner, T. P., and Weinstein, P. P. (1972b). *Z. Zellforsch. Mikrosk. Anat.* **126,** 17–24.
Branton, D., and Deamer, D. W. (1972). "Membrane Structure." Springer-Verlag, Berlin and New York.
Cassada, R. C., and Russell, R. L. (1975). *Dev. Biol.* **46,** 326–342.
Cook, G. M. W., and Stoddart, R. W. (1973). "Surface carbohydrates of the Eukaryotic Cell." Academic Press, New York.
Deas, J. E., Aguilar, F. J., and Miller, J. H. (1974). *J. Parasitol.* **60,** 1006–1012.

DeGrisse, A. T. (1977). "De ultrastruktuur van het zenuwstelsel in de kop van 22 soorten planten-parasitaire nematoden, behorende tot 19 genera (Nematoda: Tylenchida)." Rijksuniversiteit, Ghent, Belgium.

Dick, T. A., and Wright, K. A. (1973). *Can. J. Zool.* **51,** 187–196.

Dick, T. A., and Wright, K. A. (1974). *Can. J. Zool.* **52,** 245–250.

Dropkin, V. H., and Bird, A. F. (1978). *Int. J. Parasitol.* **8,** 225–232.

Dwyer, D. M. (1977). *Exp. Parasitol.* **41,** 341–358.

Fujimoto, D., and Kanaya, S. (1973). *Arch. Biochem. Biophys.* **157,** 1–6.

Gutman, G. A., and Mitchell, G. F. (1977). *Exp. Parasitol.* **43,** 161–168.

Harrison, B. D., Robertson, W. M., and Taylor, C. E. (1974). *J. Nematol.* **6,** 155–164.

Himmelhoch, S., and Zuckerman, B. M. (1978). *Exp. Parasitol.* **43,** 208–214.

Himmelhoch, S., Kisiel, M. J., and Zuckerman, B. M. (1977). *Exp. Parasitol.* **41,** 118–123.

Hollis, J. P. (1962). *Nature* (London) **193,** 798–799.

Hudson, R. J., and Kitts, W. D. (1971). *J. Parasitol.* **57,** 808–814.

Ibrahim, I. K. A. (1967). *Proc. Helminthol. Soc. Wash.* **34,** 18–20.

Ibrahim, I. K. A., and Hollis, J. P. (1973). *J. Nematol.* **5,** 275–281.

Ibrahim, I. K. A., Joshi, M. M., and Hollis, J. P. (1978). *Proc. Helminthol. Soc. Wash.* **45,** 233–238.

Johnson, P. W., and Graham, W. G. (1976). *Can. J. Zool.* **54,** 96–100.

Johnston, M. R. L., and Stehbens, W. E. (1973). *Int. J. Parasitol.* **3,** 243–250.

Kondo, E., and Ishibashi, N. (1978). *Appl. Entomol. Zool.* **13,** 1–11.

Krieg, C., Cole, T., Deppe, U., Schierenberg, E., Schmitt, D., Yoder, B., and von Ehrenstein, G. (1978). *Dev. Biol.* **65,** 193–215.

Lee, D. L. (1965). *Parasitol.* **55,** 173–181.

Lee, D. L. (1970). *J. Zool. (London)* **161,** 513–518.

Lee, D. L., and Atkinson, H. J. (1976). "Physiology of Nematodes." Macmillan, New York.

Lehninger, A. L. (1975). "Biochemistry." Worth Publ., New York.

Lubinsky, G. (1931). *Z. Parasitenkd.* **3,** 775–779.

Lumsden, R. D. (1975). *Exp. Parasitol.* **37,** 267–339.

McBride, O. W., and Harrington, W. F. (1967). *Biochemistry* **6,** 1484–1498.

McKinnon, G. A., and Lubinsky, G. A. (1966). *Can. J. Zool.* **44,** 1090–1091.

Manter, H. W. (1929). *J. Parasitol.* **16,** 101.

Marks, C. F., Thomason, I. J., and Castro, C. E. (1968). *Exp. Parasitol.* **22,** 321–337.

Martinez-Palomo, A. (1978). *J. Parasitol.* **64,** 127–136.

Neville, A. C. (1975). "Biology of the Arthropod Cuticle." Springer-Verlag, Berlin and New York.

Poinar, G. D. (1973). *Proc. Helminthol. Soc. Wash.* **40,** 37–42.

Poinar, G. O., and Hess, R. (1976). *IRCS Med. Sci. Libr. Compend.* **4,** 296.

Poinar, G. O., and Hess, R. (1977). *Exp. Parasitol* **42,** 27–33.

Popham, J. D., and Webster, J. M. (1978). *Can. J. Zool.* **56,** 1556–1563.

Riddle, D. L. (1978). *J. Nematol.* **10,** 1–16.

Riding, I. L. (1970). *Nature (London)* **226,** 179–180.

Rutherford, T. A., Webster, J. M., and Barlow, J. S. (1977). *Can. J. Zool.* **55,** 1773–1781.

Siddiqui, I. A., and Viglierchio, D. R. (1977). *J. Nematol.* **9,** 56–82.

Singh, R. N., and Sulston, J. E. (1978). *Nematologica* **24,** 63–71.

Stewart, T. B., and Godwin, H. J. (1963). *J. Parasitol.* **49,** 231–234.

Taylor, C. E., and Robertson, W. M. (1969). *Ann. Appl. Biol.* **64,** 233–237.

Taylor, C. E., and Robertson, W. M. (1970a). *Ann. Appl. Biol.* **66,** 375–380.

Taylor, C. E., and Robertson, W. M. (1970b). *J. Gen. Virol.* **6,** 179–182.

Taylor, C. E., and Robertson, W. M. (1975). *In* "Nematode Vectors of Plant Viruses" (F. Lamberti, C. E. Taylor, and J. W. Seinhorst, eds.), Vol. 2, pp. 253–276. Plenum, New York.

Weinberg, M., and Keilin, M. (1912). *C. R. Seances Soc. Biol. Ses Fil.* **73**, 260–262.

Wright, K. A. (1976). *In* "The Organization of Nematodes" (N. A. Croll, ed.), pp. 71–105. Academic Press, New York.

Wright, K. A., and Hope, W. D. (1968). *Can. J. Zool.* **46**, 1005–1011.

Zuckerman, B. M., Himmelhoch, S., and Kisiel, M. (1973). *Nematologica* **19**, 109–112.

Zuckerman, B. M., Kahane, I., and Himmelhoch, S. (1979). *Exp. Parasitol.* **47**, 419–424.

10

Nematode Sense Organs

K. A. WRIGHT

Department of Microbiology and Parasitology
Faculty of Medicine, University of Toronto
Toronto, Ontario M5S 1A1, Canada

I. Introduction 237
II. Cuticular Sense Organs 239
   A. Basic Anatomy of a Cuticular Sense Organ 239
III. Internal Sensory Receptors 278
   A. Cephalic Internal Receptors 278
   B. Photoreceptors 279
   C. Internal Receptors of the Body Wall 280
   D. Internal Receptors of the Alimentary Tract 282
IV. Conclusions 285
   A. Some General Characteristics of Nematode Sense Organs 285
   B. Comparison of the Sensory Complement of Secernentian and Adenophorean Nematodes 290
   C. Specialization of Sense Organs in Plant Parasites 291
   D. Specialization of Sense Organs in Animal Parasites 291
   E. Internal Receptors 292
   References 293

I. INTRODUCTION

Nematodes have long been recognized to have peripherally located sense organs. These comprise modifications of the cuticle as papillae, pores, or setae associated with an underlying nerve process. However, their generally small size precluded any in-depth understanding of either structure or function. As recently as 1971 a review of nematode anatomy (Hirschmann, 1971) considered the nature and function of nematode sense organs within only three and a half pages. Only 5 years later, McLaren (1976a) required 70 pages to review the same

NEMATODES AS BIOLOGICAL MODELS
VOLUME 2

subject, primarily because of the recent contributions from electron microscopic studies. Although most of these studies were of animal parasites, similar studies of plant-parasitic species followed quickly, and in 1975 two major papers were published dealing with the free-living microbial feeder, *Caenorhabditis elegans* (Ward *et al.*, 1975; Ware *et al.*, 1975).

This nematode has been extensively studied as a model system to investigate developmental processes, and, since it is small, it has been feasible to reconstruct with great accuracy the total cellular composition of various parts of its anatomy. These studies in turn allow us to reappraise others, especially those of the larger animal parasites where cell identities are often harder to trace. They have also shown that nerves are associated with internal tissues of the body in manners suggesting that they may monitor internal functions or detect external stimuli capable of penetrating body tissues. It therefore seems important to recognize two classes of sensory organs (1) cuticular or peripheral sense organs, and (2) internal sensory receptors.

Most information has accumulated on cephalic cuticular sense organs. These comprise a single pair of laterally positioned amphids, a circle of six inner labial sense organs, six outer labial sense organs, plus four submedian cephalic sense organs arranged in radial symmetry around the mouth. Since it is to be expected that these sense organs will be important in feeding behavior, and since head structures show specific morphological adaptations to a wide range of feeding processes, the cephalic sense organs will be treated ecologically in order to adumbrate any functional–anatomical patterns related to free-living, plant- or animal-parasitic life styles.

Classically, nematodes have been divided into two major taxonomic groupings, the classes Secernentia and Adenophorea. Secernentian nematodes occur primarily as free-living microbivores in terrestrial environments, as parasites of plants, or as parasites of animals. Adenophorean nematodes are primarily marine with only a few groups occuring as parasites of plants or animals. The arrangement of sense organs in these two classes will also be considered.

Other cuticular sense organs occur on the ventral surface of the male's tail, where they no doubt function during mating, whereas secernentian nematodes have a single lateral pair of sense organs in the tail termed phasmids. A number of types of cuticular sense organs have also been identified in the body wall. These are most prominent as setae in adenophorean marine nematodes.

Internal sensory receptors have been identified in the body wall and associated with the alimentary tract.

Although the nervous system of the nematodes has generally been considered to be relatively simple, our increasing understanding of the structure and function of sense organs must make us realize both the complexity and diversity of the sensory control mechanisms of these seemingly simple animals.

II. CUTICULAR SENSE ORGANS

A. Basic Anatomy of a Cuticular Sense Organ

This section establishes the basic composition of sensory units associated with the cuticle, and the terminology to be used in this chapter. Justification of this scheme and variations from it will emerge from subsequent sections.

Each sensory unit is composed of three components: (1) cuticle, (2) sensory dendrites, and (3) processes of nonneuronal cells (Fig. 1). The cuticle plays a major role in the functional anatomy of the sense organ since it normally forms the interface between the nematode and its environment. By correlation to other groups of animals, especially arthropods, it is generally agreed that a gap or pore in the cuticle overlying the sensory cell process is evidence that stimulatory molecules have direct access to the nerve cell (chemoreception). (The absence of a pore through the cuticle does not preclude chemoreception, although in that case some evidence of modification of the cuticle allowing increased permeability would be expected.) In contrast, sense organs suspected of being mechanosensitive usually show modifications of the cuticle that would facilitate mechanical triggering of the sensory cell. This may include the elevation of the cuticle to form a prominent papilla, or internal modifications of the cuticle ensuring more efficient transfer of mechanical deformation to the sensory cell.

The neural component consists of the tip of one or more dendrites of sensory neurons. The cell bodies of the neurons from cephalic sense organs, and presumably others, occur in ganglia of the central nervous system, whereas long afferent dendrites connect the receptor to the perikaryon. The tip of the dendrite is usually somewhat enlarged and often contains vesicles, banded ciliary rootlets, and, rarely, mitochondria. Arising from the tip of the dendrite is one or more process derived, but now extensively modified, from a cilium. This process will be referred to as the *dendritic process*. The term cilium, or modified cilium, is rejected since even in their simpler forms the processes show significant departures from the functional anatomy of kinocilia, or even sensory cilia of some animals that essentially retain the anatomy of functional cilia (Bedini *et al.*, 1973, 1975; Bonar, 1978). The term dendritic process also has an earlier history of use in sensory physiology of arthropods. The dendritic process contains, almost exclusively, longitudinally aligned microtubules. In its basal region the microtubules are grouped into a circlet of doublet pairs resembling axonemal doublets of kinocilia. Dense material may occur between the microtubules in the tip of the dendritic process in a fashion similar to that characterizing the "tubular body" of arthropod mechanoreceptors (McIver, 1975). The presence of this material is considered to be an indicator of mechanosensitivity. In contrast, simple ending processes lacking dense material and associated with a cuticular pore are seen as candidates for chemoreception (Slifer, 1970). Thus the architectural relationships of cuticle and dendritic process are undoubtedly crucial for the functioning of the sense organ.

Two nonneuronal cells are associated with each sense organ (Fig. 1). They have been adequately characterized only for sense organs of the cephalic region. Their cell bodies lie at the nerve ring level, whereas long, thin cell processes extend to the body wall where they expand and wrap around the neuronal components of the sense organ. The expanded peripheral tip of one of the cells is inserted into the hypodermis of the body wall and is probably responsible for formation of the specialized cuticular component of the sense organ. It has been referred to as a "support cell," "cap cell," or "socket cell." *Socket cell* will be used here to reflect its probable role in forming the cuticular "socket" into which the neuronal units fit. The tip of the second nonneuronal cell encloses or ensheaths the tips of the sensory dendrites. It has been called the "sheath cell," "pocket cell," or "gland cell." *Sheath cell* will be used here to emphasize its major role in ensheathing the receptor region of the dendrites, thereby providing the appropriate environment for their operation. Junctional complexes occur between dendrites and sheath cells. The extracellular space so enclosed by the sheath cell is then sealed from the extracellular spaces of the rest of the nematode and is here designated the *receptor cavity*.

1. Anterior Cuticular Sense Organs

a. Free-Living, Terrestrial Nematodes. Nematodes belonging to the order Rhabditida (class Secernentia) are the major species occuring in terrestrial environments where they feed principally on microbial flora. Although they were once seen as prime candidates for "the primative nematodes," DeConinck, (1965) argued that the origin of nematodes should be sought among the adenophorean marine forms. Thus the Rhabditida are more commonly now seen as the primary terrestrial stock from which the major secernentian animal parasites have evolved [see Chabaud's phylogenetic scheme in Anderson *et al.* (1974)].

Caenorhabditis elegans can be considered a typical rhabditid. Ward *et al.* (1975) used computer assistance to reconstruct its anterior sense organs and their connections to the central nervous system, whereas Ware *et al.* (1975) independently used cinematographic techniques to do the same. It is rewarding to find such close agreement in these two studies. However, fine differences in description and/or presentation of findings do occur. What is presented here is a synthesis of the two studies.

The six *inner labial papillae* are all of the same type. They each comprise two sensory dendrites (Fig. 2B). The single dendritic process of one of these dendrites penetrates through the cuticle so that its tip lies at the opening of a pore located at the tip of a papilliform projection of the cuticle. Characteristically, the

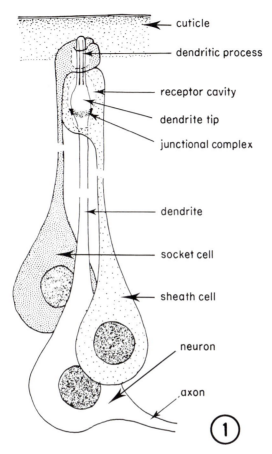

Fig. 1.   Basic composition of cuticular sense organs.

dendritic process undergoes two right-angle bends in penetrating the cuticle before entering the tip of the dendrite. The base of the dendritic process includes a
circlet of five to seven doublet microtubules. The double form of some of the
microtubules may extend nearly to the tip of the dendritic process. The dendrite
tip contains a cluster of lucent vesicles and very small remnants of a banded
rootlet. This dendrite shows characteristics of a chemoreceptor. The dendrite
forms junctional complexes with the surrounding sheath cell. The dendritic process of the second dendrite enters the cuticle, and curves so that its tip lies at right
angles to the first, but remains enclosed in cuticle. Microtubules run through the
process while at its tip, dense material occurs between the tubules. The base of
the dendritic process includes a circlet of seven doublet microtubules that continue to various levels into the dendritic process. The expanded dendrite tip

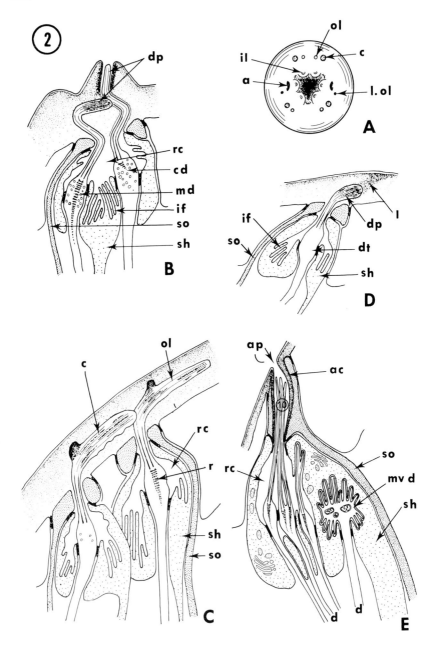

Fig. 2. Cephalic cuticular sense organs of *Caenorhabditis elegans*. (A) Distribution of sense organs over the head. (B) Inner labial sense organ. (C) Outer labial and cephalic sense organs. (D) Lateralmost outer labial sense organ. (E) Amphid—the circled number indicates the number of

contains vesicles and a 4 to 7 μm long banded rootlet. The dendrite forms junctional complexes with the enclosing sheath cell. This dendrite shows characteristics of a mechanoreceptor. The socket cell encloses the tips of the dendritic processes just below the cuticle (Fig. 3). The sheath cell surrounds the tips of the dendrites and encloses a small receptor cavity (Fig. 4). In the more anterior part of the cavity, processes may project from the inner surface of the sheath cell, whereas more posteriorly, complex infoldings of the inner surface of the sheath cell occur (Figs. 2B, 4, and 5).

These sense organs are clearly bifunctional with a presumptively chemosensitive dendrite exposed to the environment and a presumptively mechanosensitive dendritic process embedded in the cuticle. The arrangement of the two dendritic processes invites speculation on their mode of operation. They are housed under a papilliform elevation of cuticle, suggesting that mechanical collisions with this papilla will trigger the mechanoreceptor. What precludes accidental mechanical stimulation of the chemosensitive process? Two features may be significant. The outermost cuticular canal enclosing the chemosensory dendritic process is surrounded by denser cuticle that may reinforce this part of the canal, and protect the dendritic process. During a deformation of the cuticle that triggers the mechanoreceptor, the bends of the chemosensory process may allow its depression without sufficient compression to result in mechanical triggering. It should be noted that the bends lie in the inner regions of the cuticle, which are probably more "fluid" and deformable than the cortex.

Lying in the submedial axis are four *outer labial sense organs* that each contain a single dendrite with a single dendritic process. The thin dendritic process projects into the cuticle and runs anteriorly for about 2–2½ μm, lying parallel to the cuticle surface (Figs. 2C and 3). A dense branchlike process projects outward into the cuticle from the dendritic process. There is probably no external papilliform projection of cuticle associated with this sense organ. Dense material occurs as elongate bridges between microtubules along the length of the dendritic process. Apically there are only four microtubules, while basally there may be up to eight. A ring of five to seven doublet microtubules occurs in the base of the dendritic process. The dendrite tip contains a 3 μm long striated rootlet. Both socket and sheath cells occur, and extensive infoldings of the inner

---

dendritic processes entering the amphidial canal. Key to abbreviations: a, amphid; ac, amphidial canal; ap, amphidial pore; ax, axon; c, cephalic sense organ; cd, chemosensory dendrite; cu, cuticle; d, dendrite; dc, dense cuticle; dp, dendrite process; dt, dendrite tip; E, esophagus; f, filament-containing cytoplasm; g, gubernaculum; if, infoldings of membrane; il, inner labial sense organ; l, "ligamentous" connection to cuticle cortex; l.ol, lateralmost outer labial sense organ; md, mechanosensitive dendrite; mv, microvillus; mvd, microvillus-bearing dendrite; ol, outer labial sense organ; p, papilla; pg, pigment granule; r, rootlet; rc, receptor cavity; ret, reticular membrane infoldings; sg, secretory granule; sh, sheath cell; so, socket cell; sp, spicule; spm, spicule protractor muscle; spp, spicule pouch; srm, spicule retractor muscle; v, Golgi-derived vesicles.

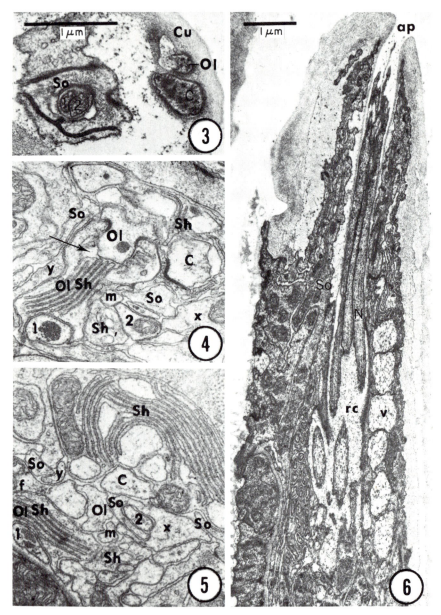

Fig. 3–5. Serial cross sections through inner labial, outer labial, and cephalic sense organs of *Caenorhabditis elegans* (×26,000). See the legend to Fig. 2 for the key to abbreviations. (From Ward *et al.*, 1975, reproduced by permission.)

Fig. 3. A section through the tip of the inner labial sense organ's socket cell; only the dendritic processes of the outer labial and cephalic sense organs appear.

sheath-cell membrane open into the small receptor cavity. The sense organ has characteristics of a mechanoreceptor, and could record pressures on the cuticle surface.

The two remaining outer labial sense organs lie just ventral to the midlateral line. They are of a distinctly different type than those described above. They comprise a single dendrite with a single dendritic process that is shorter and more bulbous, with more dense material between microtubules (Figs. 2D and 7). The tip of the dendritic process is surrounded by material that appears to connect it to the cortex of the cuticle. The dendrite tip lacks a rootlet. Socket and sheath cells are similar to those described previously. Since the tip of the dendritic process is connected to the cortex of the cuticle, it might be triggered in a fashion similar to the presumed stretch receptors of *Nippostrongylus brasiliensis* (see Section II, A, 1). Thus they may detect shear stresses in the cuticle.

Four *cephalic sense organs* lie close to the submedian outer labial sense organs just lateral to the submedian line. Their single dendritic process projects into the cuticle about 2½ μm and lies parallel to the surface (Figs. 2C and 3). There is a short dense branch projecting outward from it, and the dendritic process is enlarged, containing several microtubules with dense material organized as cores or rods between the microtubules. Basally, the dendritic process contains a circlet of six to eight doublet microtubules. Doublet microtubules extend about 1 μm into the dendritic process. The socket and sheath cells resemble those already described. The receptor cavity is the smallest of the sense organs. They appear to be mechanoreceptors. In view of the slight differences (Fig. 2C) in their form from the submedian outer labial receptors they may have somewhat different sensitivities to similar stimuli. Ward *et al.* (1975) have shown that in males these sense organs include another dendritic process arising from a second sensory dendrite. The dendritic process extends through the cuticle to a pore opening at the apex of a small papilla. It may function as a chemoreceptor during mate selection.

Amphids are the largest and most complex of the cephalic sense organs. They open via a prominent cuticular pore (about 0.6 μm). The body cuticle continues inward from the pore for a length of about 4–5 μm, forming the cuticular amphidial canal (Figs. 2E and 6). The amphidial canal contains 10 dendritic

Fig. 4. Section at the level of the dendrite tips. Arrow notes receptor cavity and membrane infoldings of the sheath cell are evident. 1 and 2, Dendrites of the inner labial sense organ; x and y, unciliated presumed internal receptor neurons; m, ciliated presumed internal receptor neuron.

Fig. 5. Section below the level of the receptor cavities, showing the extensive infoldings of sheath cell membranes.

Fig. 6. Longitudinal section through the amphid of *Caenorhabditis elegans* ($\times$14,000). The matrix of the receptor cavity is assumed to be released from vesicles (v) in the sheath cell cytoplasm. (From Ward *et al.*, 1975, reproduced by permission.)

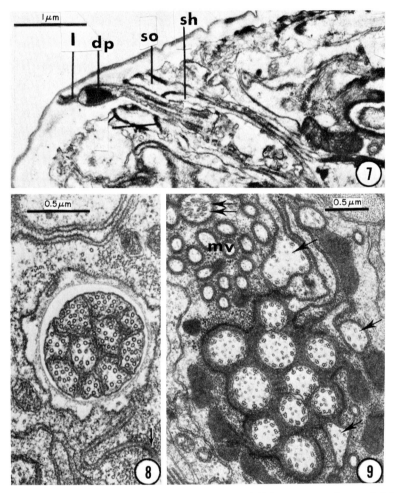

Fig. 7. Longitudinal section through the lateralmost outer labial sense organ of *Caenorhabditis elegans* (×24,000). See the legend to Fig. 2 for the key to abbreviations. (From Ware *et al.*, 1975, reproduced by permission.)

Fig. 8. Cross section through the amphidial canal of *Caenorhabditis elegans* showing 10 dendritic processes containing single microtubules about 1 μm below the pore (×45,000).

Fig. 9. Cross section of the 10 dendritic processes of the amphid in the region with doublet microtubules. (×31,000). The dendritic process of the microvillus-bearing dendrite is noted by double arrows. Dendrite processes that project into the sheath cell of the amphid are noted by arrows. Mv, Microvillus. (From Ward *et al.*, 1975.)

processes (Fig. 8). There is a slight expansion of the amphidial canal just within the pore so that there is some space between the tips of the dendritic processes; however, toward the inner end of the canal the processes are more closely packed. The amphidial canal is enclosed by the tip of the socket cell. Internal to the amphidial canal, the dendritic processes gradually diverge. Dendritic processes range from about 6 to 7.5 μm long. In the base of each of the dendritic processes is a circlet of nine doublet microtubules, whereas a few single tubules occur in the center of the circlet (Fig. 9). Doublet microtubules extend to within about 1 μm of the tip of the dendritic processes. The ten dendritic processes arise from the tips of eight sensory dendrites; two dendrites each possess two dendritic processes, whereas the remaining six dendrites have only one process each (Fig. 10). Dendrite tips contain a cluster of vesicles and a long banded rootlet. The tips of the sensory dendrites occupy a receptor cavity of considerable size formed by the sheath cell. Where the sheath cell connects to the socket cell, it forms a continuation of the amphidial canal as the cell membrane is reinforced by the presence of dense material on its external surface. The adjacent cytoplasm contains many dense filaments, perhaps tonofilaments, that may insert into the membrane. The cytoplasm of the sheath cell contains vesicles with flocculent contents similar to material in the receptor cavity (Fig. 6). The vesicles may be derived from Golgi dictyosomes in the sheath cell. The cell body of the sheath cell, located in the lateral ganglion at the nerve ring level, also contains endoplasmic reticulum, Golgi, and vesicles, suggestive of synthesis of material in the cell body. The interconnecting process of the sheath cell between the cell body and receptor is larger than in the other sense organs. The part of the sheath cell enclosing the receptor cavity is quite extensive, accounting for a major volume of the tissues on either side of the nematode's head. In addition to the eight amphidial dendrites described above, three other dendrites enter the receptor cavity and give rise to dendritic processes that project back into the sheath cell (Fig. 9). One bears two dendritic processes (Fig. 10). Each includes an axonemal pattern of nine doublet microtubules. In addition, another dendrite enters the dorsal region of the amphidial sheath cell from outside but does not emerge into the receptor cavity. Its expanded tip gives rise to a single dendritic process with doublet microtubules and about 50 microvilli projecting both anteriorly and posteriorly that are interdigitated with the sheath cell (Figs. 2E, 9, and 10).

Amphids have long been considered to be chemorecepetors because of the amphidial pore. It is likely that the 10 dendritic processes that enter the amphidial canal are chemosensitive and function to record the external environment. However, inferences drawn only from their structure should be restricted to statements about broad levels of sensitivity such as chemo- versus mechanoreception. Extensive experimental studies have been carried out to correlate the modified chemotactic behavior of mutants with abnormalities in sense-organ structure. Thus, Lewis and Hodgkin (1977) have found that 8 of 13 mutant strains showed

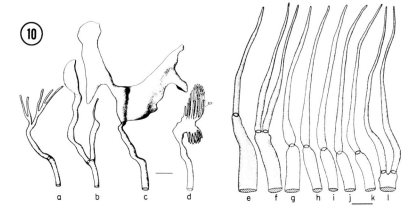

Fig. 10. Computer reconstructions of dendrites of the amphids of *Caenorhabditis elegans*. Dendrites a–e have dendritic processes that project into the sheath cell in the receptor cavity of the amphid. Dendrite d bears many microvilli projecting both anteriorly and posteriorly. It does not enter the receptor cavity of the amphid. Dendrites e–l have dendritic processes that enter the amphidial canal. (From Ward *et al.*, 1975, reproduced by permission.)

abnormalities in the construction of the cephalic sense organs. Defects occurred only in dendrites or the dendritic processes of the amphid and the presumed chemosensory dendrite of the inner labial sense organs. Mutant E 1126 showed defects only in the presumed chemosensory dendrites of the inner labial sense organs. This mutant avoids Cl⁻ ions, which are attractive to the wild-type strain. Mutant E 1066 showed abnormalities only in the dendritic processes that enter the amphidial canal, and were abnormally lethargic in bacterial cultures. Mutants with more severe damage to the other dendrites of the amphid as well showed aversion to Na⁺ ions, which are normally attractive. Similar abnormalities were not found in mutants with locomotor coordination defects. Although it might be hoped that such correlations of aberrant behavior in mutants to abnormal sense-organ development might allow the detailed identification of functions for individual sensory dendrites, Ward (1977) has expressed some misgiving, in view of the extent of anatomical derangement found in the mutants. These studies, however, can be construed to give support to the general conclusions drawn from anatomy.

b. Plant-Parasitic Nematodes. Nematodes of the orders Tylenchida and Dorylaimida are highly successful as parasites of plants. They feed from plant cells by stylets that are inserted into plant tissues, and they show specific orientation and attraction behavior to their host plants. In the Tylenchida a wide range of parasitic associations is exhibited, from species occurring exclusively in the soil and feeding only on the periphery of roots to the sedentary and endoparasitic

species of the Heteroderidae. Second-stage larvae of the endoparasitic species occur in the soil and migrate between plants. The Dorylaimida, however, include species that are predators of soil meiofauna, as well as species parasitic to plants. All stages of the ectoparasites migrate between plants. Thus, although these plant-parasitic species share the soil environment with microbial feeding nematodes, they also show specialized behavior patterns related to feeding. It may be expected that their sensory structures will show both similarities and variations from the free-living nematode pattern.

Six *inner labial sense organs* occur in all of the tylenchid nematodes described. These open by fine pores through the cuticle. In *Macrotrophurus arbusticola, Ditylenchus dipsaci, Tylenchulus semipenetrans, Heterodera glycines,* and *Meloidogyne incognita,* the pores of the inner labial sense organs lie on the outer surface of the oral region about 0.5 μm or less from the oral opening (Fig. 11). In *D. dipsaci* the pores occur on papilla-like elevations. The lateralmost inner labial sense organs of *Radopholus similis,* however, open just within the rim of the buccal cavity, whereas the rest open on the outer surface through small papillae. In *Rotylenchus robustus* and *Macroposthonia rustica* the pores of all of the inner labial sense organs open just within the buccal capsule (Fig. 11) (De-Grisse *et al.,* 1974; Baldwin and Hirschmann, 1973, 1975).

The composition of the *inner labial sense organs* has been well described in *Meloidogyne incognita* and *Heterodera glycines.* In second-stage larvae of *M. incognita* they consist of two dendrites (Fig. 12), each with a single dendritic process (Endo and Wergin, 1977). The dendritic process of one extends into the fine cuticular pore that opens to the exterior. The second ends simply just below the cuticle. In adult males, apparently both dendritic processes extend into the pore. Five doublet microtubules occur in the dendritic processes, and they may extend various distances into them. A small receptor cavity occurs around the tips of the dendrites, and the membrane of the sheath cell bears closely spaced infoldings that open to the receptor cavity (Fig. 13). The inner labial sense organs of males of *H. glycines* are very similar, with two fine dendritic processes entering the cuticular pore (Baldwin and Hirschmann, 1975).

Six *outer labial sense organs* have been described in most, but not all, tylenchs. The single dendritic processes of these end in the cuticle. The lateralmost of the outer labial sense organs is described as reduced in *Radopholus* and *Rotylenchus* (DeGrisse *et al.,* 1974). Dendrites of the outer labial sense organs lie close to dendrites of the cephalic sense organs and wrap around them to various extents (Fig. 11). Only the four submedian outer labial sense organs were found in the cuticle of *H. glycines,* although the dendritic processes of the lateral outer labial sense organs were located deeper in the head tissues (Baldwin and Hirschmann, 1975). Each submedian outer labial sense organ of *Heterodera* contains two dendrites. Only one dendritic process enters the cuticle and ends with a dense tip. Eight doublet microtubules were

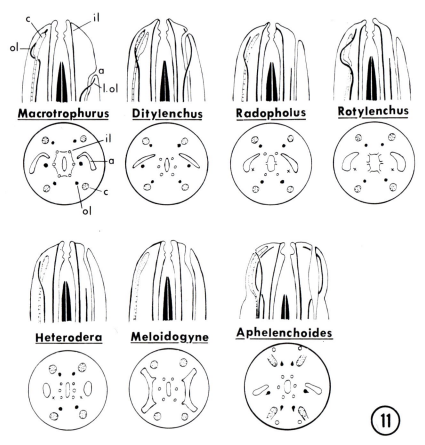

Fig. 11. Longitudinal and *en face* views of some plant parasites showing the location of the dendritic processes of the cephalic cuticular sense organs. In longitudinal views, the left side represents a submedian plane showing inner labial, outer labial, and cephalic sense organs; the right side is a lateral plane showing inner labial and lateralmost outer labial sense organs and the amphid. Longitudinal views show how dendritic processes of outer labial sense organs wrap around cephalic processes as they progress through the cuticle. The fine tips of outer labial processes are not shown in longitudinal views but their ending points are noted in the *en face* views. Lateral outer labial dendrites marked by x do not reach cuticle. See the legend to Fig. 2 for the key to abbreviations. (After DeGrisse *et al.* 1974; Baldwin and Hirshmann, 1975; Endo and Wergin, 1977; and DeGrisse, *et al.*, 1979.)

found that project for various distances into the dendritic process. No single microtubules were found. No outer labial sense organs occur in second-stage larvae or males of *M. incognita* (Endo and Wergin, 1977), or of *Tylenchulus semipenetrans* (Natasasmita and DeGrisse, 1978a).

Four *submedian cephalic sense organs* occur in all tylenchs (Fig. 11). Their single enlarged dendritic process ends in the cuticle and contains dense intermicrotubular material (Figs. 11 and 12). They have been well described in *M. incognita*. In second-stage larvae there may be four to six doublets plus one or two single microtubules in the base of the dendritic process (Endo and Wergin, 1977). In adult males, seven doublets have been described (Baldwin and Hirschmann, 1973). A receptor cavity is present with infoldings of the internal sheath cell membrane.

The cephalic cuticular sense organs of *Aphelenchoides fragariae* (a member of the Aphelenchoidea, the second major tylench group) include six inner labial, six outer labial, and four cephalic sense organs (Fig. 11). However, the cephalic sense organs include two dendrites; one is an enlarged mechanoreceptor, whereas the other opens to a pore on the outer surface of the head (DeGrisse *et al.*, 1979).

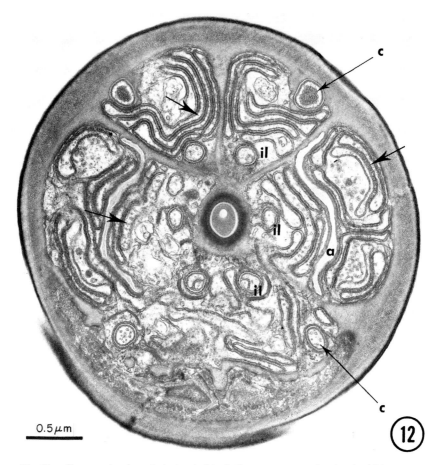

Fig. 12. Cross section through the head of *Meloidogyne incognita* showing dendritic processes of cephalic cuticular receptors and irregular processes of internal receptors (arrows) (×30,300). (From Endo and Wergin, 1977, reproduced by permission.)

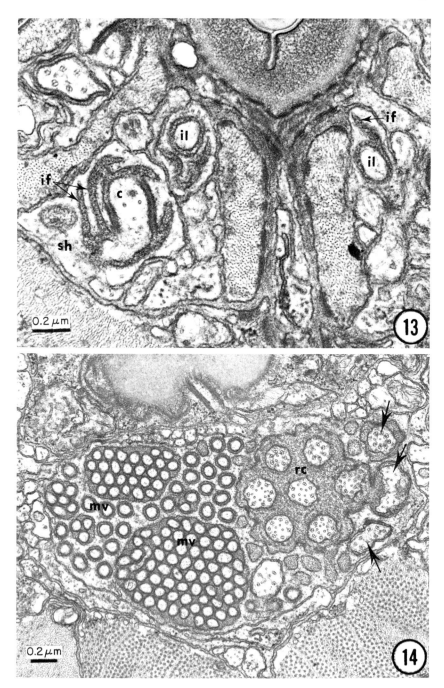

Inner labial, outer labial, and *cephalic sense organs* occur in the dorylaimids *Aporcelaimellus* spp. and *Xiphinema americanum* (Lippens *et al.,* 1974; Wright and Carter, 1980). Their composition is probably simialr in both. In *X. americanum* all have pores opening through small papillae on the surface of the cuticle (Figs. 15 and 16). Both inner and outer labial sense organs contain four dendritic processes, each arising from a single dendrite. The dendritic processes contain a ring of nine doublet microtubules and the dendrite tips contain banded rootlet material. Cephalic sense organs contain two or three dendritic processes and dendrites. Both socket cell and sheath-cell processes may be present, but they have not been traced to their cell bodies. Sheath cells may contain a few small secretory globules, but no infoldings occur in the membrane of the receptor cavity.

In general, *amphids* in tylenchid nematodes closely resemble those of *Caenorhabditis elegans.* The amphidial canal is characteristically expanded just within the amphidial pore. In *Meloidogyne* this forms an I-shaped pouch (Wergin and Endo, 1976). The shape and position of the amphidial openings in other species are somewhat variable (Fig. 11). In *Meloidogyne* second-stage larvae and males, and in *Heterodera* males, seven dendritic processes enter the amphidial canal (Wergin and Endo, 1976; Baldwin and Hirschmann, 1973, 1975). Only a single dendritic process per dendrite has been described. Dendrite tips have banded rootlets. Six to eight doublet microtubules occur in a circlet with up to five internal single tubules in the basal level of the dendritic processes. In *Meloidogyne* this pattern extends for about 1 μm, beyond which the dendritic process is expanded two or threefold, and only single microtubules occur. However, in *Heterodera* males some of the doublet tubules extend considerably further. Other tylenchs have similar amphids, although *Tylenchulus semipenetrans* has only six dendritic processes in the amphidial canal (DeGrisse and Natasasmita, 1975). In addition to dendritic processes entering the amphidial canal, up to five processes have been found in the receptor cavity projecting into the sheath cell. One ventrally located dendrite tip also bears one or two short dendritic processes and a large number of microvilli that project both anteriorly and posteriorly and are interdigitated with the sheath-cell membrane (Fig. 14). There are about 200 microvilli on this dendrite in second-stage larvae of

Fig. 13. Section through inner labial and cephalic sense organs of *Meloidogyne incognita* showing reduced receptor cavities with infoldings (if) of sheath-cell membrane ($\times$57,000). See the legend to Fig. 2 for the key to abbreviations. (From Endo and Wergin, 1977, reproduced by permission.)

Fig. 14. Section through enlarged receptor cavity of amphid of *Meloidogyne incognita* containing seven dendritic processes that enter the amphidial canal, microvilli of the specialized microvillus-bearing dendrite projecting into the sheath cell, and accessory dendritic processes that also project into the sheath cell (arrows) ($\times$35,000). See the legend to Fig. 2 for the key to abbreviations. (From Wergin and Endo, 1976, reproduced by permission.)

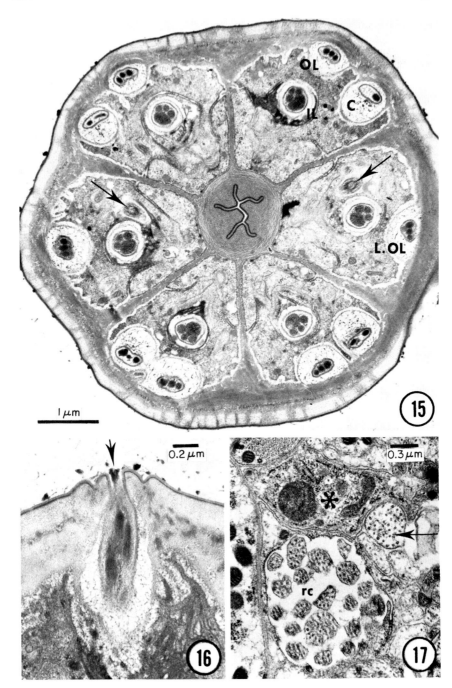

15

16

17

Meloidogyne (Wergin and Endo, 1976), 350 in males (Baldwin and Hirschmann, 1973), but only 75 in males of *Heterodera* (Baldwin and Hirschmann, 1975), and small numbers in other tylenchs studied (DeGrisse *et al.,* 1974). (These values compare to 50 microvilli on the comparable dendrite in *Caenorhabditis elegans.*) Both socket and sheath cells are undoubtedly present, although the locations of their cell bodies have not been plotted. The sheath cell contains some secretory product, indicating that material is released into the receptor cavity.

Although the individual components of the tylench amphids thus resemble those of *C. elegans,* there is one major anatomical difference. The very prominent microvillus-bearing dendrite is exposed to the receptor cavity in tylenchs, while it is not in *C. elegans.*

Amphids of the dorylaim nematodes are considerably different. Ultrastructural observations are available only for *Aporcelaimellus* spp. (Lippens *et al.,* 1974) and *Xiphinema americanum* (Wright and Carter, 1980). The amphidial canal of these species in considerably wider, resulting in a less compact distribution of the dendritic processes. There are also many more dendritic processes per amphid and more dendrites may possess more than one dendritic process.

In *X. americanum* the invagination of cuticle that forms the amphidial canal and is associated with the socket cell is relatively short in relation to the total length of the receptor cavity (Fig. 18) (Wright and Carter, 1980). The sheath-cell cytoplasm that encloses the dendritic processes is a thin layer containing occasional globules that appear to be released into the receptor cavity. However, no other extensive developments of the cytoplasm occur. Nineteen dendritic processes (Figs. 17 and 18) arise from 14 dendrites. One of these sensory units has a short dendritic process in the dorsal aspect of the receptor cavity. Its dendrite emerges through the sheath cell and lies external to the amphid some distance anterior to the level at which the remaining 18 dendritic processes enter dendrites. The basal parts of the dendritic processes contain a well-developed axonemal pattern of nine doublet tubules, whereas banded rootlets occur in the dendrite tips. At this level the sheath-cell cytoplasm extends internally and connects to an enlarged cell body with a prominent nucleus (Fig. 18). The cytoplasm here contains some granular endoplasmic reticulum, mitochondria, and Golgi units, but no secretory granules. The cytoplasm continues posteriorly an unde-

Fig. 15.   Cross section through the head of *Xiphinema americanum* showing distribution of the cephalic cuticular receptors and two dendritic processes of internal receptors (arrows) ($\times 16,500$). See the legend to Fig. 2 for the key to abbreviations.

Fig. 16.   Longitudinal section of an inner labial sense organ of *Xiphinema americanum* showing its pore (arrow) through the cuticle ($\times 34,000$).

Fig. 17.   Section through the amphid of *Xiphinema americanum* showing 18 dendritic processes in the receptor cavity. Arrow notes the long dendritic process of the second internal receptor. Asterisk notes cytoplasm of the cell body of the amphidial sheath cell ($\times 25,000$). See the legend to Fig. 2 for the key to abbreviations.

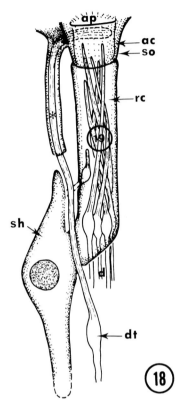

Fig. 18. The amphid of *Xiphinema americanum* showing its short amphidial canal, but long receptor cavity, and anterior location of sheath cell body. Asterisk notes the long dendritic process of an internal receptor. Circled number indicates the number of dendritic processes in the receptor cavity. See the legend to Fig. 2 for the key to abbreviations.

termined distance, but dendrites are no longer ensheathed by the cell. Nuclei of these sheath cells are the most anterior of a group of nuclei that occur in lateral and median cords and may represent hypodermal cells. Thus the sheath-cell nuclei occur about one third the distance from the head to the nerve ring.

c. Animal-Parasitic Nematodes. Nematodes of the orders Strongylida, Oxyurida, Ascaridida, and Spirurida, all belonging to the Secernentia, have evolved as highly specialized and successful parasites of animals. Recent studies of their cuticular sensory structures have been largely restricted to cephalic sense organs. Considerable variation in the functional anatomy of these sense organs is to be expected since major specializations of their cephalic anatomy have occurred to allow for the diverse feeding habits of these groups. Recent studies of cephalic "papillae" have not always related them to the nomenclatorial schemes devised for cephalic sense organs of free-living nematodes, but rather stress identification of functional types.

Two types of mechanoreceptors were identified in *Nippostrongylus brasiliensis* (Wright, 1975). These should be recognized as comprising the inner labial, outer labial, and cephalic sense organs (Fig. 19). Inner labial sense organs (type I) are composed of a single dendrite whose dendritic process projects into the anterior cephalic cuticle and is connected to the cuticular cortex by a ligament-like strand of material (Fig. 19C). The dendritic process contains microtubules and intervening dense material. Toward the base of the dendritic process, doublet microtubules appear until at its base there is a circlet of nine. The dendrite tip includes a banded rootlet. It was suggested that this sense organ serves as a "stretch receptor" by means of the ligament-like attachment of the dendritic process to the cortex of the cuticle. The cephalic cuticle in this nematode is greatly inflated, with an extensive fluid-containing median zone (Fig. 19B). The

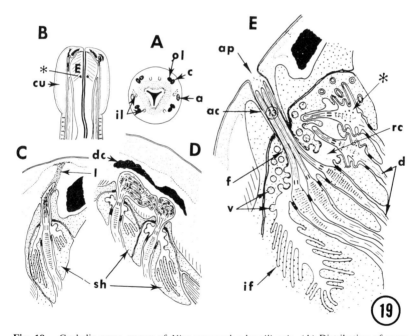

Fig. 19. Cephalic sense organs of *Nippostrongylus brasiliensis*. (A) Distribution of sense organs over the head; (B) the characteristically inflated cuticle of the head, location (*) of potentially proprioceptive internal receptors between musculature of buccal capsule and rest of esophagus; (C) an inner labial "stretch receptor"; (D) both the outer labial and cephalic sense organs with separate sheath cells and receptor cavities; (E) the general form of the amphid; the circled number indicates number of dendritic processes entering the amphidial canal and the asterisk indicates one of two accessory dendrites whose processes project into the sheath cell. See the legend to Fig. 2 for the key to abbreviations. (After Wright, 1975.)

cortex of this balloon-like cuticle is probably readily shifted on encounters with objects in the nematode's environment. Such collisions over the entire head region would be detected by the anterior inner labial "stretch receptors."

Type II sense organs consist of two dendrites, each bearing a single dendritic process (Fig. 19D). The dendritic processes extend into the cuticle and end as enlarged bulbs close to each other. The more lateral one (i.e., the cephalic sense organ) is oriented parallel to the long axis of the head, while the more median one (the outer labial sense organ) is oriented transverse the long axis of the head. A dense component of cuticle overlies both these dendritic processes and probably facilitates transmission of mechanical pressures through the inflated cuticle to the mechanoreceptors. Although these sense organs lie adjacent to each other in the cuticle, and may well function together, they have separate dendrites, receptor cavities, and separate sheath-cell processes in the body wall. They can thus be recognized as the cephalic and outer labial sense organs. Only four outer labial sense organs are therefore found. The cephalic sense organs of *Nematospiroides dubius* and *Trichostrongylus colubriformis* may be similar to those of *N. brasiliensis* (McLaren, 1976a).

The anterior sense organs of *Heterakis gallinarum* illustrate three further major structural variations (Wright, 1977). Inner labial sense organs are represented by six small dendrites whose dendritic processes contain dense intermicrotubular material and project into the cuticle at the base of a cuticular flange around the inner margin of the prominent lips (Fig. 20). These apparently mechanosensitive sense organs could detect stresses in the cuticle that may occur during the browsing style of feeding of this nematode. The large bulbous papillae situated on the outer surface of the lips each contain two enlarged dendritic processes that are arranged at right angles to each other (Fig. 20). Microtubules of these processes are interconnected by elongate strands of dense material in a quasi-geometric pattern (Fig. 21), supporting the impression that these papillae must be mechanoreceptors. The fact that these sensory processes are arranged at right angles to each other suggests that a process may be sensitive to pressure exerted from only one direction. Combining the two processes into a single papilla may ensure sensitivity to a wider range of directed pressures. Details of socket and sheath cells were not originally given. However, there appear to be two separate receptor cavities, indicating that the papilla contains both the outer labial and cephalic sense organs.

The subventral lips of *H. gallinarum* bear a further sense organ, the lateralmost outer labial sense organ. They consist of a short peg of cuticle containing a pore that encloses a single simple dendritic process (Fig. 20). A cuticular cylinder projects internally and encloses the tip of a second dendritic process. The longer process ending in the pore of the external peg could be chemosensitive, whereas the shorter process that has some dense material between its microtubules could be mechanosensitive. Presumably the

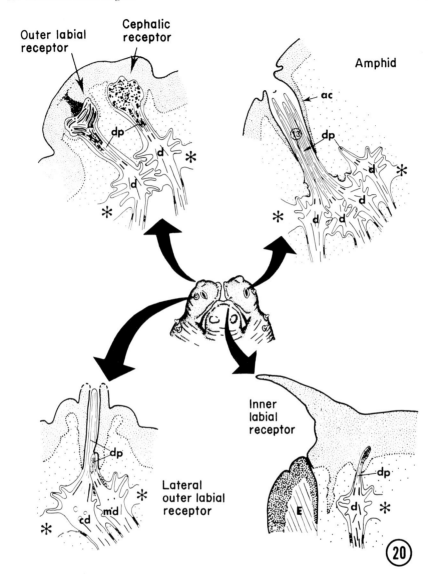

Fig. 20. Cephalic sense organs of *Heterakis gallinarum*. Asterisks note points where the receptor cavity of each sense organ opens to the highly folded membranous cords that project posteriorly
into the body. The circled number on the amphid indicates the number of dendritic processes entering
the amphidial canal. See the legend to Fig. 2 for the key to abbreviations. (After Wright, 1977.)

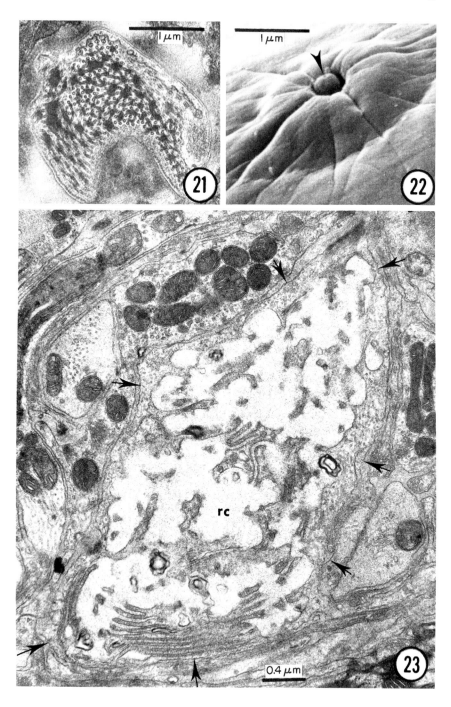

mechanoreceptor may be triggered by deflection of the peg, displacement of the cuticular cylinder, and deformation of the dendritic process. Similar sense organs can be seen on *Ascaridia galli* by scanning electron microscopy (Fig. 22: K. A. Wright, unpublished observations).

All of the cephalic sense organs of *H. gallinarum* have associated with them an extensive development of the sheath cell. This takes the form of a cord of cytoplasm containing highly infolded cell membranes (Fig. 23). The spaces within these cords are continuous with the receptor cavity. The larger of these cords (from cephalic and outer labial sense organ) extend into the body and come to lie along the outer surface of the esophagus beyond the lips, and are enclosed by hypodermal processes continuous from the dorsal and ventral hypodermal cords (see Wright 1976, 1977). The cytoplasm of the membranous cord contains microtubules, cytoplasmic filaments, and, rarely, mitochondria. There is no indication of synthetic or secretory activity.

All eight of the cephalic "papillae" of *Dipetalonema viteae, D. setariosum, Dirofilaria immitis,* and *Litomosoides carinii* apparently have the same morphology and appear to be mechanoreceptors (McLaren, 1972). A single dendritic process is enclosed by socket and sheath cell processes. There is essentially no space between the sheath cell membrane and the dendritic process, so that the receptor cavity is greatly reduced (Fig. 24). However, infoldings of the sheath cell membrane do occur. The dendrite tip produces very irregular processes that are interdigitated with the sheath cell, and gives rise to a single large dendritic process that extends into the cuticle and expands in a flattened lenticular form.

Four similar presumed mechanoreceptors were also briefly described in *Syphacia obvelata* (Dick and Wright, 1973). The base of the single dendritic process in these receptors contains many doublet microtubules, and no banded rootlets were noted in the dendrite tip.

Cephalic sense organs of the few adenophorean nematode parasites of animals that have been described differ markedly from those just described in secernentian parasites. Four anatomical types of sense organs (in addition to amphids) were distinguished on the trichuroid *Capillaria hepatica* (Wright, 1974) (Fig. 25). Since all include a pore through the cuticle, they are considered to be

---

Fig. 21.   Cross section through one of the dendritic processes in the large doublet papilla of *Heterakis gallinarum* showing distribution of microtubules and intervening dense material ($\times 20,000$).

Fig. 22.   Scanning electron microscope view of the lateralmost outer labial sense organ of *Ascaridia galli* showing the short central peg. Arrow notes an indication of a pore at the tip of the peg ($\times 20,000$).

Fig. 23.   Cross section through a membranous cord continuous with the receptor cavity of the cephalic sense organ of *Heterakis gallinarum*. The outer margin of the sheath cell is noted by arrows. The central cavity is an extension of the receptor cavity: rc, receptor cavity. ($\times 29,000$).

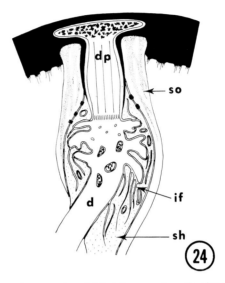

Fig. 24. A cephalic mechanoreceptor of *Dipetalonema viteae* showing the lenticular dendritic process extended into the cuticle, and the greatly reduced receptor cavity. See the legend to Fig. 2 for the key to abbreviations. (From McLaren, 1976a.)

chemosensitive. The oral opening is surrounded by six sense organs. The dorsal and ventralmost of these are simplest in form (type I), consisting of two dendrites each with a single dendritic process that extends to the cuticular pore. The lateral pair (type II) includes a third dendrite whose dendritic process lies underneath an infolding of thin cuticle. It expands as an ellipsoidal bulb and contains much dense material between microtubules—a mechanoreceptor. These six receptors (types I and II) are the inner labial sense organs. Outer labial and cephalic sense organs are represented by the type III and IV sense organs, respectively. These resemble type I, but details of the cuticular pore differ, and whereas outer labials contain two dendrites each with a single dendritic process, cephalic receptors have three dendrites. The dendritic processes of all of these sense organs contain nine doublet tubules. These sensory units were originally considered to be enclosed by hypodermal cells (Wright, 1974), since no distinctive differentiations of the enclosing cytoplasm were noted. Also, cell processes comparable to those connecting to socket or sheath-cell bodies, as later demonstrated in *Caenorhabditis elegans,* were not evident. Nevertheless, a small receptor cavity occurs but lacks infolded membranes. Anterior sense organs of *Trichinella spiralis* are apparently similar (McLaren, 1976a).

   In general, amphids of most animal parasites conform to the organization of those of *C. elegans.* Both socket and sheath cells probably occur, although the greater complexity of the cells in these larger animals often makes it difficult to identify them. Nevertheless these two cell types have been identified in *Dipetalonema viteae, Necator americanus, Syngamus trachea*, although only in *Necator* and *Syngamus* was the sheath cell traced to its nucleus (McLaren, 1972, 1974a; Jones, 1979).

The amphids of *N. brasiliensis* are characterized by having a slitlike pore opening to a cuticle-lined amphidial canal that encloses the tips of 13 dendritic processes (Fig. 19E; Wright, 1975). Near the inner end of the amphidial canal, microtubules in the dendritic processes appear as doublets until each process contains nine and a variable number of single microtubules. Each of the dendritic processes enters the tip of a dendrite in the receptor cavity. The dendrite tips have banded rootlets. In addition, there are two dendrites whose tips bear about 30–50 microvillus-like processes that indent the adjacent sheath-cell membrane. Each of these dendrites has a single dendritic process that extends through the receptor cavity and projects into the adjacent sheath cell. These dendrites resemble the microvillus-bearing dendrite of tylenchid plant-parasitic nematodes, and *C. elegans*. The sheath cell enclosing the receptor cavity in *N. brasiliensis* is extensive. It contains some vesicles of moderately dense flocculent content, resembling the contents of the receptor cavity. Further posteriorly, the receptor cavity is continuous with an extensive system of membrane infoldings that penetrate into the sheath cell cytoplasm.

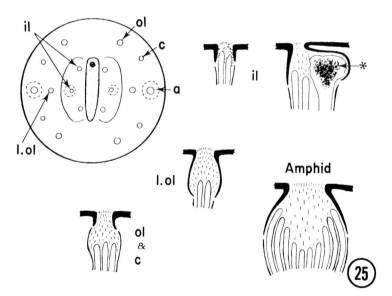

Fig. 25. Cephalic sense organs of *Capillaria hepatica*. Asterisk notes the mechanoreceptive dendritic process of the lateral inner labial sense organ. Note that all sense organs have pores opening through the cuticle. See the legend to Fig. 2 for the key to abbreviations. (After Wright, 1974.)

Amphids of *H. gallinarum* are in some ways very similar to those of *N. brasiliensis*. This species has large slitlike amphidial pores that open to an amphidial canal enclosing 13 dendritic processes (Fig. 20: Wright, 1977). The processes enter dendrites that contain only faint strands of rootlet material. The dendritic processes contain variable numbers of doublet tubules (up to 39), but none show the nine-doublet pattern of a typical axoneme. Although doublets usually occur around the periphery of the dendritic processes, as many as 10 may occur centrally as well. The tips of all of the dendrites are very irregular, producing microvilli that project into the receptor cavity or indent the adjacent sheath-cell membrane. In addition, there are two dendritic processes that project into the sheath cell. It is not clear whether these arise from one or two dendrites. The sheath cell that encloses the receptor cavity does not have many organelles (primarily microtubules), and secretory vesicles were not noted; however, very extensive areas of membrane infoldings are continuous with the membrane of the receptor cavity. These infoldings result in cords of very complexly infolded membrane that project posteriorly and are enclosed by hypodermal tissue that connects from the lateral hypodermal cords of the body wall, below the level of the nematode's lips, and is closely appressed to the outer surface of the esophagus (Wright 1976, 1977).

Amphids of filarial nematodes conform to the same general pattern of structure, but vary in some significant features. In *D. viteae, D. setariosum,* and *Dirofilaria immitis,* nine dendrites give rise to nine dendritic processes (McLaren, 1972). All dendritic processes occur within the amphidial canal, although in *D. immitis* one is notably shorter than the rest. *Litomosoides carinii* has four dendrites in one amphid, three in the other. The dendritic processes contain many doublet microtubules that appear to be associated with a lamelliform density at their base. There are no banded rootlets in dendrites of *D. viteae*. In *D. immitis* the doublet tubules occur in dense clusters. The tips of the sensory dendrites all have irregular microvillus-like projections extending from them. They also contain clusters of small clear vesicles. The socket cell is associated with the cuticle of the amphidial canal, but in addition extends over the outer end of the sheath cell. The sheath cell contains a few vacuoles but primarily has many extensive infoldings of the inner cell membrane that open to the receptor cavity. Acetylcholine esterase has been localized in association with these infoldings and the receptor cavity (McLaren, 1972).

Amphids of the hookworm *N. americanus* and the gapeworm *S. trachea* become strikingly modified at a specific stage in their life cycle. Fourteen dendritic processes occur in the amphidial canal of *N. americanus* (Fig. 26, McLaren, 1974a). Variable numbers of doublet microtubules occur in them, and dendrite tips contain banded rootlets. The amphidial canal is apparently wider than in other species so that dendritic processes are not closely packed. The most striking variation in pattern is the development of sheath cell. In infective free-living third-stage larvae, the sheath cell is small and exhibits some infoldings of the inner cell membrane similar to those seen in *Nippostrongylus,* or the filarial nematodes. Following penetration and development in the host, the amount of membrane infolding is reduced. The cytoplasm enlarges progressively and by the fourth larval stage takes on characteristics of a gland (McLaren, 1976b). The nucleus has been located at the level of the nerve ring, whereas glandular cytoplasm extends for about one fourth the length of the worm in association with the lateral hypodermal cords. The cytoplasm contains abundant granular endoplasmic reticulum, Golgi, and membrane-bound secretory granules, indicative of active protein secretion. Secretory granules appear to mature as they pass from more posterior sites of synthesis toward the receptor cavity of the amphid. Membrane-bound collecting ducts penetrate the cytoplasm in this area and ultimately open to the receptor cavity. Histochemical staining showed the presence of acetylcholinesterase in the cytoplasm of the amphidial gland and in the amphidial canal. Secretory granules of the esophageal glands also contained the same enzyme. Since it was shown that adult but not larval worms release acetylcholinesterase into an *in vitro* incubation medium, it was suggested that amphids are responsible for secretion of the enzyme (McLaren, 1974b). The large volume of the glandular part of these amphids supports such a postulate, although the enzyme may also have been released from esophageal glands.

The amphids of *S. trachea* resemble those described above with a well-developed amphidial canal surrounded by a socket cell and a receptor cavity formed by a sheath cell and enclosing the tips of twelve dendrites (Jones, 1979). Eleven of these dendrites give rise to 14 dendritic processes that enter the amphidial canal (Fig. 27). The twelfth produces a single dendritic process that projects into the sheath cell, similar to those in *Nippostrongylus* and *Heterakis.* The membrane of the receptor cavity has small infoldings that penetrate a short distance into the sheath cell cytoplasm. The cell body of the sheath cell lies posterior to the nerve ring. In young adults recovered from lungs (4–6 days after infection), the perinuclear sheath-cell cytoplasm contains some endoplasmic reticulum and Golgi bodies, but few secretory granules. By 6–7 days after infection, the adult worms migrate and attach to the tracheal mucosa. The perinuclear cytoplasm of the sheath cell then contains extensive endoplasmic reticulum (ER) and Golgi units and the more anterior part of the cell is filled with dense secretory granules. Infoldings of the receptor cavity membrane are now much more extensive and branching (Fig. 28). Presumably these serve for the release of secretory material into the receptor cavity. Nevertheless, the receptor cavity remains small. In sexually mature adults (15–25 days after infection) the number of secretory granules in the sheath cell is reduced, but the receptor cavity is greatly enlarged. Thus the major period of secretion production by the enlarged sheath cell is restricted to the phase when the worms establish themselves in their final feeding location. It may be significant that the content of the expanded

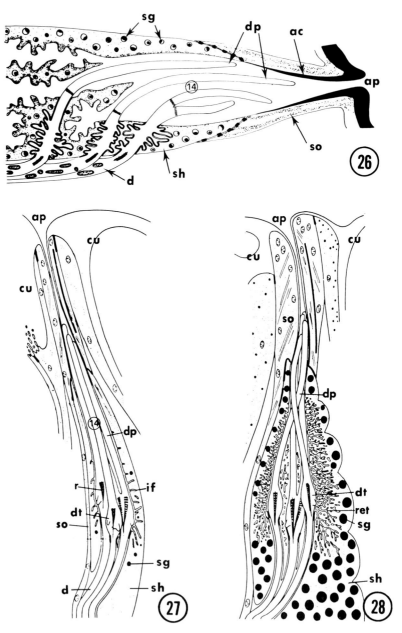

Fig. 26. Amphid of adult *Necator americanus* showing large membrane channels opening to the receptor cavity and the presence of secretory granules in the enlarged glandular sheath cell. Circled number indicates number of dendritic processes entering the amphidial canal. See the legend to Fig. 2 for the key to abbreviations. (From McLaren, 1974a, reproduced by permission.)

receptor cavity of older adults is considerably less electron dense than earlier, when the main release of secretory products occurs. Perhaps this indicates yet another functional phase of the sheath cells.

Amphids have been studied in detail in only two parasitic members of the Adenophorea-the trichuroid *Capillaria hepatica,* and the mermithid *Gastromermis boophthorae.* Both these species show notable divergences from the pattern described above.

Amphids of *C. hepatica* have small amphidial pores opening into only a small cuticle-lined chamber or amphidial canal (Fig. 25: Wright, 1974). Ten dendritic processes extend various distances in a spiral path within a long receptor cavity. At their bases the dendritic processes contain a circlet of nine doublet tubules and one to three single tubules. The processes vary in length, and each arises from a dendrite. Cell processes around the small amphidial canal are identical to adjacent hypodermal cells, but may be considered as socket-cell processes. A thin cell process surrounds the rest of the amphid and may represent the sheath cell. However, no infolding of the cell membrane or secretory granules occurs.

Observations on the amphids of the (free-living) adult *G. boophthorae* are included in this section since mermithids may be considered primarily parasites of insects; their free-living phases are nonfeeding. The amphidial canal is broad and cup shaped, and the cuticle at its base is relatively thick (Batson, 1978). Fifteen to 18 dendritic processes penetrate this basal cuticle through a narrow canal and project into the outer amphidial chamber (Fig. 29). Internally, the dendritic processes include a circlet of nine doublet microtubules just before they enter the tips of dendrites. Some dendrites give rise to more than one dendritic process. Below the cuticle of the amphidial canal are processes of hypodermal cells that may represent a socket cell, while the anterior tips of the sensory dendrites are enclosed in a receptor cavity by an extensive sheath cell. However, the dendrites run through only the anterior part of the sheath cell to the outside where they then lie in grooves along its outer surface. The anterior part of the sheath cell contains an extensive reticulum of membranes that open to the receptor cavity. More posteriorly, the extensive sheath-cell cytoplasm contains granular endoplasmic reticulum and accumulations suggestive of secretory material.

d. Free-Living Marine Nematodes. Most marine nematodes belong to the class Adenophorea and are characterized by having a more extensive or obvious complement of sense organs than their terrestrial counterparts. Many peripheral

---

Fig. 27-28. Amphids of *Syngamus trachea*, 4–6 days after infection (Fig. 27) and 6–7 days after infection (Fig. 28), showing the accumulation of secretory granules in the sheath cell and development of more reticular infoldings of the inner sheath cell membrane. Circled number indicates number of dendritic processes entering the amphidial canal. See the legend to Fig. 2 for the key to the abbreviations. (After Jones, 1979.)

sense organs are setiform. Amphids are characterized by elaborately shaped cuticular components (i.e., the amphidial canal). These produce spiral, circular, cyathiform, and "shepherd's crook" patterns that have been of great value in taxonomic studies. This suggests that several major patterns of amphid structure may ultimately be defined.

The amphids of *Oncholaimus vesicarius* consist of a "slot-shaped" amphidial pore opening into a cuticle-lined amphidial canal (Burr and Burr, 1975). This region is associated with a socket-cell process (Fig. 30). Twenty-eight to 36 dendritic processes project into the canal. These arise from only three dendrites. One dendrite has a markedly less dense cytoplasm. Nine and eight doublet microtubules plus one to three single microtubules occur in the base of the dendritic processes. The tips of the dendrites are enclosed in a receptor cavity by a sheath cell. Another dendrite occurs in a separate pouch of the sheath cell and gives rise to 10 dendritic processes that lie across the front of the esophagus where they may function in photoreception. Although both chemosensory and photoreceptive functions may thus be undertaken by the amphid, the two activities are undertaken in separate receptor cavities.

Amphids of *Tobrilus aberrans* contain the tips of 16 dendrites, each apparently with a single dendritic process (Storch and Riemann, 1973). The amphidial canal is broad, so there is ample space between tips of the dendritic processes. Although details of associated cells were not given, it appears that the amphidial canal ends abruptly and that there is a narrower diameter extension of the canal which may be the anterior part of the receptor cavity.

Detailed studies of cephalic setae of marine nematodes have not yet been published.

2. Posterior Cuticular Sense Organs

The posterior end of the nematode contains a series of ganglia and commissures connected to the dorsal and ventral nerve cords that comprise the posterior nervous system. Sense organs of the tail connect into this nerve center.

a. Phasmids. Phasmids are lateral, paired, sense organs opening by a pore through the cuticle in the tail of secernentian nematodes. They have been examined only in the filarial nematodes *Dipetalonema viteae, D. setariosum, Dirofilaria immitis,* and *L. carinii,* in the hookworm *N. americanus,* and in the guinea worm *Dracunculus medinensis.* Phasmids in filaria contain only a single dendrite bearing a single dendritic process (McLaren, 1972); in hookworms they contain two dendrites and therefore two processes (McLaren, 1976b). A socket cell is associated with a small cuticular pore and a short invagination of cuticle—a phasmidial canal. The dendritic process and dendrite tip are enclosed by a sheath-cell process, but the receptor cavity is greatly reduced. The sheath-cell membrane has complex infoldings as in the anterior sense organs. These

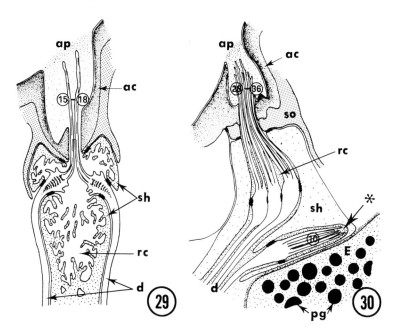

Fig. 29.   The amphid of *Gastromermis boophthorae* showing large amphidial canal and pore and arrangement of dendrites external to the central sheath cell. Circled number indicates the number of dendritic processes in the amphidial canal. See the legend to Fig. 2 for the key to abbreviations. (After Batson, 1978.)

Fig. 30.   The amphid of *Oncholaimus vesicarius* showing the large number of dendritic processes (circled number) that arise from only four dendrites. Asterisk notes the separate dendrite that may be photosensitive. See the legend to Fig. 2 for the key to abbreviations. (After Burr and Burr, 1975.)

infoldings and the receptor cavity have been demonstrated in *D. viteae* to contain acetylcholinesterase (McLaren, 1972). The phasmid of *D. medinensis* apparently encloses a much larger receptor cavity (Muller and Ellis, 1973).

b. Caudal Papillae and the Bursa of Males.   The tail of male nematodes bears prominent paired papillae on the ventral surface. They have been considered to function primarily as tactile receptors during copulation.

Males of the plant parasite *Aphelenchoides blastophthorus* have three pairs of caudal papillae; one pair adanal, the largest pair half way down the tail, and a subterminal pair (Fig. 32) (Clark and Shepherd, 1977). All of these contain two sensory dendrites, each ending with a single dendritic process. The basal part of the dendritic process contains an axoneme-like pattern of doublet microtubules. One dendritic process contains a core of dense material between single microtubules, and ends in the cuticle, whereas the second ends within a pore in the

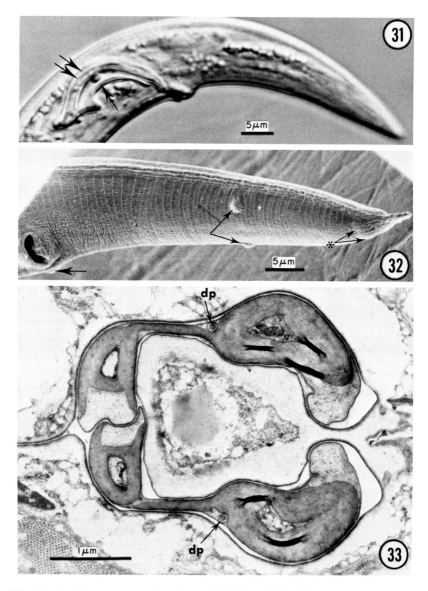

Fig. 31. Interference contrast photomicrograph of the tail of *Aphelenchoides blastophthorus*. Double arrows indicate the dorsal "limb" of the spicule and the single arrow notes the ventral "limb" (×2000). (From Clark and Shepherd, 1977, reproduced by permission.)

Fig. 32. Scanning electron microscope view of *Aphelenchoides blastophthorus* showing one adanal papilla (large arrow) just below the anus, the large median papillae (small arrows), and the location of the openings of the terminal sense organs (asterisk) (×2,400). (From Clark and Shepherd, 1977, reproduced by permission.)

cuticle (Fig. 34). The subterminal pair of sense organs each has an invagination of cuticle that encloses the tips of the dendritic processes. Sheath cells do enclose the dendritic processes and tips of the sensory dendrites, and a sizable receptor cavity, apparently with membrane infoldings, occurs. The receptor cavity and dendrite tips occur in a dorsolateral position in the tail, whereas the dendritic processes run ventrally to associate with the cuticular papilla 4 μm distant in a medioventral position.

In *D. viteae*, the papillae (three pairs precloacal, one pair adcloacal, and three pairs postcloacal) consist of large projections of the hypodermis and sensory cells into the cuticle (McLaren, 1972). Only a thin layer of cuticle overlies the apical sense organ, which consists of a single dendritic process arising from a dendrite tip (Fig. 35). Although a socket-cell process has been described, no sheath cell was identified in these early studies. The dendritic process extends into the cuticle, expands, and contains dense intertubular material as in the cephalic papillae of this species. The dendrite tip is simple, without microvillus-like processes, and no rootlet material occurs. The terminal pair of papillae differs since their dendritic processes do not expand, the dense intertubular material is distributed in an intricate reticular pattern, and the dendrite tips include rootlets. A fine canal runs from the tip of the dendritic process to the outer surface of the cuticle.

The cuticle of the male's tail is frequently extended to form lateral alae or a bursa composed of cuticle that encloses cellular rays, each with a sense organ at their tip. Although cells responsible for the formation of the bursal rays in *Caenorhabditis elegans* have been identified, and the form of the bursa has been examined by interference light microscopy (Sulston and Horvitz, 1977), detailed ultrastructural studies remain to be published. The bursa includes nine pairs of rays: the sensory tips of three pairs open to the dorsal surface of the bursa, and the remainder open ventrally (Fig. 36).

The bursa of *N. brasiliensis* serves to hold the male to the female during copulation, yet the rays consist mainly of hypodermis and associated sensory nerves, whereas muscle is restricted to their bases (Croll and Wright, 1976). Sense organs are restricted to the periphery of the bursa. They have not been characterized in detail, but each ray contains two dendrites that end in an expansion containing rootlet material and give rise to a dendritic process. Single micrographs suggest that both socket and sheath cells occur. The sheath-cell membrane lacks infoldings, no secretory granules occur, and the receptor cavity is reduced. Scanning electron microscopy (Uni *et al.*, 1977) of *Ancylostoma* spp.

Fig. 33. Cross section of the spicules of *Aphelenchoides blastophthorus* with the large dorsal "limbs" to the right and smaller ventral limbs to the left. Dendritic processes here lie external to the dense cuticle just above the "hinge" element of the spicules ($\times 25,000$). See the legend to Fig. 2 for the key to abbreviations. (From Clark and Shepherd, 1977, reproduced by permission.)

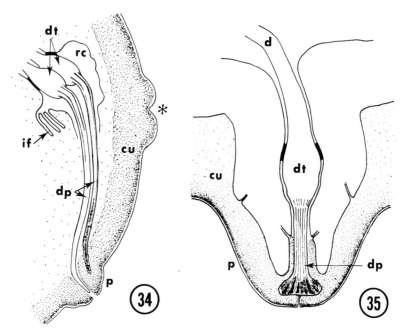

Fig. 34.   Arrangement of sensory dendrites that innervate the large median papillae of the tail of *Aphelenchoides blastophthorus*. Asterisk notes the lateral field of the tail. See the legend to Fig. 2 for the key to abbreviations. (After Clark and Shepherd, 1977.)

Fig. 35.   Diagram of caudal papillae of *Dipetalonema viteae*. Note the reduced receptor cavity. See the legend to Fig. 2 for the key to abbreviations. (After McLaren, 1972.)

shows small papillae at the inner margin of the bursa, suggesting that the receptors may be mechanosensitive.

Four sensory dendrites have been noted in each of the large postcloacal lobes of the tail of males of the trichuroid *C. hepatica,* and scanning electron microscopy has shown prominent buttonlike adcloacal papillae in *Trichuris muris* (Wright, 1978, and unpublished observations). The large postcloacal papillae of the male pinworm *Syphacia obvelata* contain a single receptor each (Dick and Wright, 1974), but their detailed anatomy remains to be further characterized.

c. Spicules.   Spicules are the principal accessory reproductive structures possessed by most male nematodes. They are formed in a dorsal outpouching of the cloacal wall (except in trichuroids where the spicule pouch is ventral), usually occur in pairs, and are protruded through the cloacal opening during copulation. Their innervation was shown by electron microscopy by Lee (1973); thus their importance as sensory structures has only recently been appreciated.

No studies have been reported of spicules of *Caenorhabditis* or other free-living nematodes, although several have been completed on tylenchid plant parasites.

Spicules of *Heterodera* spp. and *Pratylenchus penetrans* have two pores at their tip (Clark *et al.*, 1973; Wen and Chen, 1976). Dendritic processes containing single microtubules occur in the cuticle. Toward the bases of the spicule there is a core of cells including neuron processes. Through most of the spicule the core is surrounded by dense, presumably rigid cuticle that probably shields the sensory processes from mechanical impact. Since the sensory processes are associated with pores, these spicules are thought to be chemosensitive. Spicules of *A. blastophthorus* are composed of two thick "limbs" connected by a thinner "hinge" (Clark and Shepherd, 1977) (Fig. 31). A single dendritic process lies in the dorsal limb enclosed by dense cuticle. Almost halfway down the spicule, the dendritic process moves outward through the dense cuticle to lie just beneath the cuticle surface (Fig. 33). The tapering process extends toward the tip but apparently ends about 3 μm behind it. Thus the dendritic process is not very exposed to the external medium, and this raises some question regarding its efficiency as a possible chemoreceptor. Closer to the base there are spots where dense cuticle does not completely enclose the "nerve," and these were suggested to be pressure sensitive sites. Similarly, the distal end of the sensory process lies external to the rigid cuticle and hence could be mechanosensitive. The spicules of *Tylenchulus semipenetrans* each contain two sensory dendrites that enter the base or

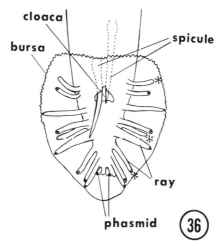

Fig. 36. The bursa of *Caenorhabditis elegans* showing its nine pairs of rays and phasmids. Asterisks note tips of rays whose sense organs open to the dorsal surface. (After Sulston and Horvitz, 1977.)

"capitulum" of the spicule (Fig. 37) (Natasasmita and DeGrisse, 1978b). Accompanying socket and sheath cell processes here enclose the expanded tips of the dendrites. The receptor cavity membrane has infoldings. Each dendrite tip gives rise to a single dendritic process that contains eight doublet axonemal tubules; however, one dendrite process is short, whereas the second extends to the tip of the spicule where there is a single pore.

In contrast to tylench spicules, the single, long, tubular spicule of the trichuroid animal parasites *C. hepatica* and *T. muris* has been shown to contain numerous distally located cuticular receptors (Wright, 1978). Pores through the cuticle are distributed along the distal part of the spicule, suggesting a chemosensitive role. Each pore is associated with a small cuticle-lined chamber that houses the tips of dendritic processes (two in *C. hepatica,* four in *T. muris*). The bases of dendritic processes contain doublet microtubules in an axonemal pattern, and the processes arise from enlarged dendrite tips. It is not clear whether a separate socket-cell process occurs, but the dendritic processes and dendrites are enclosed by sheath-cell processes. The receptor cavity is greatly reduced, but filamentous material does occur in the chamber around the tips of the dendritic processes. Although spicules of a few other animal-parasitic nematodes have been examined (e.g., McLaren, 1976a; Croll and Wright, 1976), their greater size and complexity and difficulties in sectioning have precluded detailed observations. However, a single dendritic process has been identified in the tip of the spicule of *Di-*

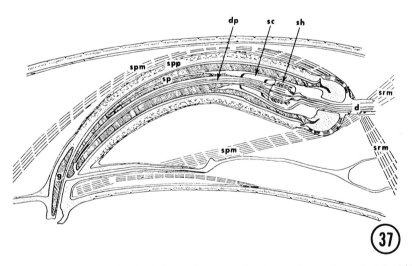

Fig. 37. The spicules of *Tylenchulus semipenetrans* showing one long and one short dendritic process within the central canal of the spicule and the receptor cavity within its anterior end. See the legend to Fig. 2 for the key to abbreviations. (From Natasasmita and DeGrisse, 1978b, reproduced by permission.

petalonema viteae, and one and perhaps a second dendritic process may occur in spicules of *Necator americanus* (McLaren, 1976a).

Other sensory units are associated with male accessory sexual apparatus. In *Aphelenchoides blastophthorus,* there are two lateral projections of the cloacal cuticle that each contain a single dendritic process (Clark and Shepherd, 1977). The tips of these projections protrude from the cloacal opening on either side of the spicule when the latter is protracted. No pores were seen in the projections. Internally the dendritic process does contain nine doublet microtubules, and a receptor cavity with an enclosing sheath cell process is evident. Such receptors could be mechanosensitive. Wen and Chen (1976) noted a pair of sensory units in the cuticle forming the large posterior lip of the cloacal opening of *Pratylenchus penetrans,* and Dick and Wright (1974) found a single "nerve" in the cuticle forming the "accessory piece" (modified cuticle of the posterior cloacal lip) of the pinworm *Syphacia obvelata.* The genital cone (gubernaculum–telamon complex) of *N. brasiliensis,* through which the spicules are protruded, contains four receptors (Croll and Wright, 1976). Each consists of a single dendritic process arising from the tip of a dendrite. The dendritic process projects into the cuticle and its expanded tip contains much dense intermicrotubular material. There is no pore through the cuticle. Although much of the genital cone is composed of dense sclerotized cuticle, the tip of the sensory process is not covered by it. These sense organs are probably mechanosensitive. A sheath-cell process is clearly present. Although the receptor cavity is small, the sheath-cell membranes bear many extensive infoldings similar to those in cephalic sense organs of this nematode. A socket cell process may also occur.

Thus, spicules might respond to a variety of stimuli, either mechanical or chemical. Laser ablation of the base of the spicule in *Panagrellus* spp. interfered with mate location via pheromones (Samoiloff *et al.*, 1973). However, Ward *et al.* (1975) suggested that chemoreceptors found only in cephalic sense organs of male *Caenorhabditis elegans* might function in mate selection. Wright (1978) pointed out that the chemosensory spicule of trichuroids might more likely sense the internal environment (receptiveness) of the female tract, since it is normally located well inside the male's cloaca and could be extruded only by artificial mechanical intervention with controlling nerves—a crude substitute for some other sensory cue. Spicules are absent in genera such as *Aspiculuris* and *Trichinella.* Adanal papillae (e.g., of the trichuroids), mechanoreceptors of the genital cone of strongyloids, or sensory units of the cloacal opening might signal close apposition of the gonopores at full coupling.

3. Somatic Cuticular Sense Organs

a. Deirids and Postlabial Papillae.   Deirids are a single pair of often papilliform sense organs lying one on either side of the body usually just posterior to the nerve ring. Those of *C. elegans* contain a single sensory dendrite with a single dendritic process that enters the cuticle, expands, and lies parallel to the cuticle's surface (Ward *et al.*, 1975). There is abundant dense material between microtubules in the dendritic process, suggesting that the unit is probably a mechanoreceptor. Both socket and sheath cell processes were identified (Ward *et al.*, 1975). McLaren (1976a) reviewed G. M. Jones's unpublished description of deirids of *Syngamus trachea* (which similarly appear to be single dendrite mechanoreceptors). She suggested that their prominent position along the body could signal the space available when nematodes penetrate restricted pores or cavities. Postdeirids may be similar sense organs located about half-way along the body.

Two pairs of postlabial papillae have been found in *Heterakis gallinarum* (Wright and Hui, 1976) and in *Ascaridia galli* (K. A. Wright, unpublished observations) in the lateral field on both sides of the body just behind the lips. These closely resemble the lateralmost outer labial papillae of the lips except that their peglike projection is longer (Fig. 38). Since they include two dendrites, one extending to a pore at the tip of the peg and the other ending just inside the internal cuticular cylinder and having dense material between its microtubules, it is likely these papillae have both chemosensory and mechanosensory capacities. Although sheath and socket processes were not described originally, it appears that a socket-cell process is associated with the outer cuticle, whereas the sheath cell surrounds the inner cuticular cylinder and also gives rise to a complex system of membrane infoldings forming a "membranous" cord (Fig. 23), as do sense organs of the head.

b. Body Pores. Dorylaimoid nematodes have numerous fine cuticular pores that open to small sensory receptors in the lateral and median hypodermal cords. In *Xiphinema americanum*, each "pore" includes four dendrites and their dendritic processes that contain a well-organized pattern of nine axonemal doublet microtubules near their base (Wright and Carter, 1980). A small socket-cell process is probably present. A sheath cell wraps around the dendrites and dendritic processes and extends posteriorly in the hypodermal cord. One such cell was traced to its cell body about 15 μm from the pore. The sheath cell encloses the sensory dendrites only at their anterior tip, and septate junctions connect den-

Fig. 39. An internal sensory-neurosecretory unit associated with a gland cell of the bacillary band of *Capillaria hepatica*. The asterisk notes infolded membranes of the gland cell that may be the site of the ion–water regulation. See the legend to Fig. 2 for the key to abbreviations. (From Wright and Chan, 1973, reproduced by permission.)

Fig. 40. A bipolar sensory cell (large asterisk) associated with a hypodermal gland cell in the body wall of *Chromadorina germanica*. Dendritic processes project close to the site of release of secretory materials below the cuticular pore. A cell identified as a glial cell is noted by small asterisk. See the legend to Fig. 2 for the key to abbreviations. (After Lippens, 1974.)

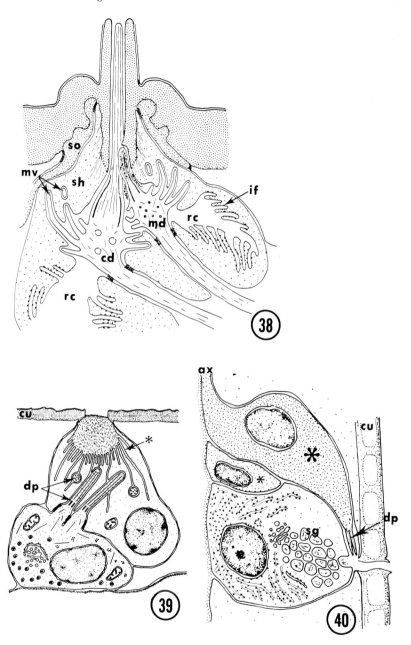

Fig. 38. Postlabial papilla of *Heterakis gallinarum*. The left-hand portion of the sheath cell extends further into the body wall as a "membranous cord." See the legend to Fig. 2 for the key to abbreviations. (After Wright and Hui, 1976.)

drites to each other and to the sheath-cell membrane. The receptor cavity is readily identified, but neither membrane infoldings nor secretions were found. Such receptors are probably chemosensitive. A single receptor was located in the ventral cord of the body wall in the esophagus region of *Capillaria hepatica* (Wright and Hui, 1976). Since this unit opens to the exterior by a pore, it may be related to the body pores of dorylaims. The pore opens to a large chamber that encloses the tips of four dendritic processes. McLaren has found similar receptors in *Trichinella spiralis* (McLaren, 1976a).

c. **Setae.**   Only one study has considered the functional morphology of setiform body sense organs of the adenophorean nematodes. In *Chromadorina bioculata,* a freshwater chromadorid nematode, setae include two dendritic processes. The seta appears to be formed of a more rigid cuticle, suggesting that it may only be bent at its base. No pores were detected in the cuticle nor was the typical dense intermicrotubular material demonstrated. Nevertheless, it was suggested that these setae may be mechanoreceptors (Croll and Smith, 1974).

III. INTERNAL SENSORY RECEPTORS

Relatively little is known about receptors that do not include the body wall cuticle as a major part of their structure. Nevertheless, it has long been recognized that the esophagus contains nerves that could include receptor units. Photoreceptors have been recognized as internal sense organs. Recent studies are beginning to document the presence of sensory units in the head tissues, in the body wall, or associated with the alimentary tract; these will be considered here as internal receptors.

A. Cephalic Internal Receptors

Nervous elements located within cephalic tissues, but clearly distinguishable from the cuticle-associated cephalic sense organs, were first recognized in the careful serial reconstruction studies of this region of *Caenorhabditis elegans* (Ward *et al.,* 1975; Ware *et al.,* 1975). There is little to indicate that these elements function as sensory receptors; however, their association with the central nervous system and the presence in them of microtubule doublets in highly suggestive.

Ward *et al.* (1975) identified two pair of these "ciliated" neurons that wrap around parts of the socket cell process of lateral inner labial sense organs (Figs. 4 and 5). They contain nine doublet microtubules about 1 μm long near their tip. One, which also contains a short banded rootlet, sends a branch to the dorsal outer labial sense organ. Four nonciliated neurons with sheetlike termini were identified. They lie just internal to the outer labial cuticular sense organs. Ware *et al.* (1975), however, identified only one pair of neuron processes, each ending in a club-shaped process with nine doublet microtubules and a banded rootlet. These lie just ventral and internal to the lateral outer labial sense organ. Ward *et al.* (1975) noted that the form of these neurons is considerably variable from animal to animal, perhaps related to age.

Flat ramifying processes have been illustrated in several tylenchid plant parasites, notably in *Meloidogyne incognita* (Fig. 12, Endo and Wergin, 1977), *Tylenchulus semipenetrans* (Natasasmita and DeGrisse, 1978a), and *Aphelenchoides fragariae* (DeGrisse *et al.*, 1979). Indeed, they are probably characteristic of all tylenchids (A. T. DeGrisse, personal communication). In *Tylenchulus* they have been shown to connect to four submedian dendrites that contain six to seven microtubule doublets. None have been found to have sheath cells or a receptor cavity associated with them.

Two pairs of dendrites occur in the dorylaim plant parasite *Xiphinema americanum* (Wright and Carter, 1980). These closely resemble the form of other sensory dendrites but do not associate directly with cuticle. One pair occurs just below the cuticle and hypodermis and just lateral to the lateral inner labial sense organs. The dendrite tip is enlarged and gives rise to a short dendritic process containing nine doublet microtubules (Fig. 15). The dendritic process expands and contains a granular, moderately dense material, but few microtubules. The second pair of internal dendrites occurs with the dendrites of the amphid. The dendrite tip lies posterior to the tips of the amphidial dendrites and a single long dendrite process containing a typical nine-doublet axonemal pattern of microtubules arises from it (Fig. 18). The dendritic process passes close to the cell body of the amphidial sheath cell and progresses anteriorly to the amphidial pore. It is enclosed by a process of the anterior hypodermis at its tip. At the level of the amphidial pore, the dendritic process turns through 90° to lie along the outer lip of the amphidal slit.

No comparable internal cephalic receptors have been identified in animal parasites.

B. Photoreceptors

Although phototropic behavior has been demonstrated in a variety of nematodes, the sensory units functional in photoreception have not been clearly identified. Nevertheless, many marine nematodes have pigment bodies associated with the anterior esophagus that have generally been thought to be eyespots, or ocelli. These appear to have several morphologies. Generally pigment granules are found in cells of the esophagus. In *Oncholaimus vesicarius* they occur in the radial muscle cells (Burr and Burr, 1975), while in *Deontostoma californicum* they occur in marginal cells of the esophagus (Siddigui and Viglierchio, 1970a). The pigments of *Araeolaimus elegans,* however, are located outside the esophagus (Croll *et al.*, 1975). Sensory elements have been identified with some certainty in only two nematodes. In *O. vesicarius,* one of the amphidial dendrites produces a group of 10 dendritic processes that lie in a pouch of the sheath cell separate from the amphidial receptor cavity (Fig. 30). Since these 10 dendritic processes are oriented medially in front of the pigmented esophagus, it has been suggested that they may be responsible for demonstrable photoreceptive behavior (Burr and Burr, 1975; Burr 1979). In contrast, the pigment-containing esophagus of *D. californicum* forms a cup that cradles an ovoid cell body containing many closely packed membrane lamellae (Siddiqui and Viglierchio, 1970b). This cell may be a modified bipolar neuron whose axon enters the lateral nerve bundle. Multilaminated membrane bodies, not clearly associated with a cell body or neurons, have been noted adjacent to the eyespot pigment in *Enoplus communis, Chromadorina* sp., and *Araeolaimus elegans* (Croll *et al.*, 1975). Pigment granules have been identified by histochemistry or spectroscopy as melanin. They probably serve to shade the photoreceptor unit, allowing directional discrimination. No true lens occurs. Pigment identified as hemoglobin, and apparently contained in semicrystalline form in the hypodermis of the anterior body wall, constitutes the "chromotrope" of the head of the mermithid *Mermis nigrescens*. The free-living adults of this species do show phototaxis, and changes in oviposition rate with light changes.

C. Internal Receptors of the Body Wall

The bacillary band of the trichuroid nematodes consists of a large number of lateral hypodermal gland cells embedded in the hypodermal cords, which open to the exterior through cuticular pores. It has been suggested that they may function in osmotic and/or ionic regulation in the nematode (Wright, 1963). Toward the anterior level of the esophagus a small number of these cells have a single sensory cell associated with them (Wright and Chan, 1974). This cell lies along the internal surface of the body wall exposed to the pseudocoelom (Fig. 39). Four to six short processes project into the hypodermal gland cell, and from each a dendritic process arises. They contain a complex of nine doublet tubules basally, but only single tubules at their tip. The dendritic processes project between the infolded cell membrane of the hypodermal gland cell so that they lie in the extracellular channels formed by the complexly infolded apical membrane. They may sense ionic gradients that form here during ionic or water regulation according to the standing gradient theory. The cell body contains some endoplasmic reticulum and Golgi, but primarily large numbers of dense-cored vesicles. These might be neurosecretory products. Since no efferent cell processes were found, it was suggested that exocytosis of neurosecretions into the pseudocoelom could affect unknown target cells located elsewhere. McLaren (1976a) records the presence of similar sensory units in *Trichinella spiralis*.

A bipolar sense cell has been described (Lippens, 1974) associated with lateral hypodermal gland cells of the free-living marine nematode *Chromadorina germanica*. Hypodermal gland cells in this species contain endoplasmic reticulum and secretory products characteristic of a typical merocrine gland cell (Fig. 40). Secretory products are released through a pore in the cuticle. Adjacent to the gland cell lies a neurocyte with a short dendrite bearing two dendritic processes that extend to the cuticular pore. An axonal process was noted extending from the neurocyte and another adjacent cell was identified as a glial cell. Perhaps this receptor monitors the secretory activity of the gland. Unfortunately axonal connections are not known.

Several structures have been identified in the body wall that might serve as proprioceptors. In larval hookworms, a single dendritic process and dendrite has been identified on either side of the nematode in the hypodermis below the lateral alae at the level of the nerve ring (Smith and Croll, 1975). The processes are longitudinally aligned and may interact with the overlying cuticle via an ill-defined "associated body." They have been suggested to be proprioceptive, but as noted by McLaren (1976a) they could also be the innervation of deirids. In the same study, attention was called to the presence in the body wall of neurons with very prominent and regularly arranged microtubules. Three pairs of similar neurons in *Caenorhabditis elegans* may be responsible for touch sensitivity (Chalfie, 1979) since touch-insensitive mutants have defects in these neurons. The microtubule organization of these neurons has been analyzed by Chalfie and Thomson (1979). Fine filamentous structures have been identified in the lateral cord of some marine enoplid nematodes. They were suggested to be proprioceptors recording muscle action and/or volume or pressure changes in the nematode during locomotion (Lorenzen, 1978).

There remains yet another structure in the body wall thought to be sensory—the hemizonid. Although these structures were first noted in soil nematodes, they have also been documented in several animal parasites (Smith, 1974; McLaren, 1976a). They appear as elongate tracts running ventrolaterally in the body wall at the level of the nerve ring. Clearly, these are large nerve commissures, and Ware *et al.* (1975) have shown that in *C. elegans* they comprise axons of the amphidial neurons that emerge from the lateral ganglion, travel around the body wall, and at the ventral line return radially to the ventral ganglion. The same arrangement of nerve processes was found in adults of both *Nippostrongylus* and *Heterakis* (K. A. Wright, unpublished observations). Although the significance of such a loop is unknown, they show no developments suggestive of any sensory role.

Similar nerve tracts have been referred to as cephalids, hemizonium, and caudalids, according to their locations in the body. It is doubtful that they are sensory.

D. Internal Receptors of the Alimentary Tract

Few studies have considered how activities of the various internal organ systems might be regulated or coordinated. Since many body functions may operate via simple mechanical principles (Crofton, 1971), the need for internal regulation may seem to be minimal. Nevertheless the esophagus has a well-developed nervous system that may have both internal receptor and effector units.

The esophagus of *C. elegans* has been studied by serial-section reconstruction techniques (Albertson and Thomson, 1976). It contains 34 muscle cells, 9 marginal cells, 9 epithelial cells, 5 gland cells, and 20 neurons. Twelve neurons comprising 7 different types have tips that end just below the lumen cuticle (Fig. 41). Junctional complexes bind the neuron tip to adjacent cells. Such an attachment might result in deformation of the neuron tip when the muscles contract, thus allowing proprioception. The endings of two paired (I1 and I2) and one single interneuron (I3) occur just behind the buccal capsule, endings of pairs of

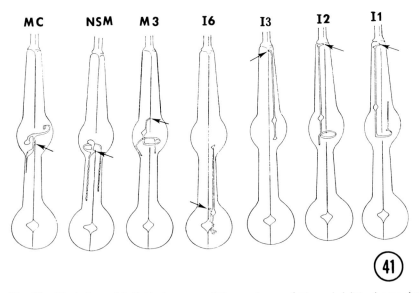

Fig. 41. Illustrations of individual neurons of the esophagus of *Caenorhabditis elegans* that have endings that may serve as proprioceptors (arrows): MC, one pair of marginal cell neurons; NSM, one pair of neurosecretory motor neurons; M3, one pair of motor neurons; I6, a single interneuron; I1 and 2, pairs of interneurons; I3, a single interneuron. (After Albertson and Thomson, 1976.)

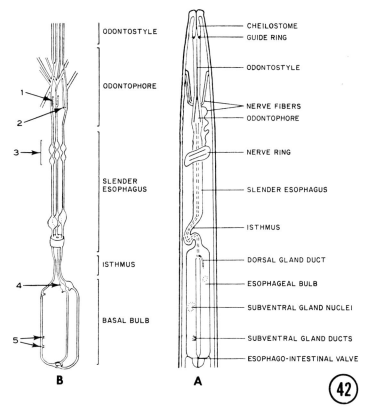

Fig. 42. The esophageal nervous system of longidorid nematodes. (A) Basic anatomy of the esophagus: (B) nerves with putative sensory sites numbered: (1) subdorsal nerve sinuses in odontophore; (2) ventral nerve sinus in odontophore; (3) "sensory" region in anterior esophagus; (4) and (5) "sensory" regions in posterior bulb. (After Robertson, 1979.)

marginal cell neurons (MC), neurosecretory motor neurons (NSM), and a motor neuron (M3) occur in the median bulb. The ending of the single interneuron (I6) occurs in the posterior bulb. The buccal capsule and median bulb thus contain more putative proprioceptors than does the posterior bulb; such a distribution may be related to the fact that food is drawn into the esophagus and collected in front of the isthmus before being passed through the posterior bulb.

Similar neuron terminals were noted between the anterior buccal muscle and the rest of the esophagus in *Nippostrongylus brasiliensis,* where they were also suggested to have a proprioceptive role (Figs. 19 and 44; Wright, 1976).

Longidorid nematodes feed by means of a long hypodermic-like stylet inserted into plant tissues. The stylet is subtended by thickened esophageal cuticle (the odontophore) that serves for insertion of stylet protractor muscles and contains

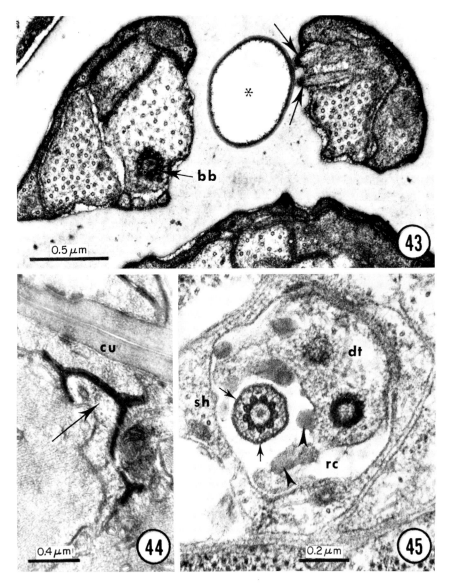

Fig. 43. Cross section of the subdorsal nerve sinuses in the odontophore of *Xiphinema diversicaudatum* showing peglike processes (arrows) from a neuron in the right sinus and a basal body in a neuron in the left sinus (bb); Asterisk notes the food canal (×44,000). (From Robertson, 1975, reproduced by permission.)

Fig. 44. Longitudinal section through the juncture of buccal capsule muscle and the rest of the esophagus of *Nippostrongylus brasiliensis* (×35,000). Arrow notes a possible proprioceptive ending of a neuron with prominent dense adherens junctions to adjacent muscle cells; cu, cuticle.

three sinuses of cell processes. (Fig. 42A). Each of these contains neurons which have peglike processes that project into thin spots in the odontophore cuticle (Robertson, 1975). In *Xiphinema diversicaudatum* a basal body lies in the neuron (Fig. 43). In *X. americanum* no basal bodies were seen (R. Carter, unpublished observations). Numerous similar peglike processes (lacking basal bodies) have also been found in *Longidorus leptocephalus* and in *X. diversicaudatum* in the anterior part of the slender esophagus, subventrally in the anterior esophageal bulb, and at two dorsal locations in the posterior esophageal bulb (Fig. 42; Robertson, 1979). Although pores do not occur through the esophageal cuticle, it is thought that these could be chemosensitive areas (especially those in the odontophore) or mechanoreceptive, detecting the flow of saliva or food, or movement of the esophagus. In addition, there are three locations where nerves were suggested to be especially subject to pressures or stresses: at the base of the odontophore, at the constricted isthmus of the slender esophagus, and at the esophageal–intestinal valve, where proprioception might occur (Robertson, 1979). Since no modifications of the neurons were noted in these locations, there is less evidence for a sensory role in these neurons, recalling the history of hemizonids.

Examination of the intestinal–cloacal junction of *Heterakis gallinarum* has shown complex modifications of the posterior cells of the intestinal epithelium (Lee, 1975). This region is surrounded by a sphincter muscle responsible for closure of the posterior end of the intestine (Fig. 46). The posterior intestinal cells contain many longitudinally aligned microtubules and, along the outer or basal cell membrane, there are numerous dense granules. Lee (1975) suggested that pressure of intestinal contents might be detected by the microtubule-containing cells, which could respond by releasing a ''sphincter muscle relaxing compound'' from the basal granules, thereby regulating defecation.

IV. CONCLUSIONS

A. Some General Characteristics of Nematode Sense Organs

Nematodes will soon be one of the better understood groups of lower invertebrates, with a simple body plan (pseudocoelomate), a well-developed central nervous system, and both peripheral and internal sense receptors. Parker (1919)

Fig. 45. Cross section of a sense organ of *Xiphinema americanum* showing one dendritic process cut through the level of the ciliary necklace (small arrows note connections of doublet microtubules to periphery), and a section of a dendrite tip showing the dense ringlike microtubule organizing center ($\times 66,000$). Arrow heads note material released from sheath cell into the receptor cavity. See the legend to Fig. 2 for the key to abbreviations.

postulated that in the evolution of sensory systems, units would develop capable of serving both as receivers and effectors of stimuli—a primitive sensory motor system. Although isolated examples of these units were first recognized in molluscs and coelenterates (Coggeshall, 1971; Westfall, 1973), it now appears that they may form an important component in the nematode's sensory system. Ward *et al.* (1975) pointed out that the mechanosensory dendrites of the inner labial sense organs of *C. elegans* make synaptic contacts directly with the anterior somatic muscle cells. Similarly, several of the proprioceptor neurons of the esophagus synapse with esophageal muscle cells, or are neurosecretory (Albertson and Thomson, 1976). Synapses originally identified in the anterior sensory nerves of the animal parasite *C. hepatica* (see Wright, 1974) may in fact be connections to somatic muscles. This nematode also has sensory–neurosecretory units associated with some cells of the bacillary band (Wright and Chan, 1973) that can be considered sensory motor units. The posterior intestinal cells of *Heterakis gallinarum,* suggested by Lee (1975) to have stretch receptor function, may also be sensory motor units, although they are presumably of endodermal rather than nervous origin.

The nematode sensory system represents an interesting level of organization between the nerve net systems of platyhelminths and the more complex central nervous systems of higher invertebrates and vertebrates. Apparently all sensory cell bodies occur in the central nervous system, whereas receptors are differentiations of their peripheral processes. Thus, sense receptors occur at the tips of long dendrites, while receptors in arthropods, for instance, are developments of peripherally located neurocytes with long axonic connections to the central nervous system (CNS). In vertebrates, the receptor units are more specialized peripheral cells with interneuron connections to the CNS.

The presence of an external proteinaceous exoskeleton (cuticle) presents the nematode with problems similar to those of other invertebrates, notably arthropods, in terms of perception of external stimuli. It is not surprising to find many parallels in the composition of the cuticular sensory receptors of both nematodes and arthropods. Fortunately, such parallels provide some assurance that functional interpretations of anatomical studies are reasonable since direct electrophysiological studies have been feasible in many arthropod systems. (Studies of sensory-defective mutants of *Caenorhabditis elegans* provide further support for functional interpretations in nematodes.)

Cuticular mechanoreceptors in both arthropods and nematodes characteristically have accumulations of dense material between microtubules of the dendritic process. The recent study by Matsumoto and Farley (1978) indicates that this material plays a major role in mechanoreception by maintaining spacing between tubules and ensuring that the surface membrane is stretched when mechanical stimuli are applied to it (Rice, 1975). However, significant differences in the

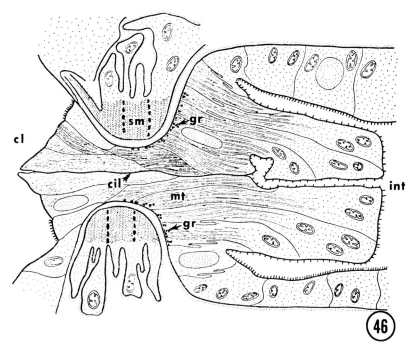

Fig. 46. The intestinal cloacal juncture of *Heterakis gallinarum* showing posterior intestina cells with many microtubules (mt) and basal granules (gr): int, intestinal lumen; cil, closed intestinal lumen; cl, cloacal lumen; sm, sphincter muscle. (From Lee, 1975, reproduced by permission.)

nature of nematode and arthropod cuticles result in differences in cuticular developments between arthropods and nematodes. Since the nematode cuticle is flexible, a specialized socket cuticle (such as arthropodan resilin) is not required (Wright and Hui, 1976); simple thinning of the cuticle may ensure flexing of a mechanoreceptor's cuticle at the right place. Mechanoreception may be mediated through nematode cuticle (1) by the penetration of the sensory dendritic processes well into the cuticle (e.g., the outer labial and cephalic sense organs of *Caenorhabditis* and tylenchid nematodes); (2) by the penetration of the entire multicellular complex of the sense organ into the cuticle, so that only a thin layer of cuticle covers the dendritic process (e.g., caudal papillae of *D. viteae*), or (3) in cuticles with an accentuated fluid median zone, by deposition of dense, presumably sclerotized proteins into the cuticle overlying the sensory dendritic processes (e.g., the outer labial and cephalic sense organs of *N. brasiliensis*).

The low permeability of nematode cuticle to compounds larger than inorganic ions necessitates cuticular pores to allow access of larger chemicals directly to sensory dendritic processes for chemoreception. Since nematodes are restricted to aqueous environments (water films for those occurring in soils or on vegeta-tion), their chemoreceptors more nearly resemble arthropod contact chemorecep-tors with single cuticular pores and simple dendritic processes.

However, the ready permeability of the nematode cuticle to water and dis-solved ions probably requires active maintenance of appropriate osmotic and ionic levels adjacent to the membrane of the receptive processes in order to allow transduction of receptor effects into propagated nerve impulses (Ward et al., 1975; Wright, 1975, 1977). It is in this activity that the sheath cell and receptor cavity probably play their major role. In insect sense organs, the receptor lymph cavity (analogous to the nematode's receptor cavity) has been shown to have high K^+ levels (Thurm, 1973; Kuppers, 1974). The degree of development of sheath cells necessary for this regulation probably depends largely upon the permeabil-ity characteristics of the cuticle and the environment occupied by the nematode. Since the cuticle generally is permeable to ions, regulation is required for both mechanoreceptors and chemoreceptors, even though only the latter open directly to the environment. Hence, the development of the sheath cell is greatly varied (e.g., the relatively simple membrane infoldings of Caenorhabditis elegans ver-sus the large membranous cords of Heterakis gallinarum). Infoldings of cell membranes with nonribosomal granules located on their cytoplasmic side, such as those found in the sheath cells, form sites where water and ions may be transported through epithelia via the standing osmotic gradient theory (Smith, 1969; Oschman and Berridge, 1971).

Sheath cells are also responsible for the production of the filamentous matrix in the receptor cavity. This may be analogous to the sensillum liquor characteris-tic of insect receptors and may function to lubricate or protect the sensory dendritic processes.

The sensory dendrite tip contains a limited number of organelles. The most prominent may be banded rootlets associated with the cilium-derived dendritic process. Rootlets may be present or absent in either mechanoreceptors or chemoreceptors. Their occurrence may be related to the level of similarity of the dendritic processes to functional cilia. That is, they generally are well developed in dendrites whose dendritic processes contain about nine doublet microtubules, whereas those processes with many doublet microtubules (e.g., filarioids, Syphacia and Heterakis) either lack rootlets or have very rudimentary strands of rootlet materials (exceptions are the terminal caudal papillae of D. viteae that do have well-developed rootlets). Their possible role in sensory reception remains unknown. Similarly, vesicles of various size and densities described in the den-drite tips are of unknown function. Perhaps their occurrence is related to the state of firing of the receptors. Mitochondria do not occur in the dendrite tips of most of the receptor units, but are characteristic of the complex dendrite with microvillus-like projections that occur in the amphids of C. elegans, tylenchids, and some animal parasites such as N. brasiliensis. This may suggest that these receptors are functionally distinct from others of the amphid. Furthermore, in *C. elegans* this neuron is not open to the receptor cavity of the amphid, whereas in the others it is.

Many authors have noted that typical basal bodies do not occur in the sensory processes of nematodes. Indeed, typical basal bodies or centrioles have not been recorded from somatic tissues, either. A few examples of small centrioles that contained only nine doublet rather than triplet tubules were located in *C. hepatica* (Wright, 1976). These doublets have a skewed orientation in the circlet, as do triplet tubules of typical centrioles. Sperm centrioles have been found to contain only single microtubules. Various numbers of doublet microtubules characterize the dendritic processes of sense organs. Most illustrations of these patterns show the doublets within the process itself. Especially in *C. elegans* and tylenchid plant-parasitic nematodes, doublet microtubules may extend nearly the entire length of the dendritic process. Toward their base the doublets show radial linkages to the cell membrane that are characteristic of the region of the ciliary membrane known as the ciliary necklace—these are regions of the ciliary axoneme, not the basal body. Gilula and Satir (1972) suggested that the ciliary necklace could have a role in controlling ciliary beat, perhaps recalling the mechanosensitivity of this region of kinocilia demonstrated by Thurm (1968). In sensory structures it could be important in the transduction of receptor effects into propagated impulses. It would then be significant that this region is conserved in nematode sense organs, even in forms with greatly increased numbers of doublet microtubules.

No organizations of tubules characteristic of basal bodies have been convincingly demonstrated in the cytoplasm of the dendrite tip. Instead, rather amorphous densities may occur. McLaren (1972) describes the many doublet microtubules of *D. viteae* as arising from a dense lamella, or of *Dirofilaria immitis* as arising from more irregular densities. Lee (1974) located a dense platelike structure in the dendrite tip of a cephalic sense organ of *Mermis nigrescens*. Serial sections through the bases of dendritic processes of *X. americanum* did not reveal basal bodies, but demonstrated a cylinder of dense material in which no microtubules could be discerned (Fig. 45; Wright and Carter, 1980). These densities may represent microtubule organizing centers that form the axonemal microtubules while microtubular basal bodies are entirely absent.

Croll (1977) and Ward (1978) have recently considered the role that sense organs may play, especially in the behavior of *C. elegans*. Ward suggests that at least nine classes of chemosensory receptor sites are required to account for the chemotaxes currently documented. These are presumably distributed over the dendritic processes of the inner labial sense organs and amphid. Gradient detection is concluded to be determined by sequential sampling by the chemoreceptors of the head as the head is moved back and forth during locomotion. Ward also points out that the position of the sensory dendritic processes with respect to the external pore of the sense organ should be considered. The levels to which dendritic processes extend into the amphidial canal and the occlusion of the canal by the processes will influence the ability of stimulatory compounds to reach receptor sites on internal dendritic processes. Nevertheless, Ward has suggested that dendritic processes of the amphid that do not enter the amphidial canal could be triggered as a result of inward diffusion of stimulants. This effect may be minimized by the restriction of the amphidial canal (e.g., note the extent of extracellular space in Fig. 8). On the other hand, the tips of these dendritic processes make intimate contact with the sheath-cell membrane of the receptor cavity by interdigitating with it. They may thereby sense either the environment of the receptor cavity or the condition of sheath-cell membranes and hence may monitor control of the receptor cavity environment (Wright, 1975). McLaren (1976b) suggested that dendritic processes in the amphidial canal might both record external stimulants and respond to the outward flow of secretions that may be formed by amphids of some animal-parasitic nematodes at special phases of their life. In such a system, it would seem difficult to provide mechanisms to allow for discrimination of external from internal stimuli (the ''stuffy nose state''?). Wergin and Endo (1976) suggested that monitoring of secretions might be the role of the microvillus-bearing dendrite of tylenchid amphids. However, the extent to which tylenchids may form secretions remains to be determined.

B. Comparison of the Sensory Complement of Secernentian and Adenophorean Nematodes

The classes Secernentia and Adenophorea represent the two major divisions of the nematodes. Secernentians occur predominantly as free-living nematodes in terrestrial habitats or as parasites of plants or animals. Adenophoreans occur predominantly in marine sediments, although some highly specialized groups have developed as parasites of plants and animals. It is interesting to note a major difference in the cuticular sensory complement of these two groups. In secernentians, amphids and phasmids are likely to be the primary receptors of chemical cues, whereas most of the remaining sense organs are mechanosensory. Development of chemoreception in these sense organs appears to vary greatly according to ecological demands. In contrast, adenophorean cuticular sense organs are apparently all chemoreceptive (excepting, perhaps, somatic setae). Furthermore, chemosensory ''papillae'' and amphids contain more sensory dendrites than do chemosensory units in secernentians. Amphids may have from 10 to 36 processes in contrast with 6 to 14 in secernentian nematodes. The processes also appear to be more openly exposed to the external environment. Even spicules of the adenophoreans (e.g., *Trichuris* and *Capillaria*) have more chemoreceptor units than those demonstrated in secernentians.

At least in *X. americanum,* the sheath-cell body has been found to lie far forward of the central nervous system, whereas those for example of *C. elegans* occur at the nerve ring level. Although membrane infoldings are characteristic of the sheath-cell membrane in receptors of secernentian nematodes, these developments are absent in the adenophorean sheath cells studied to date. Adenophoreans comprise several groups that can be recognized largely on the basis of distinctive patterns of their cuticular amphidial canal. It is likely that these will each show significant subpatterns in amphid structure when their anatomy has been examined in detail. Thus, major differences, both in patterns of amphid organization and in general sensory capacity, may occur between nematodes of the secernentian and adenophorean groups. Probably chemoreception plays a more significant role in the biology of adenophorean nematodes than in secernentians.

C. Specialization of Sense Organs in Plant Parasites

Although the plant-parasitic tylenchid nematodes have a complement of sense organs that generally resemble those of *C. elegans,* they do show some specific modifications, presumably associated with their mode of feeding or of host location. Inner labial sense organs appear to be solely chemosensitive. The pores of some or all of these sense organs in some species (e.g., *Radopholus, Rotylenchus,* and *Macroposthonia*) occur just inside the buccal capsule. The lateralmost outer labial receptors seem to be the most variable since they are apparently reduced (i.e., they do not reach the cuticle) in *Radopholus, Rotylenchus,* and *Heterodera.* It is noteworthy that the more highly sedentary of the tylenchid endoparasites, such as *Tylenchulus semipenetrans* and *Meloidogyne incognita,* have the greatest reduction in cephalic mechanoreceptors (i.e., all six outer labial sense organs are absent). They also have the most highly developed microvillus-bearing dendrite in the receptor cavity of the amphid. The complex interdigitation of these microvilli with the sheath-cell membrane prompted Baldwin and Hirschmann (1973, 1975) and Wergin and Endo (1976) to consider this region a gland. However, the adjacent cytoplasm does not show an extensive complement of endoplasmic reticulum or Golgi that would be required to form protein or carbohydrate secretory products. Although the functional significance of this complex dendrite/sheath-cell arrangement is still unknown, it is probably not related to glandular activity.

D. Specialization of Sense Organs in Animal Parasites

It is not surprising that animal parasites show considerable departures from the pattern of cephalic sense organs of *Caenorhabditis elegans.* Taxonomic studies have for a long time discussed fusion or loss of various sense organs. In general, the most obvious is the very close apposition of outer labial and cephalic sense organs and their housing in a single large papilla, as in ascarids and *Heterakis* or even *Nippostrongylus*. In fact, this is only a minor change since both sense organs apparently retain independent receptor cavities. The lateral outer labial sense organs may be absent as in *Nippostrongylus*, or may have developed into distinctly different receptor types such as the combined chemomechanoreceptor of *Heterakis*. Among the secernentian animal parasites, all inner labial sense organs appear to be mechanoreceptors.

Few adenophorean parasites have yet been studied. However, it may be significant that those species that have a highly chemosensory capacity are primarily parasites of internal organs or develop essentially as intracellular parasites (Lee and Wright, 1978; Wright, 1979).

Perhaps the most striking variation of any of the sense organs is that found in amphids of strongyloid parasites where the sheath cell body enlarges and undertakes active secretion. In *Necator*, this occurs throughout the adult stages (McLaren, 1976b), whereas in *Syngamus* it occurs only when the transition is made from migrating juvenile worms to attached adults (Jones, 1979). The secretory product of these cells has not been identified with certainty, although both acetylcholinesterase and anticoagulants have been suggested (McLaren, 1976b). The role of a secreted acetylcholinesterase is uncertain, although it has been suggested that it could modify intestinal motility and hence make it easier for nematodes to remain in place in the intestine (i.e., the "biochemical holdfast," Lee, 1969; Ogilvie and Jones, 1971). Alternatively, esterase might modify the permeability of host membranes, releasing nutrients for the parasites (Lee, 1970). Regardless, this phase of amphid activity should be seen as a specialized activity independent of other sheath cell roles such as osmotic–ionic regulation of the receptor cavity or formation of the normal receptor cavity matrix. Does the secretory activity interfere with the sensory activity of the amphid, or does it provide a mechanism for changing the sensitivity of amphids for a subsequent style of host–parasite interactions?

E. Internal Receptors

Little can, or should, be said at this stage about internal receptors. Our understanding of their structure is rudimentary and our understanding of their physiology is even more elementary. Perhaps future studies of mutant strains of *C. elegans* will enlighten us. At the most, we can now identify structures that are candidates to function as regulators of internal function, through proprioception of body wall, esophageal, or rectal function, monitoring of body-wall glands or ion–osmoregulatory cells, or perhaps chemoreception of food. Internal receptors could also receive exogenous stimuli if the stimulus can penetrate tissues of the animal. The most obvious of these are the photoreceptors. Internal cephalic receptors of *C. elegans* and tylenchoid plant parasites could be sensitive to light or other forms of electromagnetic radiation.

The study of *Caenorhabditis elegans* as a model system has already provided much basic information on the nature and probable function of nematode sense organs. Nevertheless, significant variations have also been found in nematodes with other basic lifestyles (plant and animal parasites) and between secernentian and adenophorean nematodes. The future extends the challenge of determining more specifically how these sensory units function to coordinate the various phases of the nematode's biology. Undoubtedly, this model system will continue to contribute prominently.

REFERENCES

Albertson, D. G., and Thomson, J. N. (1976). *Philos. Trans. R. Soc. London Ser. B.,* **275,** 299–325.

Anderson, R. C., Chabaud, A. G., and Wilmott, S. (1974). CIH "Keys to the Nematode Parasites of Vertebrates," Vol. 1. Commonwealth Agricultural Bureaux, Farnham Royal, England.

Baldwin, J. G., and Hirschmann, H. (1973). *J. Nematol.* **5,** 285–302.

Baldwin, J. G., and Hirschmann, H. (1975). *J. Nematol.* **7,** 40–53.

Batson, B. S. (1978). *Tissue and Cell* **10,** 51–61.

Bedini, C., Ferrero, E., and Lanfranchi, A. (1973). *Tissue & Cell* **5,** 359–372.

Bedini, C., Ferrero, E., and Lanfranchi, A. (1975). *Tissue & Cell* **7,** 253–266.

Bonar, D. B. (1978). *Tissue & Cell* **10,** 153–165.

Burr, A. H. (1979). *J. Comp. Physiol.* **134,** 85–93.

Burr, A. H., and Burr, C. (1975). *J. Ultrastruct. Res.* **51,** 1–15.

Chalfie, M. (1979). (personal communication)

Chalfie, M., and Thomson, J. N. (1979). *J. Cell Biol.* **82,** 278–289.

Clark, S., and Shepherd, A. M. (1977). *Nematologica* **19,** 243–247.

Clark, S., Shepherd, A. M., and Kempton, A. (1977). *Nematologica* **23,** 103–111.

Coggeshall, R. (1971). *Tissue & Cell* **3,** 637–648.

Crofton, H. D. (1971). *In* "Plant Parasitic Nematodes" (B. M. Zuckerman, W. F. Mai and R. A. Rohde, eds.), Vol. 1, pp. 83–113. Academic Press, New York.

Croll, N. A. (1977). *Annu. Rev. Phytopathol.* **15,** 75–89.

Croll, N. A., and Smith, J. M. (1974). *Nematologica* **20,** 291–296.

Croll, N. A., and Wright, K. A. (1976). *Can. J. Zool.* **54,** 1466–1480.

Croll, N. A., Evans, A. A. F., and Smith, J. M. (1975). *Comp. Biochem. Physiol. A* **51,** 139–143.

DeConinck, L. (1965). *In* "Traité de Zoologie" (P. -P. Grassé, ed.), pp. 586–600. Masson, Paris.

DeGrisse, A., Natasasmita, S., and B'Chir, M. (1979). *Rev. Nématol.* **2,** 123–144.

DeGrisse, A., and Natasamita, S. (1975). *Meded. Fac. Landbouwwet Rijksuniv. Gent.* **40,** 489–495.

DeGrisse, A. T., Lippens, P. L., and Coomans, A. (1974). *Nematologica* **20,** 88–95.

Dick, T. A., and Wright, K. A. (1973). *Can. J. Zool.* **51,** 197–202.

Dick, T. A., and Wright, K. A. (1974). *Can. J. Zool.* **52,** 179–182.

Endo, B. Y., and Wergin, W. P. (1977). *J. Ultrastruct. Res.* **59,** 231–249.

Gilula, N. B., and Satir, P. (1972). *J. Cell Biol.* **53,** 494–509.

Hirschmann, H. (1971). *In* "Plant Parasitic Nematodes" (B. M. Zuckerman, W. F. Mai, and R. A. Rohde, eds.), Vol. 1, pp. 11–63. Academic Press, New York.

Jones, G. M. (1979). *J. Morphol.* **160**, 299–322.

Kuppers, J. (1974). *In* "Mechanoreception" (J. Schwartzkopff, ed.), pp. 387–399. Westdeutscher Verlag, Landen, Germany.

Lee, D. L. (1969). *Parasitology* **59**, 29–39.

Lee, D. L. (1970). *Tissue & Cell* **2**, 225–231.

Lee, D. L. (1973). *J. Zool.* **169**, 281–285.

Lee, D. L. (1974). *J. Zool.* **173**, 247–250.

Lee, D. L. (1975). *Parasitology* **70**, 389–396.

Lee, T. D. G., and Wright, K. A. (1978). *Can. J. Zool.* **56**, 1889–1905.

Lewis, J. A., and Hodgkin, J. A. (1977). *J. Comp. Neurol.* **172**, 489–510.

Lippens, P. L. (1974). *Z. Morphol. Tiere* **79**, 283–294.

Lippens, P. L., Coomans, A., DeGrisse, A. T., and Lagasse, A. (1974). *Nematologica* **20**, 242–256.

Lorenzen, S. (1978). *Zool. Scr.* **7**, 175–178.

McIver, S. B. (1975). *Annu. Rev. Entomol.* **20**, 381–397.

McLaren, D. J. (1972). *Parasitology* **65**, 507–524.

McLaren, D. J. (1974a). *Int. J. Parasitol.* **4**, 25–37.

McLaren, D. J. (1974b). *Int. J. Parasitol.* **4**, 39–46.

McLaren, D. J. (1976a). *In* "Advances in Parasitology" (B. Dawes, ed.), Vol. 14, pp. 195–265. Academic Press, New York.

McLaren, D. J. (1976b). *In* "Organization of Nematodes" (N. A. Croll, ed.), pp. 139–161. Academic Press, New York.

Matsumoto, D. E., and Farley, R. D. (1978). *Tissue & Cell* **10**, 63–76.

Muller, R., and Ellis, D. S. (1973). *J. Helminthol.* **47**, 27–33.

Natasasmita, S., and DeGrisse, A. (1978a). *Meded. Fac. Landbouwwet. Rijksuniv. Gent* **43**, 769–777.

Natasasmita, S., and DeGrisse, A. (1978b). *Meded. Fac. Landbouwwet. Rijksuniv. Gent* **43**, 779–794.

Ogilvie, B. M., and Jones, V. E. (1971). *Exp. Parasitol.* **29**, 138–177.

Oschman, J. L., and Berridge, M. J. (1971). *Fed. Proc. Fed. Am. Soc. Exp. Biol.* **30**, 49–56.

Parker, G. H. (1919). "The Elementary Nervous System." Lippincott, Philadelphia, Pennsylvania.

Rice, M. J. (1975). *In* "Sensory Physiology and Behavior" (R. Galun, P. Hillman, I. Parnas and R. Werman, eds.), pp. 135–165. Plenum, New York.

Robertson, W. M. (1975). *Nematologica* **21**, 443–448.

Robertson, W. M. (1979). *Nematologica* **25**, 245–254.

Samoiloff, M. R., McNicholl, P., Cheng, R., and Balakanich, S. (1973). *Exp. Parasitol.* **33**, 253–262.

Siddiqui, I. A. and Viglierchio, D. R. (1970a). *J. Nematol.* **2**, 274–276.

Siddiqui, I. A. and Viglierchio, D. R. (1970b). *J. Ultrastruct. Res.* **32**, 558–571.

Slifer, E. H. (1970). *Annu. Rev. Entomol.* **15**, 121–142.

Smith, D. A. (1969). *Tissue & Cell* **1**, 443–484.

Smith, J. M., (1974). *J. Nematol.* **6**, 53–55.

Smith, J. M., and Croll, N. A. (1975). *Int. J. Parasitol.* **5**, 289–292.

Storch, V., and Riemann, F. (1973). *Z. Morphol. Tiere* **74**, 163–170.

Sulston, J. E., and Horvitz, H. R. (1977). *Dev. Biol.* **56**, 110–156.

Thurm, U. (1968). *Symp. Zool. Soc. London* **23**, 199–216.

Thurm, U. (1973). *In* "Mechanoreception" (J. Schwartzkopff, ed.), Westdeutscher Verlag, Landen, Germany.

Uni, S., Iseki, M., and Takada, S. (1977). *Jpn. J. Parasitol.* **26**, 157–167.

Ward, S. (1977). *Soc. Neurosci. Res. Symp.* **2**, 1–26.

Ward, S. (1978). *In* "Taxis and Behavior" (G. L. Hazelbauer, ed.), pp. 143–168. Chapman & Hall, London.

Ward, S., Thomson, N., White, J. G., and Brenner, S. (1975). *J. Comp. Neurol.* **160,** 313–338.

Ware, R. W., Clark, D., Crossland, K., and Russell, R. (1975). *J. Comp. Neurol.* **162,** 71–110.

Wen, G. Y., and Chen, T. A. (1976). *J. Nematol.* **8,** 69–74.

Wergin, W. P., and Endo, B. Y. (1976). *J. Ultrastruct. Res.* **56,** 258–276.

Westfall, J. A. (1973). *J. Ultrastruct. Res.* **42,** 268–282.

Wright, K. A. (1963). *J. Morphol.* **112,** 233–259.

Wright, K. A. (1974). *Can. J. Zool.* **52,** 1207–1213.

Wright, K. A. (1975). *Can. J. Zool.* **53,** 1131–1146.

Wright, K. A. (1976). *J. Nematol.* **8,** 92–93.

Wright, K. A. (1977). *J. Parasitol.* **63,** 528–539.

Wright, K. A. (1978). *Can. J. Zool.* **56,** 651–662.

Wright, K. A. (1979). *J. Parasitol.* **65,** 441–445.

Wright, K. A., and Carter, R. (1980). To be published.

Wright, K. A., and Chan, J. (1973). *Tissue & Cell* **5,** 373–380.

Wright, K. A., and Chan, J. (1974). *Can. J. Zool.* **52,** 21–22.

Wright, K. A., and Hui, N. (1976). *J. Parasitol.* **62,** 579–584.

Index

A

Acanthonchus duplicatus
  cuticle, 215
  ion regulation, 153
Acrobeloides buetschlii, respiration, 129
Acrobeloides nanus, dauer larvae, 201
Adenophorea, 267
  definition, 238
  sense organs, 290-291
Adenylate energy charge, 114-115
Aerobic metabolism, 166, 173
Age pigment, 8-11
  Caenorhabditis briggsae, 8-10, 90
  Caenorhabditis elegans, 8-9
  characteristics, 8-11
  Panagrellus redivivus, 8-9
  retardation, 9, 90
  Turbatrix aceti, 9
Age synchrony
  density gradient, 7
  DNA synthesis inhibitors, 6-7, 30-34
  glass bead, 6, 30, 83
  heat shock, 7, 35-36
  screening, 7-8, 30-31, 34-35
  single nematode, 6, 83
Aging-changes, 3-46
  age pigment, 8-11, 90
  axenic culture, 4-5
  behavioral changes, 14-19
  cultural conditions, 4-5, 32
  cuticle, 22-25
  electron dense aggregates, 10, 20
  environmental factors, 14-32
  enzymes, 29-46
  fecundity, 13-14

model, advantages, 4
monoxenic culture, 4-5, 32
nutrition effect, 16-18, 74
osmotic fragility, 20-22, 90, 156
sex-related changes, 11-14
specific gravity, 20-22, 90
surface charge, cuticle, 22-23
Aldolase, age related changes, 37-43
Alimentary tract, internal receptors, 282-285
Amphid
  acetylcholinesterase, 292
  Caenorhabditis elegans, 242-248
  Nippostrongylus, 257
  Tylenchidae, 253-256
Anaerobic metabolism, 136-137, 166, 173
Ancylostoma caninum
  excretion, 155
  lipid content, 125
  permeability, body wall, 156
  protein metabolism, 185
Ancylostoma tubaeforme
  excretion, 155
  oxygen consumption, 123
Anguina tritici
  anhydrobiosis, 195, 225-226
  cuticle, 225-226
  dessication
    physiological changes, 205-207
    survival, 160, 212, 204
Anhydrobiosis
  induction, changes during, 204-206
  revival, changes following, 206-207
Aphelenchoides blastophthorus
  caudal papillae, 269-272
  spicule, 270, 273
Aphelenchoides fragariae

Aphelenchoides fragariae (*cont.*)
  cephalic sense organs, 250–251, 279
  labial sense organs, 250
Aphelenchoides ritzemabosi, fatty acids, 167
Aphelenchoides rutgersi
  amino acid requirements, 53–54
  axenic culture, 48, 50
  excretion, 171–172
Aphelenchus avenae
  carbohydrate metabolism, 178
  cryptobiosis, 169
  dessication, physiological changes, 204–206
  energy regulation, 188–189, 198
  enzymes, 198
  glycerol production, 179
  glycogen reserves, 169–170
  life cycle, 197–198
  lipid utilization, 136, 169–170
  monoxenic culture, 48, 50, 196–197
  osmoregulation, 148, 150, 199
  osmotic fragility, age changes, 156
  oxygen consumption, 105
  permeability, body wall, 155–156
  respiration, 123, 130, 198, 206
  sex ratios, 197
  toxicology model, 199
Aporcelaimus
  amphid, 255
  oxygen consumption, 105
Ascaridia galli
  lipid content, 125
  oxygen consumption, 105, 129
  postlabial papillae, 276
Ascaris
  cuticular lesions, 227
  cytochromes, 111–114
  development, changes in glycogen content,
    168–169
  energy regulation, 186–188
  fatty acid, beta oxidation, 184
  gluconeogenesis, 183
  glucose metabolism, 173–174
  glycogen content, 168
  glyoxylate cycle, 180–181
  mitochondria, 110
Ascaris lumbricoides
  anaerobic metabolism, 136–137
  excretion, 154–155
  hatching, 157–158
  heme incorporation, 74

  hemoglobin, 132–134
  lipid content, 125
  osmoregulation, 149–152
  oxygen consumption, 105, 128–130
Ascaris suum
  cuticle structure, 214
  cuticular lesions, 227–230
  epicuticle, 219
  hatching, 157
Aspiculuris tetraptera
  mitochondria, 110
  osmoregulation, 150
Assay, toxicants
  carcinogenic effects, 96
  developmental parameters, 82–90
  enzyme activity, 90–91
  mutagenic effects, 91–95
Axenic culture, 47–77
  aging changes, 4–5
  axenization, 72
  basal medium, 48–56
  contaminants, 72–74
  definition, 47
  growth factor, 56–63
  nutrient absorption, 63–71
  sterility check, 72
  Turbatrix aceti, 4

B

Basal medium, formulation, 48–56
Behavior
  age changes, 14–19, 32
  chemotaxes, 289–290
  feeding, 16–19
  movement, 15–16
  oviposition, 18–19
Biospace experiments, proposed, 25–26
Body size, oxygen demand, related, 116–119
Body wall, permeability, 155–156, 214
Bursaphelenchus lignicolus
  age pigment, 9
  dauer larvae, 226
  osmotic fragility, age related, 156

C

Caenorhabditis
  carbohydrate reserves, 169–170
  lipid utilization, 136
  oxygen consumption, 105

Caenorhabditis briggsae
  age pigment, 8-10
  aging changes, 6-26
  amino acid synthesis, 172
  axenic culture, 48-56
  cuticle structure, 216-217
  cytochromes, 111
  dauer larvae, 200
  energy regulation, 188-189
  epicuticle, 220-222
  excretion, 171-172, 201
  glycerol production, 178-179
  glycocalyx, 221-222
  glyoxylate cycle, 180
  life span, 4
  oxygen consumption, 122, 129-130
  permeability, body wall, 155-156
  permeability changes, 22-24
  phosphogluconate pathway, 182
  protein metabolism, 185
  requirement
    amino acid, 52-54
    lipid-related, 54-56
    vitamin, 52
  species identification, 53-54
  surface carbohydrates, 23-24
  surface charge, 22-23
  ultrastructure, aging, 10
  vitamin E, 90
Caenorhabditis elegans
  age pigment, 8-9
  age synchrony, 30-36
  aging changes, 4-5, 11, 20-24
  alimentary tract, internal receptors, 282-283
  amphid, 242-248, 255
  axenic culture, 33-34, 84
  body wall internal receptors, 281
  bursa, 271, 273
  cephalic internal receptors, 278-279
  cuticle structure, 217-219
  cuticular sense organ, 240-248
  dauer, 200, 217-219, 226
  deirids, 275-276
  epicuticle, 220
  esophagus, cell types, 282
  feeding behavior, 16, 19
  gene, essential number, 92
  glyoxylate cycle, 181-182
  labial sense organs, 240-246
  mitochondria, isolated, 91

  molting, 232-233
  mutant, sense organ, 247-248
  nutritional requirements, 48-63
  oxygen consumption, 120, 126, 130
  phasmid, 273
  ultrastructure, 11, 218, 244, 246
Capillaria hepatica
  amphid, 267
  body pores, 278
  caudal papillae, 272
  cephalic sense organ, 261-263
  spicule, 274
Carbohydrate, 168
  dessication, changes associated with, 204-206
Centrophenoxine, 9, 20-21, 90
Chemoreceptor, 287-289
  amphid, 245-248
  Caenorhabditis elegans, 243
  definition, 239
Chromadorina bioculata, setae, 278
Chromadorina germanica
  body wall, internal receptors, 281
  sensory cell, 276-277
Collagen, 216
Cooperia punctata
  gluconeogenesis, 183
  lipid content, 125
  protein metabolism, 185
Cryptobiosis, 194
  Anguina tritici, 195
  Ditylenchus dipsaci, 195
Cuticle
  acid mucopolysaccharides, 221
  age changes, 22-25
  axenic culture, 48
  blisters, 89
  chemical composition, 215
  dauer larvae, 226
  development, 232
  epicuticle, 219-225
  glycocalyx, 221-222
  invaginations, 223-225
  microorganism interactions, 227-232
  nomenclature, 216-219
  permeability, 24-25, 155-156
  sense organ, 239
  structure, 214-225
  surface carbohydrates, 23-24, 221

Cuticle (*cont.*)
  surface charge, 22–24, 221–222
  wrinkling, 24
Cyclic nucleotides, *Panagrellus redivivus,* 5
Cytochrome oxidase inhibitors, 91
Cytochromes, 109–114, 166, 174–175, 177–
  178, 199

D

Dauer larvae
  cuticle morphology, 200, 217–219, 226
  epicuticle, 220–233
  formation, 74, 84, 200–201, 226
  oxygen consumption, 122
Deirids, definition, 275
Dendritic process
  Caenorhabditis elegans, 240–248
  definition, 239
  diagram, 241–242
Deontostoma
  cuticle, 215
  osmoregulation, 150
  photoreceptors, 280
DNA synthesis inhibitors, 6–7, 30–34, 87, 90
  disadvantages, 6–7, 32–33
  pheromone, inhibition of production, 88
Dessication survival, 160–161
  theoretical considerations, 202–204
Developmental parameters, 19–20
  fecundity, 87–89, 51
  generation time, 13, 32, 51, 87
  length, 30, 83–87
  maturation, 51
  molting, 84–87
  survival, 34
  volume, 5, 84
Dictyocaulus viviparus, carbohydrate
  metabolism, 175
Dipetalonema setariosum
  amphid, 264
  cephalic sense organ, 261
  phasmid, 268
Dipetalonema viteae
  amphid, 264
  caudal papillae, 271–272
  cephalic sense organ, 261–262
  phasmid, 268–269
  spicules, 274–275
Diplogaster stercorarius, dauer larvae, 201

Dirofilaria immitis
  amphid, 264
  cephalic sense organ, 261
  phasmid, 268
Ditylenchus dipsaci
  carbohydrate metabolism, 176–178
  cephalic sense organ, 250
  dessication, physiological changes, 205–206
  dessication survival, 160, 195–196
  excretion, 171
  fatty acids, 167
  labial sense organ, 249–250
  oxygen consumption, 105, 126
Ditylenchus myceliophagus
  amino acid excretion, 171
  dessication, physiological changes, 205–206
  dessication survival, 195
Ditylenchus triformis
  carbohydrate metabolism, 176–178
  cytochromes, 111, 178
  excretion, 171–172
  fatty acids, 167
Dopaminergic neurons, 17–19
  Aphelenchus avenae, 199
Dormancy, 194
  oxygen consumption, 121–122
Dorylaimus obtusicaudatus, respiration, 123
Dracunculus medinensis, phasmid, 268–269

E

Ecdysis, *see* Molting
Electron transport, 110–114, 166
Enoplus brevis
  growth curve, 119
  hemoglobin, 132–136
  osmoregulation, 150–152
  oxygen consumption, 105, 123, 129, 131
  permeability, body wall, 155–156
  volume measurement, 144–146
Enoplus communis
  growth curve, 119
  hemoglobin, 132–136
  osmoregulation, 150–152
  oxygen consumption, 105, 123, 129, 131
  permeability, body wall, 155–156
  volume measurement, 144–146
Embden–Meyerhof pathway, 173, 176
Endotokia matricida, 36, 88–89

Energy metabolism
    end products, 171–173
    energy reserves, utilization, 168–170
    energy storage molecules, 167–168
    fatty acids, β-oxidation, 183–184
    gluconeogenesis, 183
    glycerol production, 178–179
    glycolysis, 173–178
    glyoxylate cycle, 179–182
    phosphogluconate pathway, 182
    protein metabolism, 185
    regulation, 185–189
    tricarboxylic acid cycle, 173–178
Enolase, age-related changes, 37–44
Enzymes, 90–91
    aging
        altered, 36–45
        heat sensitivity, 38
        specific activity changes, 36–37
    altered, formation, 40–41
    Aphelenchus avenae, 198
    detection, altered, 36–37
    mammalian–nematode comparison, 38–39
    unaltered, 39
Esophagus, *Caenorhabditis elegans*
    cell type, 282
    neurons, 282–283
Eustrongylides ignotus
    anaerobic metabolism, 136
    lipid content, 125
Excretion, 117, 153–155, 171–173
    amino acids, 171–172
    ammonia, 172
    urea, 172–173

F

Fatty acids, 167–168
    β-oxidation, 183–184
    metabolic end products, 171
Feeding, age changes, 16–18
Ferritin, 69–71
    cationized, 22–23, 221–222
Formaldehyde-induced fluorescence, 17–19

G

Gastromermis boophthorae, amphid, 267, 269
Globodera pallida
    dormancy, lipid utilization, 121–122
    hatching factor, 157

Globodera (Heterodera) rostochiensis
    calcium content, 146
    hatching factor, 157–160
    osmoregulation, 148–151
    oxygen consumption, 115, 129
    respiration, 123
    volume measurement, 144–146
Glucose-6-phosphate dehydrogenase, age re-
        lated changes, 37
Glycerol production, 178–179
Glycocalyx, *C. briggsae*, 24, 221–222
Glycogen
    changes during development, 168–169
    content, 168
Glycolysis, 173–178
Glyoxalate cycle, 179
Growth factor
    chick embryo extract, 50
    Escherichia coli, 59
    heated liver extract, 50, 57
    heme, 56–63
    sterol, 56–63
    yeast extract, 58

H

Haemonchus contortus
    axenic culture, 76
    carbohydrate metabolism, 174
    cytochromes, 111, 174
    hemoglobin, 133
    oxygen consumption, 104–105
Hatching
    factor, 156–160
    ionic influences, 156–160
    osmotic influences, 156–160
Helicotylenchus dihystera, dessication survival,
        204
Heme
    absorption, 63–71
    nutritional requirement, 56–63
Hemoglobin, 104, 109, 132–136, 215
Heterakis gallinarum
    amphid, 264
    cephalic sense organs, 258–261
    internal receptors, 285
    postlabial papillae, 276–277
    ultrastructure sense organs, 260
Heterodera glycines
    cephalic sense organ, 250
    labial sense organ, 249–250

Heterodera oryzae
   anaerobic metabolism, 136
   oxygen consumption, 126
Heterodera schactii, hatching, 157–158
Hymenolepis diminuta, oxygen consumption, 128
Hymenolepis nana, axenic culture, 76

I

Inorganic ion content, estimation, 146–147, 161
Intestinal epithelium, ultrastructure, 10

L

Laser ablation, 81–82, 95
   spicules, 275
Lipid
   concentration, 167
   depletion
      dormancy, 121–122
      starvation, 123–126
   dessication, changes associated, 204–206
Litomosoides carinii
   amphid, 264
   cephalic sense organ, 261
   oxygen consumption, 105, 130
Longidorus leptocephalus, internal receptors, 285
Longidorus macrosoma, 285
Longevity
   Anguina tritici, 195
   Ditylenchus dipsaci, 195
   increases, 19–20, 90
   male–female, 11
   nutrition effect, 16–18, 74
   parental age, 12
   virgin–nonvirgin, 11–12

M

Macroposthonia rustica, labial sense organ, 249
Macrotrophurus arbusticola
   cephalic sense organ, 250
   labial sense organ, 249–250
Malnutrition, effect, 74–76
Mass culture, 30–35
Mechanoreceptor, 286–287
   cephalic sense organs, 245
   definition, 239

labial sense organs, 243–245
Nippostrongylus, 257–258
Meloidogyne incognita
   amphid, 252–254
   cephalic internal receptors, 279
   cephalic sense organ, 250–252
   egg shell, 215–216
   excretion, 171–172
   labial sense organ, 249–251
   lipid utilization, 170
   ultrastructure, 251–252
Meloidogyne javanica
   cuticle, 223–225
   hatching, 158
   lipid utilization, 121, 170
   mitochondria, 110
   molting, 233
   respiration, 123, 129
Mermis nigrescens
   cuticle permeability, 214
   cuticle structure, 214–215
Mermis subnigrescens, hemoglobin, 132
Mesodiplogaster lheritieri
   dauer larvae, 201
   respiration, 132
Metabolism, end products, 171–173
Mitochondria
   enzymes, 109
   function, 109–114
   structure, 109
Molting, 232–233
   block
      by sterol deficiency, 88
      injury to nerve ring, 86
   timing, toxicant assay, 84–86
Monhystera disjuncta
   osmoregulation, 150
   volume regulation, 148
Mononchus papillatus, oxygen consumption, 120, 123
Monoxenic culture, 74
   aging changes, 4–5
Movement
   age changes, 15–16
   in soil, 160
   oxygen consumption, related, 121–122
Muscle
   hemoglobin, 132–133
   ultrastructure, 17

Mutagens, 91–93
  ethyl methanesulfonate, 91–92
Mutant, 81
  lethal, 92
  reproductive system, 87
  sense organs, 247–248
Mutation frequency
  Caenorhabditis elegans, 92
  Panagrellus redivivus, 92–93

N

Necator americanus
  amphid, 264, 266
  oxygen consumption, 127
  phasmid, 268
  spicule, 275
Nematode model
  aging, 3–44, 89–90
  assay
    carcinogens, 96
    mutagens, 91–93
    pharmaceuticals, 89–90
    radiation effects, 93–95
    toxicants, 82–91, 199
  cuticle studies, 233–234
  evaluation, 44–45, 81–82, 166, 202
  nutrition, 47–79
Nematodirus filicollis, oxygen consumption,
  105
Nematospiroides dubius
  cephalic sense organs, 258
  growth curve, 119
  oxygen consumption, 105
Neoaplectana bibionis, dauer larvae, 201
Neoaplectana carpocapsae
  axenic culture, 48, 50
  sterol requirement, 59
Neoaplectana glaseri
  amino acid requirements, 53
  axenic culture, 48, 50
Nervous system
  esophageal neurons, 282
  molting, control, 86, 95
  sensory neuron, 239
Neurotransmitter, 17–19
  Aphelenchus avenae, 199
  in amphids, 262, 292
  in phasmid, 269
Nippostrongylus brasiliensis

amphid, 257, 263, 265
  bursa, 271–272
  carbohydrate metabolism, 175
  changes in carbohydrate reserves, 169
  cuticle, 215
  energy regulation, 188
  internal receptors, alimentary tract, 283–284
  lipid content, 124–125
  mitochondria, 110
  molting, 233
  oxygen consumption, 104–105, 108–109,
    126–127, 130
  pheromone, 88
  protein metabolism, 185
  sensory receptors, 245
  spicules, 275
  sterol requirements, 54
Nippostrongylus muris
  excretion, 155
  permeability, body wall, 156
Nutrition, 48–77
  absorption, 63–71
  aging effect, 16–18, 74
  culture medium, basal, 48–56
  malnutrition effects, 75–76
  requirements
    amino acid, 52–54
    carbohydrate, 56
    growth factor, 56–63
    heme, 56–63
    lipid, 54–56
    sterol, 50, 88
    vitamin, 52

O

Osmoregulation, 149–160
Osmotic fragility, age changes, 20–22, 90, 156
Obeliscoides cuniculi, carbohydrate
  metabolism, 175
Oncholaimus vesicarius
  amphid, 268–269
  photoreceptors, 280
Onchocerca volvulus, epicuticle, 220, 225
Osmoregulation, mechanisms, 149–158
Osmotic pressure, estimation, 146–147
Oxygen
  availability, 127–129
  demand, factors influencing, 116–127
  diffusion into nematodes, 102–109

Oxygen (*cont.*)
  diffusion, theoretical relationships, 104–108
  low level, effects, 129–130
  partial pressure, 102–109, 173
Oxygen consumption
  factors influencing, 116–127
  maximum theoretical, 105
  modifying factors, 104
Oxygenases, 115–116
Oxidative phosphorylation, 114

P

Panagrellus redivivus
  age pigment, 8–9
  aging changes, 5, 12–14
  axenic culture, 58, 50, 84
  carbohydrate metabolism, 177–178
  carbohydrate reserves, 170
  culture contaminant, 72
  dessication, physiological changes, 205–206
  energy regulation, 188–189
  fatty acids, 167
  fatty acid, β oxidation, 184
  fecundity, 88
  glycerol production, 178–179
  glyoxylate cycle, 180
  growth, 83–87
  ionizing radiations, effects, 93–95
  lipid requirement, 55
  longevity, 11, 90
  molts, 86, 88
  osmoregulation, 148–152
  osmotic fragility, 21
  oxygenases, 115
  oxygen consumption, 105
  pheromone, 88
  phosphogluconate pathway, 182
  radiation effects, 93–95
  respiration, 123
  steroid requirement, 88–89
  vitamin E, 90
Panagrellus silusiae
  axenic culture, 50, 62
  cytochromes, 111
  dessication survival, 201
Panagrolaimus davidi, osmoregulation, 150
Panagrolaimus rigidus, growth curve, 119
Parascaris equorum, cuticular lesions, 227
Paratrichodorus pachydermus, virus interaction, 230–231

Pelodera chitwoodi
  oxygen consumption, 120
  respiration, 123
Pelodera punctata, respiration, 129
Pelodera strongyloides
  dauer larvae, 201
  excretion, 173
  respiration, 130
Pharmaceuticals, aging, effect, 9, 20–21
Phasmids, 268–269
Pheromone, 87
  inhibition, 88
  Panagrellus, 275
  Rhabditis pellio, 14
Phosphogluconate pathway, 182–183
Photoreceptors, 279–280
Plectus cirratus, oxygen consumption, 120
Plectus granulosus, respiration, 123
Plectus parietinus, generation time, 201
Plectus rhizophilus, dessication survival, 201–202
Pontonema vulgare
  growth curve, 119
  oxygen consumption, 105
  respiration, 123
Porrocaecum decipiens, lipid content, 125
Pratylenchus penetrans
  excretion, 171
  fatty acids, 167
  spicules, 273
Pratylenchus scribneri, cytochromes, 111
Procaine, 21, 90
Protein turnover, *Turbatrix aceti,* 41–43
Protostrongylus stilesi, surface carbohydrates, 221

R

Radiation effects
  ionizing, 93–95
  laser, 75
  microwave, 95
  outer space, 25–26
  X-ray, 93–95
Radopholus similis
  cephalic sense organ, 250
  labial sense organ, 249–250
Receptor cavity
  definition, 240
  diagram, 241–242

Reesimermis nielseni, cuticle permeability, 214
Reproduction, age related changes, 12–14
Respiration, 101–141
  rate, body size related, 116–119
Rhabdias bufonis
  carbohydrate metabolism, 176
  energy regulation, 188
Rhabditis anomala, glyoxylate cycle, 180
Rhabditis dubius, dauer larvae, 201
Rhabditis pellio, pheromone, 14
Rhabditis terrestris, osmoregulation, 150–151
Rhabditis tokai, longevity, 12
Ribosomes, aging changes, 43
Romanomermis culicivorax
  cuticle permeability, 214
  cuticle structure, 214
  respiration, 129
Rotylenchus robustus
  cephalic sense organ, 250
  labial sense organ, 249–250

S

Scutellonema cavenessi, dessication survival,
  204
Secernentia, 256, 268, 290
  definition, 238, 240
  sense organs, 290–291
Sense organs, 237–295
  amphid, 245–248, 253–256
  body pores, 276–278
  caudal papillae, 269–272
  cephalic, 245, 250–253
  cephalic internal receptors, 278–279
  components, 239–241
  cuticular, 239–278
  deirids, 275–276
  internal receptors, 278–285
  internal receptors
    alimentary tract, 282–285
    body wall, 280–282
  labial, 240–246, 249–250
  male bursa, 269–272
  phasmids, 268–269
  photoreceptors, 279–280
  postlabial papillae, 276
  setae, 278
Serotonin, 17–19
  Aphelenchus avenae, 199
Setae, 278

Setaria cervi, carbohydrate metabolism, 176
Sex related changes, during aging, 11–14
Sheath cell
  definition, 240
  diagram, 241–242
Socket cell
  definition, 240
  diagram, 241–242
Specific gravity, age changes, 20–22, 90
Stephanurus dentatus, cuticular lesions, 227
Strongyloides fulleborni, axenic culture, 76
Strongyloides ratti
  anaerobic metabolism, 136
  carbohydrate metabolism, 175
  changes in carbohydrate reserves, 169
  fatty acid, β oxidation, 184
  glyoxylate cycle, 180
  lipid content, 125
Strongylus brevicaudata, carbohydrate
  metabolism, 176
Strongylus edentatus, cuticular lesions, 227, 229
Strongylus equinus, cuticle structure, 214
Superoxide dismutase, age-related changes,
  37–38
Syngamus trachea
  amphid, 265–266
  hemoglobin, 134
Syphacia muris, carbohydrate metabolism, 175
Syphacia obvelata
  cephalic sense organ, 261
  cuticle, 223

T

Target theory, 94
Tobrilus aberrans, amphid, 268
Toxocara mystax, oxygen consumption, 129
Tricarboxylic acid cycle, 173–178
Trichinella spiralis
  body wall, internal receptors, 281
  carbohydrate metabolism, 174
  cephalic sense organ, 262
  cytochromes, 111–113
  lipid content, 125
  mitochondria, 110
Trichodorus christiei, oxygen consumption, 105
Trichostrongylus colubriformis
  cephalic sense organs, 258
  oxygen consumption, 128
Trichostrongylus retortaeformis, hatching, 157

Trichuris muris, spicules, 274
Turbatrix aceti
  age pigment, 9
  age synchrony, 7, 29–36
  axenic culture, 4, 48
  carbohydrate metabolism, 176–178
  carbohydrate reserves, 170
  cytochromes, 111
  dessication, physiological changes, 205–206
  energy regulation, 188–189
  enzyme
    aging, 36–44
    detection of altered, 36–37
  fatty acids, 167
  fatty acid, β oxidation, 184
  fecundity patterns, 13
  glycerol production, 178–179
  glyoxylate cycle, 180–182
  heme requirement, 56
  ionizing radiations, 93
  longevity, 11
  permeability changes, 22
  phosphogluconate pathway, 182
  sterol requirement, 55–56
  ultrastructure, 17, 223
Tylenchorhynchus claytoni, fatty acids, 167
Tylenchorhynchus martini, swarming disease,
  230, 232
Tylenchulus semipenetrans
  amphid, 253
  cephalic internal receptors, 279
  labial sense organ, 249
  lipid utilization, 170
  spicules, 273–274

U

Ultrastructure
  bacterial lesions, 228
  Caenorhabditis briggsae age pigment, 11

Caenorhabditis elegans
  age pigment, 11
  amphid, 246
  cuticle, 218
  intestine, 70
  labial sense organs, 244, 246
  cuticle surface charge, 23
  dauer cuticle, 218
Meloidogyne incognita, 251
Meloidogyne javanica, 224
  sense organs, 244, 246, 251, 260
  spicules, 270
Turbatrix aceti, age changes, 15–17

V

Videotape analysis, feeding rate, 16
Vitamin E, longevity increase, 19, 31, 90
Volume change, method of estimation, 144–
  146
Vulval contraction, age changes, 18–19

X

Xiphinema americanum
  amphid, 255–256
  body pores, 276
  cephalic sense organs, 253–254, 279
  internal receptors, 284–285
  labial sense organs, 253–254
  ultrastructure, 254
Xiphinema diversicaudatum
  internal receptors, alimentary tract, 284–285
  virus interaction, 230–231
Xiphinema index, ion regulation, 154